TERMINAL TRANSFERASE
IN IMMUNOBIOLOGY
AND LEUKEMIA

ADVANCES IN EXPERIMENTAL MEDICINE AND BIOLOGY

Recent Volumes in this Series

TERMINAL TRANSFERASE IN IMMUNOBIOLOGY AND LEUKEMIA

Edited by

Umberto Bertazzoni
CEC Official at
CNR Institute of Biochemical and Evolutionary Genetics
Pavia, Italy

and

F. J. Bollum
Uniformed Services University of the Health Sciences
Bethesda, Maryland

PLENUM PRESS • NEW YORK AND LONDON

Library of Congress Cataloging in Publication Data

Main entry under title:

Terminal transferase in immunobiology and leukemia.

(Advances in experimental medicine and biology; v. 145)
"Proceedings of an EMBO workshop on terminal transferase in immunobiology
and leukemia, held May 28–31, 1981, in Elba, Italy" — Verso t.p.
Includes bibliographical references and index.
1. Deoxyribonucleate nucleotidyltransferase — Congresses. 2. Leukemia — Diagnosis
— Congresses. 3. Immunocompetent cells — Congresses. I. Bertazzoni, Umberto, 1937–
.II. Bollum, F. J. (Frederick James) III. European Molecular Biology Organi-
zation. IV. Series. [DNLM; 1. DNA nucleotidyltransferases — Congresses. 2. Lympho-
cytes — Analysis — Congresses. 3. Leukemia — Analysis — Congresses. W1 AD559 v. 145/
QU 141 T319 1981]
QP606.D46T47   616.99'4190756   82-3691
ISBN 978-1-4684-8931-6   ISBN 978-1-4684-8929-3 (eBook)   AACR2
DOI 10.1007/978-1-4684-8929-3

Proceedings of an EMBO workshop on Terminal Transferase in
Immunobiology and Leukemia, held May 28–31, 1981, in Elba, Italy

© 1982 Plenum Press, New York
Softcover reprint of the hardcover 1st edition 1982
A Division of Plenum Publishing Corporation
233 Spring Street, New York, N.Y. 10013

PREFACE

This book contains the proceedings of a conference devoted to the study of the structure and function of Terminal deoxynucleotidyl Transferase (TdT) and its utilization as biochemical marker in immunobiology and leukemia, held in Elba, Italy on May 28-31, 1981. The enzyme has been known to nucleic acid biochemists for more than 20 years and has proved to be an excellent tool for making deoxypolymers, labeling DNA fragments, and adding homopolymer tails to restriction endonuclease fragments from DNA.

Since the discovery of its peculiar tissue distribution, normally restricted to the thymus and bone marrow, and of its abnormal occurrence in human leukemic lymphoblasts, TdT has become one of the most widely used markers in the study of lymphocyte differentiation and in the classification of hematopoietic neoplasia. The subject seemed most appropriate for a meeting where molecular and cellular biologists, immunologists and hematologists could convene for the first time to discuss both basic research and the clinical aspects of the problem.

Among the goals achieved by this workshop was the sharing of information about the enzymology of TdT, biochemical and immunological methodology and the correlation of TdT with other markers in the diagnosis of leukemia. The remarkable accordance of results, presented here by five independent hematological institutions, from analysis of TdT in several thousands of leukemic patients, marks the importance of this enzyme as a diagnostic and prognostic tool in these diseases. The discussion about the ontogeny of lymphocytes was of particular interest, since terminal transferase represents a very specific intracellular marker for prelymphocytes and appears to be associated with the development of cells in B and T lineage.

The conference was attended by 45 researchers from Australia, Austria, France, Germany, Israel, Italy, Japan, Switzerland, United Kingdom and USA. The enthusiastic participation of all the attendants, the high quality of their presentations, the uninterrupted discussion and the friendly atmosphere all contributed towards the success of the meeting. We believe that the application of terminal transferase in the diagnosis of leukemia is an outstanding contribution from basic

science to clinical medicine and stems from fruitful cooperation between distant disciplines. This book reflects that experience. It also
provides a reference for information scattered in many different jour-
nals and gives an up-to-date account of the new and exciting developments in the terminal transferase story.

                              Umberto Bertazzoni
                              F. J. Bollum

ACKNOWLEDGEMENTS

This International Workshop, sponsored by EMBO, was also made possible by generous contributions from:

        ITALIAN RESEARCH COUNCIL (CNR)
        ENTE VALORIZZAZIONE ISOLA D'ELBA
        ISTITUTO FARMACOLOGICO SERONO
        GRUPPO LEPETIT
        ISTITUTO SIEROTERAPICO SCLAVO
        BOEHRINGER ITALIANA

CONTENTS

SECTION I

STRUCTURES AND PROPERTIES OF TERMINAL DEOXYNUCLEOTIDYL TRANSFERASE

SECTION II

DEVELOPMENT OF IMMUNOLOGICAL REAGENTS FOR DETECTION OF
TERMINAL DEOXYNUCLEOTIDYL TRANSFERASE

SECTION III

USE OF TERMINAL DEOXYNUCLEOTIDYL TRANSFERASE AS A MARKER
FOR STUDIES ON DIFFERENTIATION OF LYMPHOCYTES

CONTENTS

SECTION IV

TERMINAL TRANSFERASE AS A MARKER FOR MALIGNANT LYMPHOID CELLS

TERMINAL TRANSFERASE: PAST TO PRESENT

F. J. Bollum

Biochemistry, USUHS

4301 Jones Bridge Road, Bethesda, MD 20814

HISTORICAL INTRODUCTION

It has been said that cells do not make proteins for irrelevant reasons. This means that every protein has a specific function. If we fail to understand that function the flaw is in human nature, not in Nature itself. A corollary of this logic is that if a unique kind of cell produces a unique protein, then that unique protein must be related to a special function of the unique class of cells. Terminal deoxynucleotidyl transferase (TdT) is such a protein, existing only in a limited population of lymphoid cells. The purpose of this meeting is to consider the phenomenology associated with the occurrence of this protein, and hopefully deduce Nature's reason for producing it.

The events that lead to the discovery of terminal transferase were not so logical. Terminal transferase was an accidental finding during the purification of eukaryotic DNA polymerases. The nucleotide-polymerizing enzymes were discovered in the middle 1950's, at a time when oxidative metabolism held sway in biochemical research. Molecular Biology and Immunobiology were not even in a well-defined infancy, being more at the stage of conception --- at least at the molecular level. Grunberg-Manago and Ochoa discovered polynucleotide phosphorylase by reinterpretation of experiments designed to study oxidative phosphorylation in bacteria. Success in the polymerization of ribonucleotides probably encouraged Kornberg, Bessman, Lehman, and Simms in their discovery of E. coli DNA polymerase I in 1956.

1

Fumbling Around with Eukaryotic DNA Polymerases

I began work on eukaryotic DNA polymerases in 1957, and after the initial work on the inducible rat liver DNA polymerase, I moved to calf thymus gland as a more favorable source for large scale puri- fication. At the University of Wisconsin I was surrounded by "oxi- dizers" who seemed to be able to isolate any complex reaction and study its detail by using artificial electron acceptors. The sim- plicity and colorfulness of oxidative metabolism was attractive to me and I wondered whether simple nucleotide acceptors could be used to simplify DNA polymerase studies. I wrote to H. G. Khorana in Vancou- ver, who was preparing chemically synthesized oligodeoxynucleotides, and asked for some samples to test as deoxynucleotide acceptors.

Two years later, after my move to Oak Ridge, the samples arrived as dried chromatographic spots on filter paper in an envelope. I eluted the spots and at the end of the useful lifetime of a set of P^{32}-dXTP's I mixed $d(pT)_3$, $d(pT)_4$, $d(pT)_5$, $d(pT)_6$ and $d(pT)_7$ with various dXTP's and the crude DNA polymerase preparations I was study- ing. The reaction mixtures were run out on paper chromatograms, for a holistic look at any possible products, and the chromatograms were exposed to X-ray film for a couple of days. I spent the next week looking at those films while making a new set of P^{32}-dXTP's. There were low molecular weight products on the films, but I don't think I ever succeeded in convincing my colleagues at Oak Ridge of that. They preferred to interpret the result as an artifact of spotty vision!

Terminal Addition Becomes Reality

When the new batch of P^{32}-dXTP's was ready I repeated the ex-

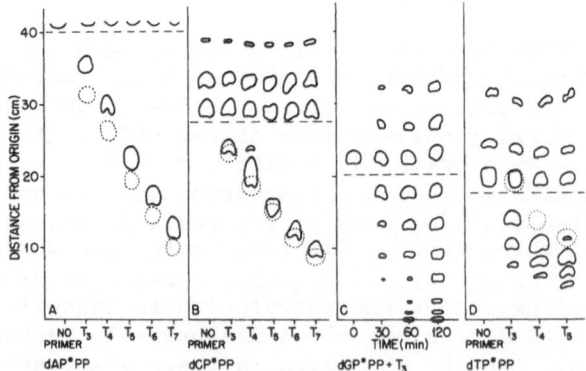

Figure 1. Paper Chromatographic Detection of Terminal Transferase. Deoxynucleoside triphosphates are above the dashed line, oligonucleotide initiators (dashed spots) and products (solid lines) are below the dashed line.

periment and the result was undeniable. New compounds were produced
(Figure 1, taken from (1)), and later characterized (2) as 3'-addi-
tions to the oligodeoxynucleotide initiator. Eleanor Groeniger
subsequently demonstrated that DNA polymerase and "end-addition"
were separate reactions and Masahiko Yoneda physically separated
the two activities on hydroxylapatite and Sephadex® G-100 (Figure 2).

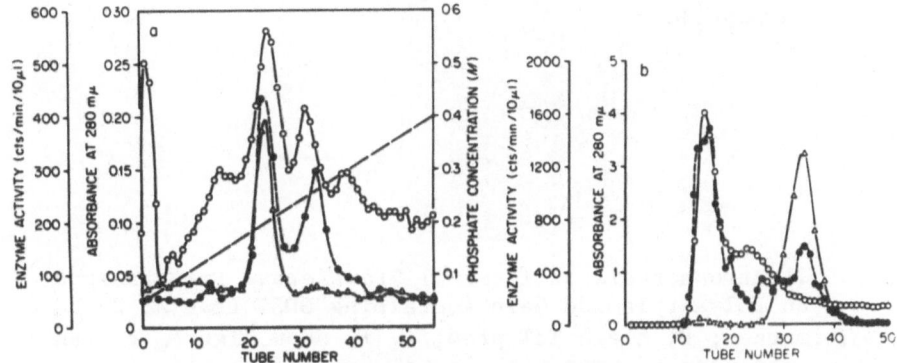

Figure 2.   Separation of Terminal Transferase and DNA polymerase
            Activity. (Reproduced from ref. 3). Separation on
            hydroxylapatite (r.) and Sephadex® G-100 (1.).   DNA
            polymerase (-●-), terminal transferase (-Δ-),
            and absorbancy at 280 nm (-○-).

     Independent studies at Yale also demonstrated the presence of
a "non-replicative" activity in extracts of calf thymus gland (4).
Work continued there into the late 60's, but then they lost interest.
I continued to work on terminal transferase because I thought it
might have some specific function in thymus gland and because it
certainly had uses for polydeoxynucleotide synthesis (5) and modification. TdT was viewed at that time as possibly a degraded form
of replicative DNA polymerase, and several investigations claimed to
demonstrate "end-addition" activity in miscellaneous tissues. But
I like to polymerize deoxynucleotides, so I persisted.

     When I moved to the University of Kentucky in 1965, I took the
protein fractions containing DNA polymerase and TdT with me. We
continued to study the properties of terminal transferase (6) and
make polydeoxynucleotides with these fractions, which were about
10% pure. In 1970 Lucy Chang produced homogenous enzyme (7) just
before her Christmas sabbatical and the new SDS acrylamide gel procedure (8) was essential for the demonstration of the asymmetric two
peptide structure of the native protein. Figure 3 shows the α and
β peptides of a typical (not quite homogeneous) preparation. This
enzyme preparation has been repeated in many laboratories, but the
true structure of TdT is still in the process of refinement.

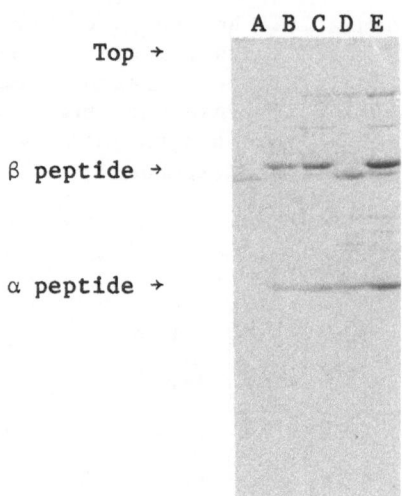

Figure 3.   Electrophoresis of Terminal Transferase Preparations
            on Polyacrylamide Gels Containing SDS. Lane A, 24 Kd
            marker; B, clean TdT prep; C, E, show high M_W contamin-
            ants; and D, prep with 24 Kd β peptide

Fumbling Around in Biology

In these days of analytical nihilism we tend to forget the value
and power of work on isolating rare proteins in tangible amounts.
Lucy Chang's purification and demonstration of TdT as a pure and
separate substance gave confidence in further research, and set
the stage for current biological studies. She then opened the
curtain by demonstrating that TdT was unique to thymus, and not a
common degradation product of DNA polymerase in cells proceeding from
a proliferative to a differentiative phase (9). The enzyme was unique
to thymus in all species where we were able to find this organ.
Presence in bone marrow was demonstrated in 1974 by Mary Sue Coleman
using assays of increased sensitivity (10).

The unique location of TdT in thymus and bone marrow burst into
the real world when McCaffrey et al. detected high levels of enzyme
in human leukemias (11). The need for biological studies, at the
tissue and the cellular level was obvious. The cellular studies
became possible with the development of antibodies to TdT, and the
results obtained from these investigations are quite striking.

Development of Enzyme Assays for TdT

Demonstration of the existence of a deoxynucleotide polymerizing activity in partially purified DNA polymerase preparations was
only marginally interesting in 1960. Purification of eukaryotic DNA

polymerase was a more immediate problem then, and besides the oligo-
nucleotide-initiated reaction did not work well in crude extracts.
Thus terminal transferase could be a degraded form of DNA polymerase --- definitely not interesting. Alternatively, of course, it
might be possible to develop new assays that would permit an examination of the biological distribution of the enzyme.

By 1967 many properties of the terminal transferase reaction
had been explored and we set out to examine the tissue distribution
of the enzyme. The first procedure used was rather indirect.
Reasoning that degradative enzymes or inhibitors might interfere
with the reaction at low activity levels, we fractionated ammonium
sulfate concentrated extracts on 5-20% sucrose gradients and assayed
all fractions for DNA polymerase activity and terminal transferase,
still using oligodeoxynucleotide initiators. This procedure was
applied to a wide variety of tissues in a variety of eukaryotic species. The results gave confidence that the enzyme activity did exist
in crude extracts of certain tissues and no others. The specific
tissue was thymus gland, and it was present in the thymus of all animals that we looked at, including chickens. Confident that we were
working with a bona fide and somewhat unique activity, we purified
the enzyme to homogeneity (7).

During the process of assaying literally thousands of sucrose
gradient fractions we sought more direct assays for terminal transferase --- not requiring such elaborate fractionation. We thought
that nuclease action was our principal problem since we could get no
convincing evidence for tissue inhibitors. This problem could be
circumvented by using a short polynucleotide initiator that could
survive nibbling by exonuclease and splitting by endonuclease and
still be long enough to bind to the enzyme. Earlier work had shown
that polymers of deoxyguanylate were resistant to nuclease degradation (12), so we used dGTP as the substrate to be polymerized.
The assay reaction was then:

$$n\text{-}dG^*TP + d(pA)_{50} \xrightarrow[M^{++}]{E} d(dA)_{50}(p^*G)_n + nPP_1$$

which could be scored by measuring acid-insoluble radioactivity by
batch work-up on glass fiber disks (13). These reaction conditions
are surprisingly specific and give very low backgrounds in negative
tissues and a good progress curve in positive samples. In our experience, use of 0.2 M potassium cacodylate buffer and Mg^{++} gives the
best results and other buffers, especially TRIS, may be found to be
inhibitory.

There are several interesting sidelights on the development of
this specific assay for terminal transferase. Other investigators
decided that activated DNA was a suitable substitute for the single-

chained polynucleotide initiator. As a result they found a rather
wide distribution for TdT activity. We have not confirmed these results with the more specific assay. In our assays on sucrose gradient
fractions (7,9), we assayed for DNA polymerase with activated DNA and
terminal transferase with short single-stranded polydeoxynucleotides.
Tissue extracts that did not respond in the terminal transferase
assay did show some response in the assay using activated DNA. By
careful examination of the sedimentation patterns Lucy Chang decided
that the activated DNA response in the low molecular weight region of
the gradient was a new DNA polymerase, since it did not sediment at
the same rate as terminal transferase and did not respond to the same
assay conditions. This observation lead to the demonstration (14,15)
and purification of DNA polymerase-β (16).

During 1970-1974 we used our specific terminal transferase assay
(with and without sucrose gradient fractionation) to assay hundreds
of different tissue extracts for activity. The major new finding was
the presence of small amounts of enzyme in bone marrow extracts (10),
in certain leukemic patient samples (10,11,17) and a restricted set of
lymphoblastoid cell lines (18). A specific point of contention at this
time was whether both B- and T-cell lines contained the enzyme. Using
specific assays we examined about 100 myeloma cell lines provided by
Michael Potter at NIH and all were negative (18). There were some
curious positive cases in Abelson virus-induced tumors, and these are
now thought to be pre-B lines. The generality that developed was that
terminal transferase was present in pre-T and possibly pre-B lines,
but that was difficult to prove at the gross tissue level by enzyme
assay.

Development of Specific Antibody

The availability of a pure enzyme should have made preparation
of antibody to terminal transferase a rather simple matter. It took
us a rather long time to do this, however, and we were finally successful only by using repeated injections of cross-linked antigen
(19). The procedure we use has now been successful in at least 20
rabbits so we know that it works. The antibody we prepared in rabbits has a broad species cross-reactivity and has been used to examine TdT distributions in systems as diverse as humans (20) and
chickens (21).

The antibody we prepared in rabbits against calf thymus terminal transferase is used as an immunoadsorbent column-purified
immunoglobulin. The greatest use of this material has been for
studies of cell populations containing terminal transferase using
indirect immunofluorescence. With fluorescence microscopy it is
relatively easy to survey cell populations from a variety of tissues.
These studies provided immediate confirmation of the enzyme assays
carried out earlier: Cells that contain TdT in adult animals are
present in thymus (60%, mostly cortical thymocytes) and to a much

lesser extent in bone marrow (1-5%, in bone marrow lymphoblasts).
In thymus the terminal transferase seems to be distributed in nucleus
and cytoplasm, while in bone marrow localization is exclusively
nuclear.

Leukemias and Lymphomas

Immunofluorescent staining for terminal transferase is ideally
suited for clinical studies since our procedure can be applied directly to dried bone marrow and blood films. The results obtained
by immunofluorescence again confirm the results obtained by enzyme
assay. In acute lymphoblastic leukemia at diagnosis 40-95% of the
lymphoblasts are TdT^+. In the acute phase of chronic myelogenous
leukemia a finding of 20-50% TdT^+ blasts indicates lymphoblastic proliferation. Rare acute myelogenous leukemias will also show elevated
levels of TdT^+ cells. Lymphoblastic lymphomas contain elevated levels
of TdT^+ cells in lymph nodes, where these cells are not normally
found.

At the present time TdT analysis in leukemia and lymphoma patients is mostly useful for diagnosis, but diagnosis is the prelude
to therapy. Looking for TdT^+ cells in spinal fluid, testicular bi-
opsis and pleural effusions during therapy seems a rather obvious
extension of diagnosis to careful management. In several studies
that have followed TdT^+ in bone marrow cells during therapy and
remission, the measurement did not provide a dependable early indication of relapse. As a matter of fact, determination of TdT^+ cells
is said to be rather variable during therapy and remission. But
given the sensitive nature of the bone marrow to the chemotherapeutic
agents used, this does not seem unsusual. Perhaps closer attention
to patient status and drug therapy at the time of determination might
provide a better understanding of this problem.

In monitoring patients on therapy or in remission it would appear
that TdT determinations should be part of a multiparameter analysis
applied to bone marrow aspirates. The multiparameter approach does
provide sensitive detection of abnormal phenotypes (22). Several of
the clinical studies in this Volume will discuss these ideas in more
detail.

Ontogeny of TdT^+ Prelymphocytes

The most striking use of the cytochemical test for TdT^+ cells has
been in the study of the development of the lymphoid system. In
rodents TdT^+ cells arise first in the thymus in the 14-16 day embryo.
No other organs contain positive cells at this stage of development.
Shortly after birth the bone marrow population appears and reaches its
stable level within 10 days after birth (23). The thymus and bone
marrow populations are stable during early life (Fig. 4) and attenuate
only with thymic involution and age.

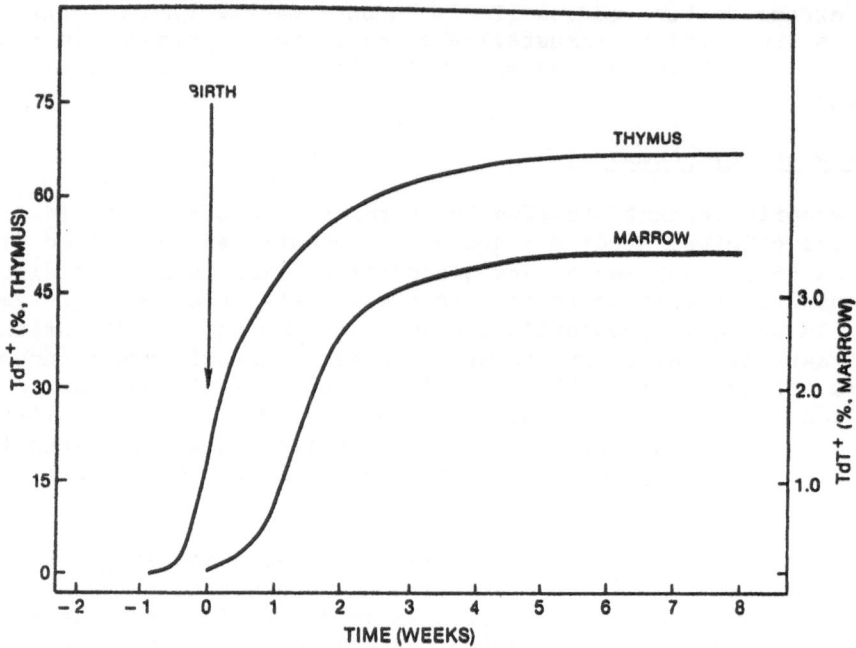

Figure 4.   Stable Populations of TdT$^+$ Cells.

In the early postnatal period some very suggestive transient populations can be detected. At 10 days after birth a very sharp wave of TdT$^+$ cells appears in the circulation. In lung, liver and spleen waves of positive cells appear and disappear with different kinetic curves (Fig. 5). These are the transient populations of TdT$^+$ cells. The meaning of these transients remains a mystery, but they may mark the migration of precursor populations from the primary lymphoid organs into secondary sites. For example, the transient TdT$^+$ population in peripheral blood arises in the thymus since it does not occur in homozygous nude mice (24). All of the other transients are seen in nudes so they must arise from non-thymic sources, presumably bone marrow. Immunological marker studies on neonatal mice and rats confirm these sources of transients.

A simple interpretation of the transient populations would be that TdT$^+$ marks an early prelymphocyte that populates all of the secondary lymphoid organs, becoming the parents of functional T and B cells (Fig. 6, taken from 25). Our present studies show blood, lung, liver and spleen to be subject to this immigration --- all organs of well-known immunologic and hematologic functions. Closer study may show skin, gut and lymph nodes to be subject to similar population migration. After involution of thymus and bone marrow, or under hematologic stress, these secondary sources may serve as continuing sources for the prelymphocyte population.

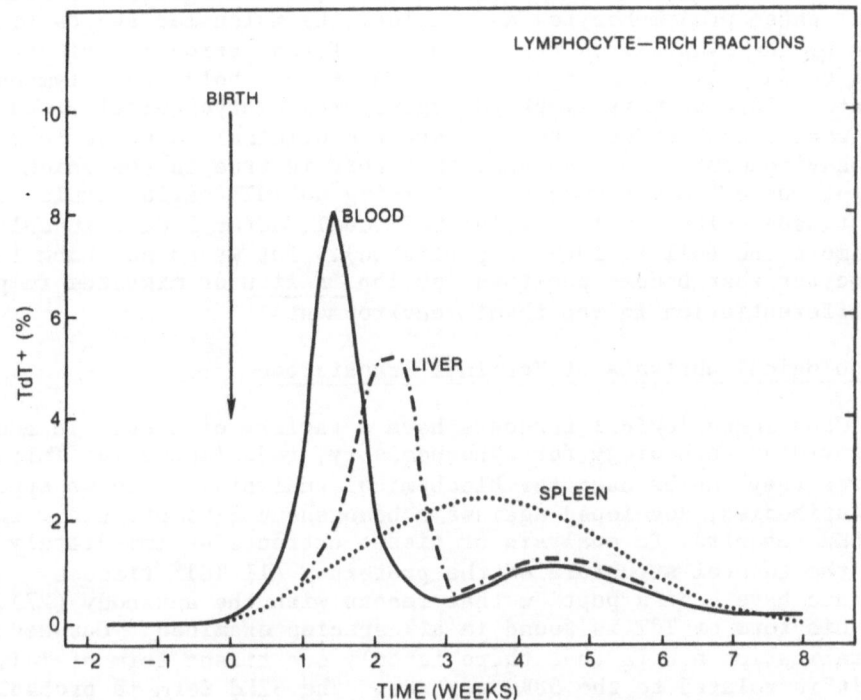

Figure 5. Transient Populations of TdT$^+$ Cells.

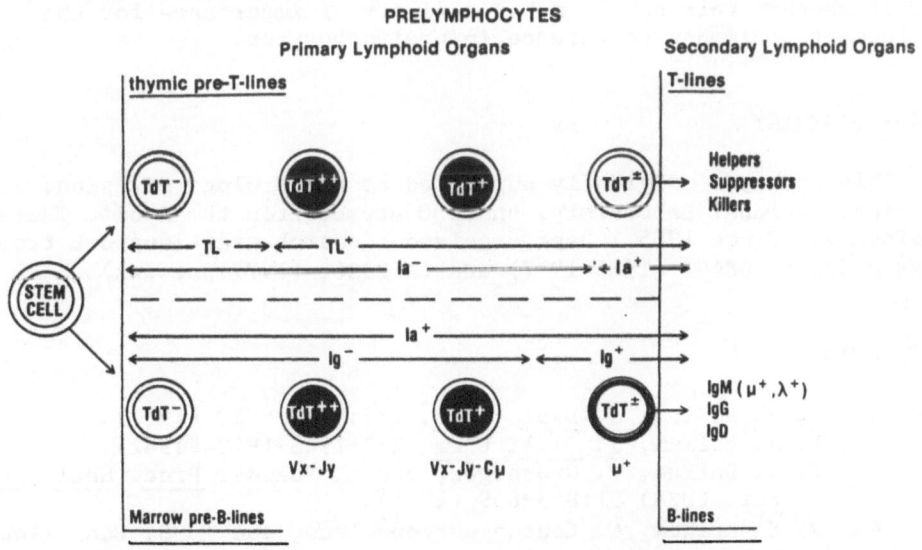

Figure 6. Differentiation of Prelymphocytes.

In the future, we will need to find out the ultimate source
of all these prelymphocytes and explore the molecular events brought
about by the lymphokines that induce differentiation in primary
and secondary lymphoid organs. Current theory holds that lymphocytes
migrating into primary lymphoid organs from embryological sites ---
yolk sac, neural crest, etc. --- are the original settlers in new
microenvironments. We can show that this is true in the chick
embryo, where 8 day thymuses, containing no TdT⁺ cells, explanted
into tissue culture will develop TdT⁺ cells after 5 days in culture
(Sugimoto and Bollum, 1978, unpublished). But we do not know if the
thymocytes that become positive develop in situ or migrated in prior
to differentiation in the thymic environment.

Immunological Analysis of Terminal Transferase

Good immunological reagents have a variety of uses. In addition
to providing technology for cytochemistry, radioimmune and ELISA procedures they can be used for biochemical analysis. When we applied
our antibodies, developed against a homogenous 32Kd protein with 24Kd
and 8Kd subunits, to analysis of tissue extracts we immediately clarified the general structure of the protein. All TdT⁺ tissues
examined have a 58Kd peptide that reacts with the antibody (27),
and this form of TdT is found in all species examined. Out best
generalization now is that there is only one tissue form of TdT,
and it is related to the 58Kd peptide. The 32Kd form is probably
an artifact produced during isolation.

These findings now provide a unifying view of TdT structure
throughout those animals that have an immune system. The future will
tell us whether this new form has activity of importance for the
development of immune competence in prelymphocytes.

ACKNOWLEDGEMENT

This work was originally supported at the Biology Division,
Oak Ridge National Laboratory, under Contract from the Atomic Energy
Commission. Since 1965 I have received research grant support from
USPHS grant CA 08487 (1965-1977) and CA 23262 (1977-present).

REFERENCES

1. F. J. Bollum, J. Biol. Chem. 235:PC18-PC20 (1960).
2. F. J. Bollum, J. Biol. Chem. 237:1945-1949 (1962).
3. F. J. Bollum, E. Groeniger, and M. Yoneda, Proc. Natl. Acad.
    Sci. (USA) 51:853-859 (1964).
4. J. S. Krakow, C. Coutsogeorgopoulous, and E. S. Cannellakis,
    Biochem. Biophys. Res. Commun. 5:477-481 (1961).

5. F. J. Bollum in: "Procedures in Nucleic Acid Research"
    G. Cantoni, ed., Harper and Row, New York, pp 577-583
    (1966).

6. K. Kato, J. M. Goncalves, G. E. Houts, and F. J. Bollum,
    J. Biol. Chem. 242:2780-2789 (1967).

7. L. M. S. Chang and F. J. Bollum, J. Biol. Chem. 246:909-916
    (1971).

8. A. L. Shapiro, E. Viñuela, and J. V. Maizel, Jr., Biochem.
    Biophys. Res. Commun. 28:815-819 (1967).

9. L. M. S. Chang, Biochem. Biophys. Res. Commun. 44:124-131
    (1971).

10. M. S. Coleman, J. J. Hutton, P. DeSimone, and F. J. Bollum,
    Proc. Natl. Acad. Sci. (USA) 71:4404-4408 (1974).

11. R. P. McCaffery, D. F. Smoler, and D. Baltimore, Proc. Natl.
    Acad. Sci. (USA) 70:521-525 (1973).

12. C. F. Lefler and F. J. Bollum, J. Biol. Chem. 244:594-601
    (1969).

13. F. J. Bollum, J. Biol. Chem. 234:2733-2734 (1959).

14. L. M. S. Chang and F. J. Bollum, J. Biol. Chem. 246:5835-5847
    (1971).

15. L. M. S. Chang and F. J. Bollum, Biochemistry 11:1264-1272
    (1972).

16. L. M. S. Chang, J. Biol. Chem. 248:3789-3795 (1973).

17. P. S. Sarin and R. C. Gallo, J. Biol. Chem. 249:8051-8055
    (1974).

18. F. J. Bollum in: "Advances in Enzymology" A. Meister, ed.,
    John Wiley and Sons, New York, Vol. 47, pp 347-374
    (1978).

19. F. J. Bollum, Proc. Natl. Acad. Sci, (USA) 72:4119-4122
    (1975).

20. G. Janossy, J. A. Thomas, F. J. Bollum, S. Granger,
    G. Pizzolo, K. F. Bradstock, L. Wong, A. McMichael,
    K. Ganeshaguru, and A. V. Hoffbrand, J. Immunol. 125:
    202-212 (1980).

21. M. Sugimoto and F. J. Bollum, J. Immunol. 122:392-397
    (1979).

22. K. F. Bradstock, G. Janossy, F. J. Bollum, and C. Milstein,
    Nature 284:455-457 (1980).

23. K. E. Gregoire, I. Goldschneider, R. W. Barton, and F. J.
    Bollum, J. Immunol. 123:1347-1352 (1979).

24. R. Sasaki, F. J. Bollum, and I. Goldschneider, J. Immunol.
    125:2501-2503 (1980).

25. F. J. Bollum, Blood 54:1203-1215 (1979).

26. F. J. Bollum, Trends in Biochemical Sciences 6:41-43 (1981).

27. F. J. Bollum and L. M. S. Chang, J. Biol. Chem. 256:8767-
    8770 (1981).

PURIFICATION OF TERMINAL DEOXYNUCLEOTIDYL TRANSFERASE FROM PIG

THYMUS : IDENTIFICATION OF 42,000 AND 57,000 DALTON SPECIES

Tsuguhiro Kaneda, Saiko Kuroda, Osamu Koiwai* and
Shonen Yoshida*

Blood Disease Center, Nagoya National Hospital, 4-1-1
Sannomaru, Naka-ku, Nagoya 460, Japan

*Department of Biochemistry, Institute for Developmental
Research, Aichi Prefecture Colony, Kasugai 480-03, Japan

SUMMARY

    Terminal deoxynucleotidyl transferase (TdT, EC 2.7.7.31) has
been purified 4365-fold from pig thymus. It was further separated
into two molecular forms of 57,000 and 45,000 dalton by Sephadex
G-100 gel-filtration. The former sedimented at 4.2s through a
sucrose gradient, while the latter, at 3.6s. By a SDS-polyacrylamide
gel-electrophoresis, their molecular weights were estimated as
57,000 and 42,000 dalton, respectively. Thus each of the large and
small pig TdT consists of a single polypeptide of 57,000 or 42,000
dalton and has no subunit structure. These two forms were indistinguishable in antigenicity by a neutralization assay with an anti-
calf TdT antibody. The enzymological properties of 42,000 dalton-
TdT from pig thymus were very similar to those of calf TdT which
has a two-subunits structure.

INTRODUCTION

    Terminal deoxynucleotidyl transferase (TdT, EC 2.7.7.31) is
localized in thymus gland (1,2) and also in bone marrow in a low
level (3,4) in normal adult mammals. The physiological function of
TdT in thymus or bone marrow remains unknown, but a hypothesis that
TdT acts as a somatic mutator in the generation of immunological
diversity in the early stages of lymphocyte differentiation has been
proposed by Baltimore (5) and Bollum (6).
    It has been reported that TdT purified from calf thymus is
composed of two heterogenous subunits and that molecular weight of

13

a native form is 32,000 dalton (7). Recently, high molecular weight
forms of TdT (58,000-62,000 dalton) were found in human lymphoblastic
cells (8,9), murine thymus and bone marrow (10). In the present study
we first achieved a large scale purification of two forms of TdT
from pig thymus, of which molecular weights were 57,000 (TdT-57K)
and 42,000 dalton (TdT-42K). Either of them consists of a single
polypeptide.

MATERIALS AND METHODS

Purification of pig TdT : Minced pig thymus (1.7kg) was homogenized in 4 liters of TEM buffer (50mM Tris-HCl, pH 7.5, 1mM EDTA,
1mM 2-mercaptoethanol) containing 0.2M KCl and the homogenate was
centrifuged at 12,000×g for 10 min. The supernatent was diluted with
an equal volume of chilled distilled water and was kept for 2 hrs
at 4°C. After clarifying by centrifugation at 12,000×g for 10 min,
the diluted supernatent (Fraction I) was adsorbed to 500 ml of
phosphocellulose (Whatman P-1, equilibrated with TEM buffer). The
gel was washed with a 1.5 liters of TEM buffer in a column (5×26cm).
Enzyme activity was eluted from the column with a linear gradient
from 0-1.2M KCl in TEM buffer. Active fractions were collected
(Fraction II) and was subjected to $(NH_4)_2SO_4$ fractionation (30-60%).
Precipitate was collected by centrifugation and was soluted in KEMG
buffer (50mM potassium phosphate, pH 7.2, 1mM EDTA, 1mM 2-mercapto-
ethanol, 10% glycerol) containing 0.1M KCl (Fraction III). Fraction
III was applied onto Sephadex G-100 column (2.8×92cm) and the column
was eluted with the same buffer. TdT activity was eluted at approx.
30% of the column volume (Fraction IV) and was completely separated
from DNA polymerase α. Fraction IV was dialysed against TEMG buffer
(concentration of Tris-HCl was reduced to 20mM and 10% glycerol was
added) and was applied onto double-stranded DNA-cellulose column
(1.6×18cm). The column was eluted with a linear gradient from 0 to
0.3M NaCl in TEMG buffer. Activity was eluted at 40mM NaCl (Fraction
V). Fraction V was applied onto a column of hydroxylapatite (0.9×15
cm). The column was eluted with a linear gradient of 0.05-0.3M
potassium phosphate (pH 7.2) in KEMG buffer and the activity was
eluted at 0.17M with a shoulder at 0.2M (Fraction VI) (for further
analysis of these two fractions, see RESULTS). Fraction VI was
rechromatographed on a hydroxylapatite column with the same procedure
except for using KEMG buffer containing 0.1M KCl (Fraction VII).

Assay of TdT : The standard assay mixture (62.5μl) contained
50mM Tris-HCl(pH 8.3), 2mM dithiothreitol, 0.3mM $MnCl_2$, 50mM KCl,
10μg of BSA, 0.02O.D. units of oligo(dA)$_{12-18}$, and 0.1mM [^{3}H]dGTP
(40,000cpm/nmol). Activated heat-denatured calf thymus DNA (20μg)
was used as a primer in place of oligo(dA)$_{12-18}$ in the assays during
the purification of TdT. Incubation was carried out at 37°C for 30
min and acid-insoluble radioactivity was measured. One unit of enzyme
was defined as an amount catalyzing the incorporation of 1 nmol of
deoxynucleotide per hr at 37°C.

RESULTS

Separation of two molecular forms of pig TdT : The specific
activity of the most purified fraction (2nd hydroxylapatite) was
30,556 units/mg of protein, which signifies 4365-fold purification
from the crude extract. During the attempts to purify pig TdT,
multiple species of the enzyme were detected. The elution profile of
TdT from the 1st hydroxylapatite column (Step VI) was asymmetric.
The major TdT activity (HA-I) was eluted from the column at 0.17M
potassium phosphate and the minor one (HA-II) at 0.2M potassium
phosphate as a shoulder. Two fractions were applied independently
on Sephadex G-100 (2×81cm) (Fig.1A). It was revealed that both
fractions of HA-I and HA-II contained two components; a large
species of 57,000 dalton and a small one of 45,000 dalton estimated
from their elution volume from the column. The major species in HA-I

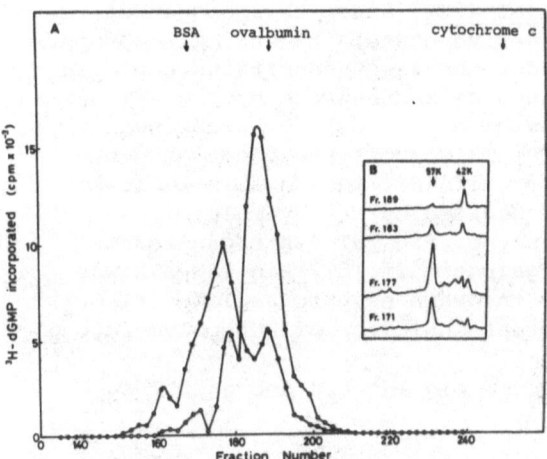

Fig. 1. Sephadex G-100 gel-filtration of pig TdT (A) and de-
sitometric scans of SDS-gels
(B). (A) Hydroxylapatite fraction I (●) and II (O) and II
(O) were independently applied
to Sepadex G-100 column (2x81
cm). Recovery of the enzyme
activity was approx.50%. (B)
Sephadex G-100 fraction no.171,
177, 183 and 189 of HA-I (●)
were subjected to SDS-PAGE. The
gel was stained by silver staining method (11). Absorbance at
600mn of each gel was measured.

Fig. 2. SDS-polycrylamide gel-electrophoresis of calf TdT (A) and pig
Tdt were subjected to SDS-PAGE ac-
cordingly to the method of Laemmli
(12). (A) Lane 1 : external markers. Lane 2 : 5µg of calf TdT.
Gel was stained by 0.25% Coomassie Brilliant Blue G-250. (B)
Lane 1 : 0.2µg of the small species of pig 0.2µg of the large
species of pig TdT (Sephadex G-
100 fraction no. 177 of HA-1, see
Fig.1A). Arrows indicate the marker proteins. Gel was stained by
silver staining method (11).

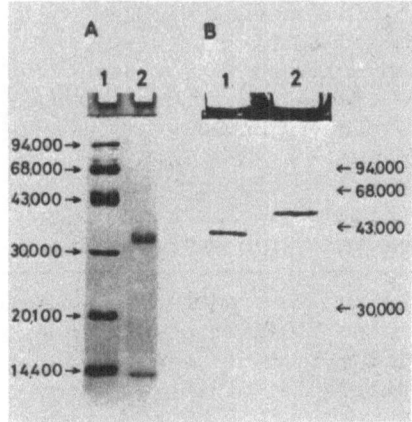

fraction is the large one, while in HA-II, the small TdT species is predominant. The peak fraction of each enzyme species was analysed by a SDS-polyacrylamide gel-electrophoresis (Fig.2). The small species of pig TdT (Sephadex G-100 fraction no.186 of HA-II, see Fig.1A) showed one band corresponding to 42,000 dalton (designated as TdT-42K). These results indicate that TdT-42K consists of a single polypeptide. On the other hand, the large species (Sephadex G-100 fraction no.177 of HA-I, see Fig.1A) exhibited a main band of 57,000 dalton (designated as TdT-57K) and a few minor bands. The major polypeptide of 57,000 dalton may be the enzyme molecule because this molecular weight agrees well with that estimated by the gel-filtration of the native form of this species. Furthermore, the amounts of 57K-polypeptide were well assorciated with the levels of TdT activity, when the fractions from Sephadex G-100 column were analysed by the SDS-gel electrophoresis (Fig.1B). Therefore, TdT-57K may also consist of a single polypeptide.

Enzymatic properties of pig TdT-42K : Since the homogenous TdT-42K can be obtained in its fully active state, its enzymatic properties were studied extensively. Under the assay conditions employed here, dGTP was most efficiently used as a substrate by pig TdT-42K among four deoxynucleoside triphosphate. The utility decreased in order of dGTP, dCTP, dTTP and dATP. This substrate preference by pig TdT-42K resembles with that of calf TdT. As shown in Table 1, the activity of pig TdT-42K was inhibited by N-ethylmaleimide or ATP to the same extents as calf enzyme. Pig TdT-42K cross-reacted well with the antibody produced against calf TdT. Pig TdT-57K was also inhibited by the antibody to the same extents as TdT-42K. On the other hand, aphidicolin, a potent inhibitor of DNA-polymerase α,

Table 1. Enzymatic properties of pig TdT-42K and calf TdT

Conditions[a]	Pig TdT-42K	Calf TdT
	pmol dGMP incorporated	
Complete reaction mixture	1136 (100%)	1143 (100%)
Primer omitted	0 (0)	0 (0)
Primer omitted + heat-denatured activated calf thymus DNA	3535 (311)	7786 (681)
Complete + 6mM NEM	91 (8)	126 (11)
Complete + 0.2mM ATP	68 (6)	91 (8)
Complete + 1.5µg of anti-calf TdT antibody[b]	182 (16)	69 (6)
Complete + 20µM aphidicolin	1088 (96)	1120 (98)

a) Assays were performed as described in MATERIALS AND METHODS using 1.0 unit of either pig TdT-42K or calf TdT.
b) The method of preparation of anti-calf TdT antibody and properties of the antibody will appear elsewhere (13).

did not inhibit TdT-42K as well as calf TdT, in agreement with the
previous report (14). Pig TdT-42K sedimented at 3.6s through a
sucrose gradient (5-20%), while calf TdT at 3.4s. The apparent Km
for dGTP using oligo(dA)$_{12-18}$ is 20μM for pig TdT-42K and 28μM for
calf TdT, which are slightly lower than those reported previously
with TdT from calf thymus or malignant human cells (9,15). Thus we
could not find any marked differnces in their enzymatic properties
between pig TdT-42K and calf TdT as far as we examined.

DISCUSSION

     The purified pig TdT-42K seems to be nearly homogenous judged
from SDS-gel electrophoresis. This enzyme species consists of a
single polypeptide of 42,000 dalton, and has no subunit structure.
On the other hand, calf TdT, which has been purified in the similar
way, showed a subunit structure consisting of 33,000 and 13,000
dalton (Fig.2A). In spite of the marked difference between pig TdT-
42K and calf TdT in their peptide structures, their substrate preferences or other enzymatic properties well resemble each other. These
results indicate that the subunit structure is not essential for
the active form of TdT.
     In the present study, a large species of pig TdT, TdT-57K, has
been found. Pig TdT-57K sedimented at 4.2s through a sucrose
gradient, while TdT-42K, at 3.6s. The preliminary experiments with
TdT-57K fraction showed that TdT-57K cross-reacted with the anti-
calf TdT antibody to the same extents as TdT-42K. Further characterization of TdT-57K is now in progress. Higher molecular weight forms
of TdT (58,000-62,000 dalton) have been reported with human lymphoblastic cells (8,9) and in murine thymus and bone marrow (10).
Recently, Nakamura et al. demonstrated that there were multiple
species of bovine TdT, including the species of 60,000 and 42,000
dalton (Nakamura,H. personal communication). Thus high molecular
forms of TdT may commonly exist in mammalian thymuses as well as in
human lymphoblastoid cells.
     Two plausible hypotheses could be proposed to explain the
multiple forms of TdT, i.e., TdT-57K and TdT-42K in pig thymus. 1)
TdT-57K is a native enzyme which is active either in vivo or in
vitro. TdT-42K is derived from TdT-57K by a proteolytic degradation
during the purification. 2) Both forms of TdT exist in vivo. Though
they show the similar enzymatic properties in vitro, they function
differently in vivo. The second possibility may be supported by the
fact that TdT-42K was also obtained in the similar quantity when
the purification was performed in the presence of protease
inhibitors (0.1mM phenylmethylsulfonylfluoride, 0.2mg/ml ovomucoid
and 2mM benzamidine).

REFERENCES

1. F.J.Bollum, Oligodeoxyribonucleotideprimed reactions catalyzed
     by calf thymus polymerase, J.Biol.Chem. 237:1945 (1962).

2. L.M.S.Chang, Development of terminal deoxynucleotidyl transferase activity in embryonic calf thymus gland, Biochem.Biophys.Res. Commun. 44:124 (1971).

3. M.S.Coleman, J.J.Hutton, P.D.Simone, and F.J.Bollum, Terminal deoxyribonucleotidyl transferase in human leukemia, Proc.Nat. Acad.Sci.USA 71:4404 (1974).

4. R.D.Barr, P.S.Sarin, and S.M.Perry, Terminal transferase in human bone-marrow lymphocytes, Lancet 1:508 (1976).

5. D.Baltimore, Is terminal deoxynucleotidyl transferase a somatic mutagen in lymphocytes ?, Nature 248:409 (1974).

6. F.J.Bollum, "Karl August Forster Lecture," Vol.14, pp.1-47, Franz Steiner Verlag, Wiesbaden (1975).

7. L.M.S.Chang and F.J.Bollum, Deoxynucleotide-polymerizing enzymes of calf thymus gland, J.Biol.Chem. 246:909 (1971).

8. F.J.Bollum and M.Brown, A high molecular weight form of terminal deoxynucleotidyl transferase, Nature 278:191 (1979).

9. M.R.Deibel,Jr. and M.S.Coleman, Purification of a high molecular weight human terminal deoxynucleotidyl transferase, J.Biol. Chem. 254:8634 (1979).

10. A.Silverstone, L.Sun, O.N.Witte, and D.Baltimore, Biosynthesis of murine terminal deoxynucleotidyltransferase, J.Biol.Chem. 255:791 (1980).

11. R.C.Switzer,III, C.R.Merril, and S.Shifrin, A highly sensitive silver stain for detecting proteins and peptides in polyacrylamide gels, Anal.Biochem. 98:231 (1979).

12. U.K.Laemmli, Cleavage of structural proteins during the assembly of the head of bacteriophage T4, Nature 227:680 (1970).

13. T.Kaneda, H.Mishima, Y.Hirota, and S.Yoshida, Immunological comparison of terminal deoxynucleotidyl transferases from various mammals in terms of calf thymus enzyme, J.Appl.Biochem. in press.

14. R.A.Dicioccio, K.Chadha, and B.I.S.Srivastava, Inhibition of herpes simplex virus-induced DNA polymerase, cellular DNA polymerase α, and virus production by aphidicolin, Biochim. Biophys.Acta 609:224 (1980).

15. K.Kato, J.M.Goncalves, G.E.Houts, and F.J.Bollum, Deoxynucleotide-polymerizing enzymes of calf thymus gland, J.Biol.Chem. 242:2780 (1967).

PURIFICATION OF TERMINAL DEOXYNUCLEOTIDYL TRANSFERASE OF 60,000 DALTON FROM MAMMALIAN THYMUS AND THYMONA BY IMMUNOADSORBENT COLUMN AND COMPARISON OF PEPTIDE STRUCTURES

Hiromu Nakamura[1], Kazushi Tanabe[2], Shonen Yoshida[3], Mutsushi Matsuyama[4] and Toshiteru Morita[1]

Departments of Experimental Radiology[1], Biochemistry[2], and Ultrastructure Research[4], Aichi Cancer Center Research Institute, Chikusa-ku, Nagoya 464, Japan, and Department of Biochemistry[3], Institute for Developmental Research Aichi Prefecture Colony, Kasugai 480-03, Aichi, Japan

INTRODUCTION

There are many discrepancies in the molecular weight of terminal deoxynucleotidyl transferase (TdT) reported, e.g. 32- and 79-kilodalton(k) forms[1-7], and also in the enzyme structure, i.e. two-subunit form (the α and β subunits)[1-3] or single polypeptide form[4-7]. However, it has not been established yet whether these differences reflect species- or tissue-specificities of the enzyme, or are due simply to the proteolytic degradation during purification process. In order to clarify this issue, we have developed a new, rapid purification method, immunoadsorbent column chromatography, which is suitable to purify TdT's from limited amounts of tissues such as rodent thymus[8]. Furthermore, the enzyme proteins from various species of mammals were compared by two-dimensional peptide mapping.

EXPERIMENTAL PROCEDURES

Prepurification of TdT

Normal thymuses were obtained from calves, rats (Wistar King A and Buffalo/Mna) and mice (C57BL/6J). Thymomas developed spontaneously in old Buffalo/Mna rats and malignant thymic lymphomas induced by irradiation of x-ray in C57BL/6J mice were also used. TdT was completely extracted from 10-100 g of the frozen tissues and partialy purified by a phosphocellulose column[8]. This fraction was used as enzyme source for subsequent immunoadsorbent column chromatography.

19

Immunoadsorbent Column Chromatography

The procedures were described in detail elsewhere[8]. A rabbit antibody (IgG fraction) against a homogeneous calf thymus TdT was rendered highly specific by elution from a TdT-conjugated column, and was immobilized on Sepharose 4B to prepare an immunoadsorbent column[8]. The column specifically adsorbed all mammalian TdT's tested. After extensive washings, nearly homogeneous TdT's were recovered from the column by a pH gradient elution (pH 4.0-2.9) at extremely high yields (ca 70% of enzyme activity and ca. 95% of enzyme protein)[8]. TdT activity was measured as previously described[9]

Gel Electrophoresis and Tryptic Peptide Mapping

Sodium dodecyl sulfate(SDS)-polyacrylamide slab gel electrophoresis was carried out as described previously[8]. Gels were stained with Coomassie blue. For peptide mapping, individual stained bands were cut out from the SDS-gel, radioiodinated by the chloramine-T method and digested with trypsin. The digested peptides were resolved by electrophoresis in the first dimension and by ascending chromatography in the second dimension as previously described[10].

RESULTS AND DISCUSSION

Molecular Weight of TdT

When the extraction and phosphocellulose steps were done in the absence of protease inhibitors and then TdT's were purified by the immunoadsorbent column chromatography, calf thymus TdT showed well-known two-subunit form, the α (10-k) and β (32-k) subunits, as shown in the 2nd lane from the left of Fig. 1. When the protease inhibitors (1mM EDTA, 2mM PMSF, 5mM benzamidine and 0.5mg/ml of ovomucoid) were used during the prepurification steps, calf thymus TdT showed several new bands with higher molecular weights while the two bands corresponding to the α and β subunits were barely detectable (Fig. 1, 3rd lane). The molecular weight of the major three bands were estimated to be 60-k, 57-k, and 42-k[8]. On the other hand, rat enzymes purified from normal thymuses and thymomas by the similar way were mainly comprised of a single polypeptide with 60-k (Fig. 1, 4th to 6th lanes). Mouse preparations from normal thymuses and malignant thymic lymphomas showed two bands with 60-k and 42-k (Fig. 1, 7th and 8th lanes). In all preparations (Fig. 1), we could not detect any trace of the 79-k polypeptide reported previously[7].

Tryptic Peptide Mapping Analyses

In order to determine possible relationships among several proteins observed upon the SDS-gels, polypeptides were analyzed by the tryptic peptide mapping. As shown in Fig. 2, upper panels A and B,

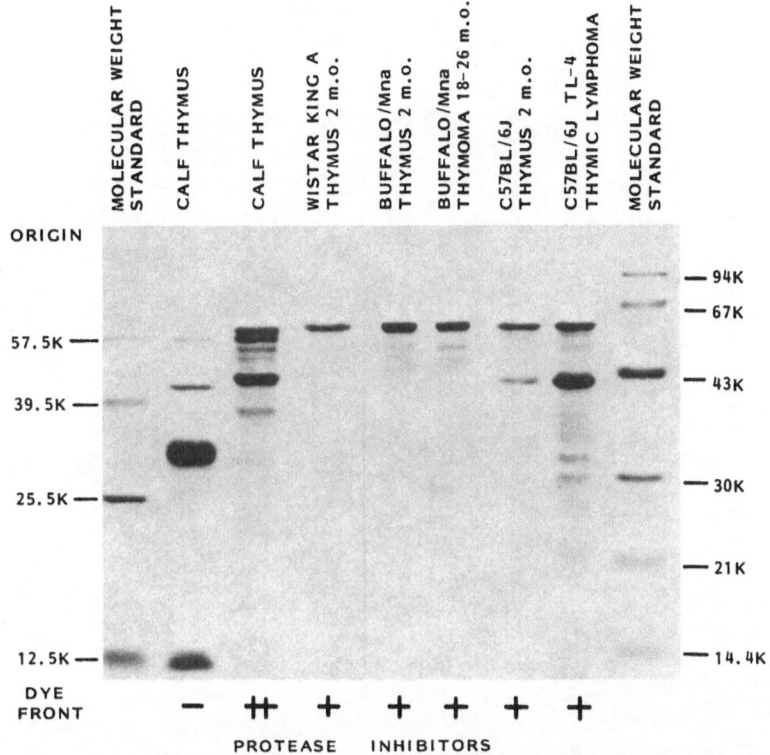

Fig.1. SDS-polyacrylamide gel electrophoretic patterns of TdT's
       purified from various tissues by the immunoadsorbent column
       chromatography.

the mapping pattern of the β subunit (32-k) was entirely different
from that of the α subunit (10-k). In contrast, the pattern of the
42-k polypeptide (Fig. 2, upper panel C) consisted of the two groups
of spots; one corresponds to those from the α subunit and another
from the β subunit. This finding was further confirmed by the mixing
experiment (Fig. 2, upper panel D. The representative spots derived
from the α subunit are indicated by arrows). The maps of both 60-k
and 57-k polypeptides obtained using high concentrations of protease
inhibitors were found to be homologous to that of the 42-k polypeptide (Fig. 2, lower panel A) except one spot indicated by an arrow
(Fig. 2, lower panel B)[8].

    The tryptic peptide maps of the 60-k components obtained from
three mammalian species revealed similar patterns; at least six spots
indicated by arrows on the rodent enzymes (Fig. 2; lower panels C and
D) were common to those of calf enzyme, while a few spots may be
specific for each species. By the peptide mapping, the other minor

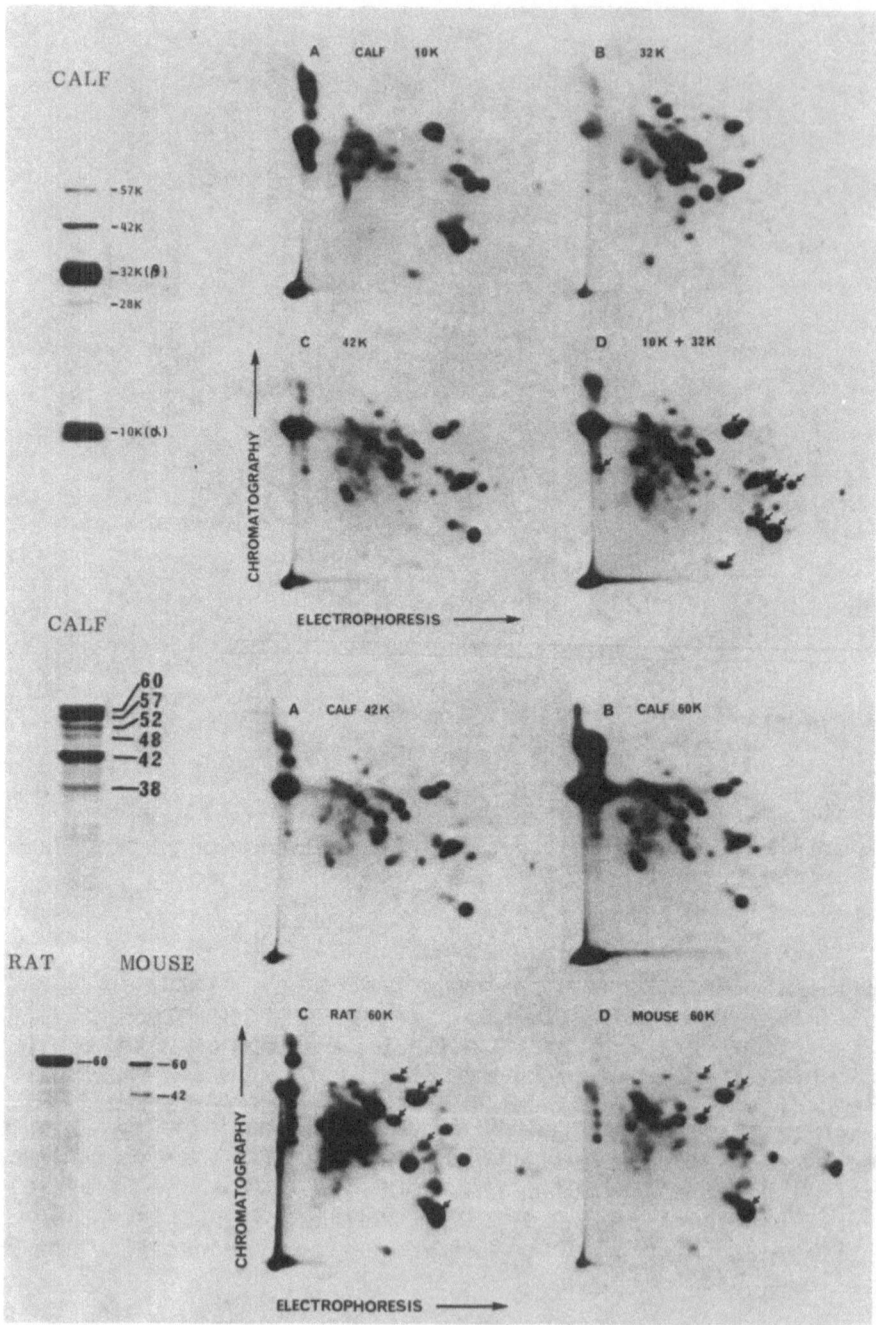

Fig.2. Tryptic peptide maps of TdT components. The polypeptides upon
        the SDS-polyacrylamide gel (left) were radioiodinated and
        analyzed by the tryptic peptide mapping (right panels)[8].

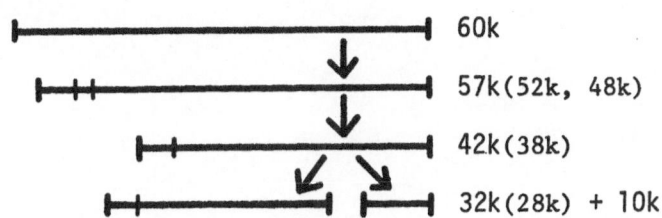

Fig.3. Possible process of TdT degradations. Minor components are
    indicated in the parentheses.

components (52-k, 48-k and 38-k) were also proved to be single polypeptide forms of TdT and the 28-k polypeptide was partially degraded
form of the β subunit[8].

    These results suggest that a single polypeptide with 60-k must
be the original form of mammalian TdT, and the α and β subunits of
the enzymes reported previously[1-3] might be proteolytic products
derived from the 60-k form, via the intermediate form of 42-k during
the purification process as shown in Fig. 3. Furthermore, the tryptic peptide maps suggest an extensive homology in amino acid sequence
among the TdT's from three species of mammals.

Acknowledgments

    We wish to thank Drs. W. Nakamura, T. Takahashi, and C. Sato
for their encouragements and for providing the mouse tissues.

REFERENCES

1. L.M.S. Chang and F.J. Bollum, J.Biol.Chem. 246: 909 (1971).
2. P.C. Kung, P.D. Gottlieb, and D. Baltimore, J.Biol.Chem. 251:
   2399 (1976).
3. F.A. Siddiqui and B.I.S. Srivastava, Biochim.Biophys.Acta 517:
   150 (1978).
4. F.J. Bollum and M. Brown, Nature 278: 191 (1979).
5. M.R. Deibel, Jr. and M.S. Coleman, J.Biol.Chem. 254: 8634 (1979).
6. A. Silverstone, L. Sun, O.N. Witte, and D. Baltimore, J.Biol.
   Chem. 255: 791 (1980)
7. D. Johnson and A.R. Morgan, Biochem.Biophys.Res.Commun. 72: 840
   (1976).
8. H. Nakamura, K. Tanabe, S. Yoshida, and T. Morita, J.Biol.Chem.
   256: in press (1981).
9. H. Nakamura, T. Morita, and S. Yoshida, Int.J.Radiat.Res. 38:
   499 (1980).
10. K. Tanabe, M. Yamaguchi, A. Matsukage, and T. Takahashi, J.Biol.
    Chem. 256: 3098 (1981).

STRUCTURE OF CALF THYMUS TERMINAL DEOXYNUCLEOTIDYL TRANSFERASE

Lucy M. S. Chang and F. J. Bollum

Uniformed Services University
of the Health Sciences
4301 Jones Bridge Road, Bethesda, MD   20814

INTRODUCTION

Terminal deoxynucleotidyl transferase (TdT), originally purified from calf thymus glands to homogeneity, was shown to have a
native molecular weight of 32,000 by equilibrium sedimentation
and two peptides of 26,500-dalton and 8,000-dalton when analyzed
by gel electrophoresis in the presence of sodium dodecyl sulfate
(SDS) (1). Rabbit antibody prepared against this homogeneous TdT
preparation (2) precipitates a single 58,000-dalton peptide from
cell extracts of human and mouse leukemia lines labeled with radioactive amino acids (3). A high molecular weight enzyme (62,000-
dalton peptide) was also purified from human leukemic lymphoblasts
(4).

When immunoreactive peptides of TdT in crude extracts were
analyzed directly on electrophoretic blots of SDS gels using the
procedure described by Towbin, Staehlin and Gordon (5), we found
a 58- to 60,000-dalton peptide to be present in all TdT containing
extracts including calf thymus, rat thymus, mouse thymus, chicken
thymus, cat thymus, human 8402 cells and mouse P1798 cells (6).
These results suggest that the homogeneous calf thymus TdT preparation is a proteolytically degraded form of the 58,000-dalton
peptide, and that TdT peptides from human, bovine, rat, chicken,
cat and mouse are immunologically related and have similar conserved polypeptide structure.

We describe in this communication a procedure to isolate high
molecular weight TdT peptides from calf thymus glands and show the
conversion of the high molecular TdT peptides to 26,000-dalton and
10,000-dalton peptides by degradation with trypsin.

PROCEDURES

Assay for TdT   TdT enzyme activity was assayed using d(pA)$_{50}$ as
initiator and ^{3}H-dGTP as substrate as previously described (7).
One enzyme unit is defined as the amount of enzyme required to
polymerize one nmole of ^{3}H-dGTP into acid-insoluble material in
one hr. Immunoreactive peptides of TdT were detected using the
immunoblot procedure described by Towbin, Staehlin and Gordon (5).
The reagents and the precise procedure used were as previously
described (6), except that rabbit antiserum to calf thymus TdT at
1 to 100 dilution in 0.1% bovine serum albumin in Tris-buffered
saline was used as the primary antiserum.

Purification of High Molecular Weight TdT from Calf Thymus Glands
High molecular TdT was purified from 2 kg of frozen calf thymus
glands. In addition to monitoring enzyme activity, enrichment of
the high molecular weight TdT peptides during purification was
monitored by the immunoblot procedure. All purification steps
were carried out at 4°.

Crude Extract: Calf thymus glands was homogenized with a
Waring Blender in 8 liters of 0.04 M potassium phosphate buffer at
pH 7.4 with 0.04 M NaCl, 1 mM phenylmethylsulfonyl fluoride and 1%
dimethylsulfoxide. The homogenate was clarified by centrifugation
for 10 min. at 5,000 rpm in the HG-4 rotor in the Sorvall RC-3
centrifuge. The supernatant was filtered through 4 layers of
cheese cloth providing 8.3 liters of crude extract.

Protamine Sulfate Precipitation of Nucleic Acid: The nucleic
acid in the crude extract was precipitated by addition of 410 ml
of 3% protamine sulfate solution to the extract with mixing,
followed by removal of the precipitate by centrifugation for 15 min.
at 5,000 rpm in the HG-4 rotor in the Sorvall RC-3 centrifuge. The
supernatant solution was filtered through 4 layers of cheese cloth
providing 7.7 liters of clear protamine sulfate supernatant.

Ammonium Sulfate Fractionation: Solid ammonium sulfate was
added to the protamine sulfate supernatant to 30% saturation, and
the solution was allowed to stand for 1 hr at 4°. The precipitate
formed was removed by mixing in 75 gm of Hyflo-Supercel (diatom-
aceous earth) per liter and then filtration through a pad of Hylfo
on S & S #410 filter paper on a Buchner funnel using gentle suction.
Solid ammonium sulfate was added to the filtrate to 55% saturation
and mixing was continued for 30 min. The solution was allowed to
stand for 2 hrs at 4°. The precipitate formed was collected by
centrifugation for 15 min. at 8,500 rpm in a GSA rotor in the Sorvall RC-5 centrifuge, and redissolved in 1 M ammonium sulfate in
0.2 M potassim phosphate at pH 7.4 and 10 mM 2-mercaptoethanol.
The redissolved ammonium sulfate fraction was clarified by centrifugation at 8,500 rpm in a GSA rotor yielding 900 ml of clarified

30-55% saturated ammonium sulfate fraction.

Isoleucine Sepharose Column: The clarified ammonium sulfate fraction was divided into two halves, and each half was loaded directly onto a 2.6 x 30 cm Isoleucine Sepharose 4B column previously equilibrated with 1 M ammonium sulfate in 0.2 M potassium phosphate at pH 7.4 and 10 mM 2-mercaptoethanol. Each column was washed with 500 ml of column buffer and protein was eluted from the column with 0.4 M ammonium sulfate in 0.2 M potassium phosphate at pH 7.4 and 10 mM 2-mercaptoethanol. The active fractions eluting from the Isoleucine Sepharose columns were pooled, producing 306 ml of Isoleucine Sepharose fraction.

Phosphocellulose Chromatography: The Isoleucine Sepharose fraction was dialyzed overnight against 2 changes of 0.075 M potassium phosphate (pH 7.2) in 10 mM 2-mercaptoethanol and 10% glycerol, and then loaded onto a phosphocellulose P-11 (Whatman) column (2.5 x 25 cm) previously equilibrated with the same buffer. The column was washed with 300 ml of the column buffer, and a one liter linear NaCl gradient from 0 to 0.5 M in the column buffer was applied. TdT activity eluted from the column as one major peak at about 0.3 M NaCl. The active fractions were pooled (340 ml) and protein in this fraction was salted out by addition of solid ammonium sulfate to 60% saturation. The protein precipitate was collected by centrifugation and redissolved in 10 ml of 0.5 M NaCl in 25 mM potassium phosphate at 7.4, 10 mM 2-mercaptoethanol containing 10% glycerol.

Gel Filtration on Sepharose 6B: The ammonium sulfate concentrated phosphocellulose fraction was fractionated on a 2.6 x 90 cm Sepharose 6B column in 0.5 M NaCl, 10 mM 2-mercaptoethanol, 25 mM potassium phosphate at pH 7.4 and 10% glycerol. The fractions containing TdT activity were pooled, and the volume was 68 ml. Protein was precipitated by addition of solid ammonium sulfate to 60% saturation, and the precipitates were redissolved in 45 ml of 50 mM Hepes buffer at pH 7.5.

Dithionitrobenzoate:Thiopropyl Sepharose Column Connected to Phosphocelluse Column: A Thiopropyl Sepharose 6B column (1 x 6 cm) was first exchanged with 5 mg per ml of 5,5'dithiobis(2-nitrobenzoic acid) (DTNB) in 50 mM sodium phosphate at pH 7.4 and 1 mM EDTA, then equilibrated with 0.2 M ammonium sulfate in 50 mM Hepes buffer at pH 7.5. The concentrated Sepharose 6B pool was loaded onto the DTNB:Thiopropyl Sepharose column. The column was washed with 50 ml of 0.2 M ammonium sulfate in 50 mM Hepes buffer at pH 7.5, followed by 50 ml of 50 mM Hepes buffer at pH 7.5. The outlet of the column was connected to a 1 x 15 cm phosphocellulose P-11 column which has been previously equilibrated with 50 mM Hepes buffer at pH 7.5. Protein linked to the Thiopropyl Sepharose column was eluted by slow desorption (3 min. per ml) with 250 ml of 25 mM dithiothreitol in 50 mM Hepes buffer and the affluent passed

directly onto the connected phosphocellulose column. After elution
of the Thiopropyl Sepharose column was complete, the two columns
were disconnected, and the phosphocellulose column was washed with
50 ml 0.05 M NaCl in 10 mM 2-mercaptoethanol, 50 mM Hepes buffer
at pH 7.5 and 10% glycerol. The phosphocellulose column was then
eluted with a 200 ml linear NaCl gradient from 50 mM to 0.5 M in
50 mM Hepes buffer at pH 7.5, 10 mM 2-mercaptoethanol and 10%
glycerol. TdT activity eluted from the phosphocellulose column as
a broad peak at 0.25 M NaCl. The pooled active fractions had a
volume of 60 ml.

Chromatography on Hydroxylapatite: After 0.6 ml of 1 M potassium phosphate at pH 7.4 was added to the phosphocellulose pool,
it was loaded directly onto a 1 x 3 cm hydroxylapatite column (9)
previously equilibrated with 10 mM potassium phosphate in 0.25 M
NaCl, 10 mM 2-mercaptoethanol and 10% glycerol. The column was
washed with 30 ml of the column buffer, then eluted with a 200 ml
linear potassium phosphate gradient from 10 mM to 75 mM in 0.25 M
NaCl, 10 mM 2-mercaptoethanol and 10% glycerol. TdT activity
eluted from the hydroxylapatite column as a broad heterogeneous
peak (Fig. 1). The active fractions were pooled (60 ml) and dialyzed against 50 mM potassium phosphate at pH 7.5, 10 mM 2-mercap-
toethanol containing 50% glycerol and stored at -20°.

RESULTS

Purification of High Molecular Weight TdT from Calf Thymus
Gland: During purification of the high molecular weight TdT from
calf thymus glands, both enzyme activity and the presence of high
molecular weight immunoreactive peptides of TdT were monitored.
When enzyme pools were made, the fractions having TdT activity but
enriched in low molecular weight (e.g., 26,000-daltons or lower)
TdT peptides were discarded. A summary of the purification scheme
is shown in Table I.

Three steps used in the purification scheme are devised to
remove low molecular weight TdT from the enzyme fraction. Previous
experiments showed that (1) purified low molecular weight TdT cannot be precipitated from the solution at 55% saturation of ammonium
sulfate, (2) the low molecular weight enzyme does not bind to
Isoleucine Sepharose, and (3) it does not bind to Thiopropyl Sepharose. These observations suggest that the peptides removed from
the native enzyme due to proteolysis are hydrophobic and are enriched in thiol groups. Even when the Isoleucine Sepharose and the
Thiopropyl Sepharose columns are included in the purification
scheme, the TdT obtained still fractionates as a heterogeneous
peak on hydroxylapatite (Fig. 1). When fractions from the hydroxylapatite column were analyzed on polyacrylamide gel in the
presence of SDS, the Coomassie Blue stained SDS gel showed that
the TdT activity eluted from the hydroxylapatite column behind a

Table 1. Purification of High Molecular Weight TdT from
Calf Thymus Glands

Enzyme Fraction	Total Protein (mg)	Sp. Ac. (nmol per hr)	Total Enzyme Units
I. Crude Extract	140,000	38	5.38×10^6
II. Protamine Sulfate Supernatant	46,000	93	4.27×10^6
III. 30-55% $(NH_4)_2SO_4$	19,000	164	3.12×10^6
IV. Isoleucine Sepharose	5,400	354	1.91×10^6
V. Phosphocellulose	367	4,930	1.81×10^6
VI. Sepharose 6B	90	16,700	1.52×10^6
VII. DTNB-Thiopropyl Sepharose: Phosphocellulose	39.3	22,700	0.89×10^6
VIII. Hydroxylapatite	9.04	98,500	0.89×10^6

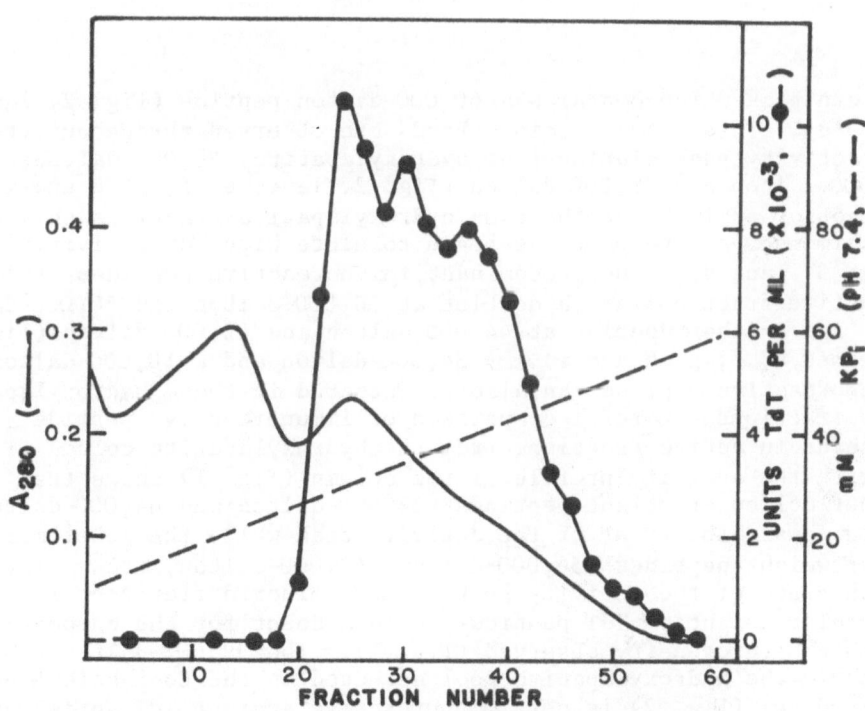

Figure 1. Chromatography of Calf Thymus TdT on Hydroxylapatite. For detail, see Procedures.

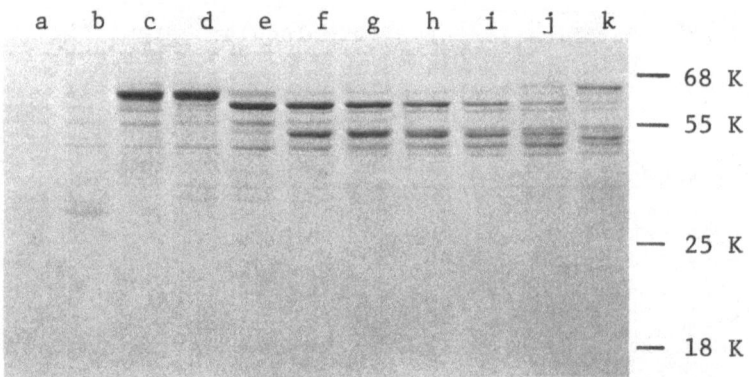

Figure 2.   Analysis of Peptides in Hydroxylapatite Fractions
            by SDS Gel Electrophoresis. Aliquots of hydroxyl-
            apatite column fractions (Fig. 1) were separated
            on a 15% polyacrylamide gel in the presence of SDS
            (8). Separated peptides were visualized by stain-
            ing with Coomassie Blue. Lanes a, b, c, d, e, f,
            g, h, i, j and k correspond to hydroxylapatite
            column fractions 5, 10, 15, 20, 24, 28, 30, 32,
            36, 40 and 44, respectively.

protein peak which contains a 60,000-dalton peptide (Fig. 2, lanes
c and d). Three major stained bands are observed throughout the
TdT activity peak eluting from hydroxylapatite, 58,000-dalton,
44,000-dalton and 39,000-dalton (Fig. 2, lanes e, f, g, h and i).
Immunoblot analysis of the same hydroxylapatide fractions showed
that immunoreactive peptides begin to elute with TdT activity
(Fig. 3, lane d). The predominant immunoreactive peptides present
in active fractions are a doublet at 58,000-dalton and 56,000-dal-
ton, and another doublet at 44,000-dalton and 42,000-dalton (Fig.
3, lanes e, f, g, h and i). A 34,000-dalton and a 10,000-dalton
immunoreactive peptide can also be detected in these hydroxylapatite fractions. Careful comparison of immunoreactive peptide
patterns in active fractions from the hydroxylapatite column (Fig.
3) with the activity profile of the column (Fig. 1) shows that
higher molecular weight peptides (58,000-dalton and 44,000-dalton)
elute toward the front of the activity peak while the lower molecular weight peptides (56,000-dalton, 42,000-dalton, etc.) elute
in the back of the activity peak. These minor differences in
molecular weight of TdT peptides could account for the chromatographic heterogeneity observed (Fig. 1). The 39,000-dalton peptide
found in the hydroxylapatite pool detected on the Coomassie Blue
stained gel (Fig. 2) is not immunoreactive against TdT antiserum,
suggesting that it is a contaminating peptide in the preparation.

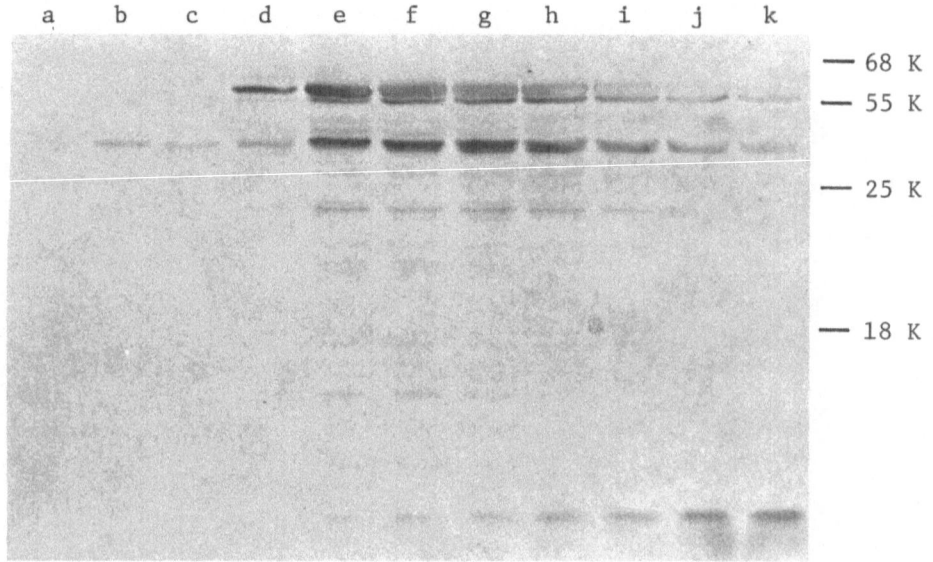

Figure 3. Immunoblot of TdT Peptides in Hydroxylapatite Column
           Fractions. A gel identical to that shown in Fig. 2
           was used for analysis of TdT peptides by the immuno-
           blot procedure. Immunoreactive peptides were visual-
           ized by immunoperoxidase staining.

     The purification procedure described starts with frozen calf
thymus glands. Low molecular weight immunoreactive peptides can
be detected in the crude extract. It is not known at the present
whether these low molecular weight forms of the enzyme have any
biological significance. It is possible that the lower molecular
weight forms arise during storage of the glands, although these
forms are also found in extracts made from fresh tissue. In any
case, continued proteolysis occurs during purification until the
enzyme has been eluted from the Isoleucine Sepharose column.
Since various molecular weight forms of TdT from calf thymus glands
are enzymatically active, the procedure described in this report
only produced an enzyme fraction enriched in the high molecular
weight (58,000-dalton and 44,000-dalton) TdT peptides. The enzyme
protein constitutes over 60% of the total protein present in the
final preparation.

     Generation of the α and β Peptides of TdT by Proteolysis:
When purified high molecular weight TdT was treated with 2 mg.per
ml of trypsin, a loss of 30% of the enzyme activity occurred after
2 min. of digestion. The activity that remains appears to be more
resistant to trypsin since about 50% of enzyme activity still remained after 10 min. of digestion. After 10 min. of digestion with
trypsin, the decline in TdT activity follows first order kinetics.

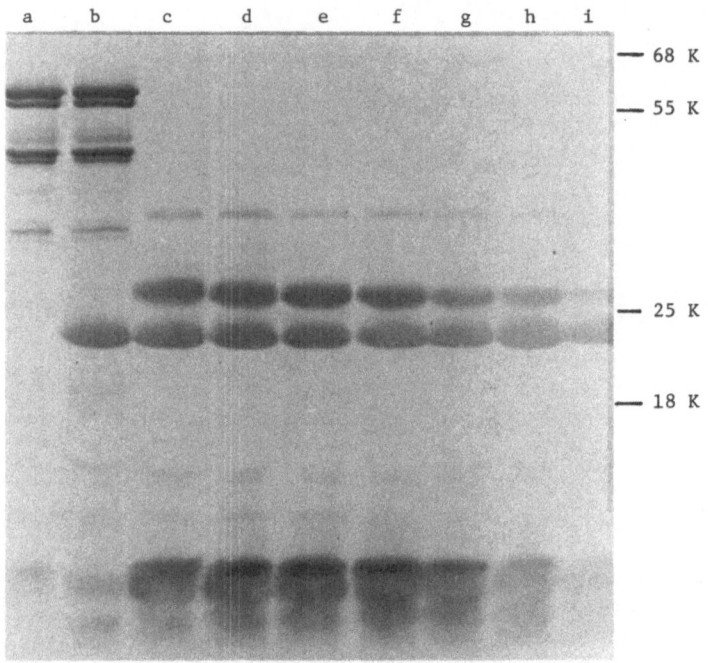

Figure 4.   Immunoblot Analysis of Calf Thymus TdT peptides After Trypsin Degradation.   Trypsin degradation of 120 μg of purified high molecular weight calf thymus TdT was carried out in a 0.5 ml reaction with
2 mg per ml trypsin in 50 mM Hepes buffer at pH 7.5.   A zero time sample (45 μl) was taken prior to addition of trypsin, and added to a
tube containing 5 μl of 20 mg per trypsin in 10mM HCl and 5 μl of 20
mg per trypsin in 10mM HCl and 5 μl of soybean trypsin inhibitor at 40
mg/ml in 50 mM Hepes buffer at pH 7.5 Degradation was started by addition of 45 μl of 20 mg per ml trypsin in 10 mM HCl to the TdT solution
kept in a 25° water bath.   At various times, 50 μl was removed from
the reaction and added to 5 μl of 40 mg per ml of soybean trypsin in-
hibtor in 50 mM Hepes buffer at pH 7.5 kpet in a tube in the ice bath.
After all samples were collected, 5 μl of each sample was used to assay for enzyme activity and the remainder of the sample was processed
for SDS gel electrophoresis and immunoblotting.   Lane a represents purified high molecular weight calf thymus TdT.   Lane b represents the
zero time sample.   Lanes c, d, e, f, g, h and i represent samples taken at 2 min., 4 min., 6 min., 10 min., 20 min., 40 min. and 60 min.
of the trypsin digestion.   The intensely stained band at 23,5000-dal-
ton is trypsin peptide and it appears to be detected by the immunoblot procedure.   The reason for the staining of trypsin is probably
due to the presence of anti-carbohydrate antibodies in the anti-calf
thymus TdT antiserum used.   When column-purified anti-TdT IgG was used
as the primary reagent in the immunoblot procedure, the trypsin peptide did not stain in the immunoperoxidase reaction.

Immunoblot analysis of trypsin-degraded calf thymus TdT peptides shows the loss of almost all of the high molecular weight immunoreactive peptide with the generation of the 26,000-dalton (α peptide) and the 10,000-dalton (β peptide) after 2 min. of digestion (Fig. 4, lane c). A 34,000-dalton peptide, not detected in the original enzyme fraction, appeared transiently during trypsin degradation. From the kinetics of the trypsin digestion (and considering the large amount of protease used in this experiment), it appears that the high molecular weight peptides are extremely sensitive to trypsin digestion, while the β-peptide is relatively insensitive (Fig. 4, lane d, e and f). Digestion of the high molecular weight TdT with 200-fold less trypsin resulted in little loss of enzyme activity, but gradual and complete conversion of the high molecular weight peptides to α and β peptides can be accomplished (data not shown). Prolonged degradation of TdT with high trypsin levels eventually results in loss of the α and β peptides (lanes g, h and i, Fig. 4).

The production of α and β peptides by trypsin suggests that generation of these peptides in thymus or during purification is caused by a serine protease. It is not clear why inclusion of phenylmethylsulfonyl fluoride in the preparation of the crude extract was not effective in inhibiting this protease activity.

DISCUSSION

The demonstration of the 58,000-dalton TdT peptide in calf thymus extract using the immunoblot technique with an antiserum raised against homogeneous low molecular weight calf thymus TdT (3) suggests that the low molecular weight form of the enzyme is derived from the 58,000-dalton form by proteolysis. Whether proteolytic degradation of TdT occurs in vivo is not known at this point. Proteolysis of calf thymus TdT in the extract does not appear to be random. Comparing the immunoreactive peptides in immunoblots of fractions in early stages of purification and in purified enzyme fractions, it is possible to order the appearance of various immunoreactive peptides. The first TdT peptide generated is the 56,000-dalton peptide, followed by the doublet at 44,000-dalton and 42,000-dalton. Most purified high molecular weight TdT preparations contain another doublet at 34,000-dalton and 32,000-dalton. Homogeneous TdT preparations frequently contain a minor 24,000-dalton peptide as well as the 26,000-dalton β peptide. The β peptide in extensively degraded TdT preparations is 24,000-dalton. The presence presence of doublets, with the smaller band of each doublet as a minor component, suggests the presence of a sensitive bond about 20 amino acids from one terminus of the 58,000-dalton form on a small fraction of TdT peptide in the extract. A possible scheme for proteolytic degradation of TdT in calf thymus extract is shown in Fig. 5.

The generation of 26,000-dalton β peptide and 8- to 10,000-dalton α peptide by trypsin suggests that the proteolysis occuring in calf thymus extract is predominantly produced by serine proteases.

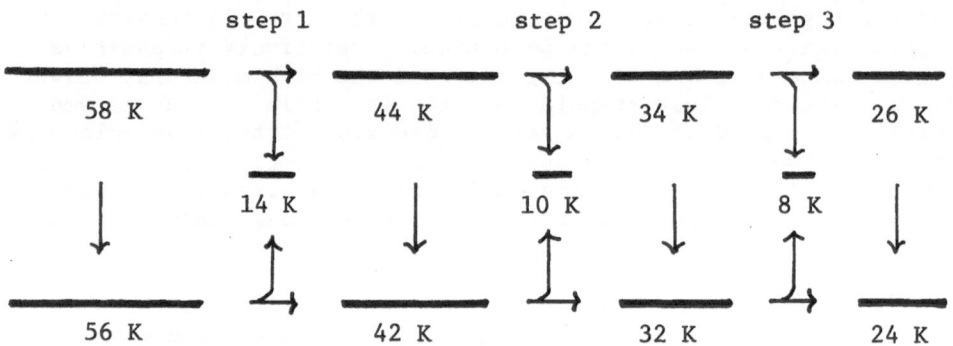

Figure 5.   Schemative Diagram of Proteolytic Degradation of Calf Thymus TdT Peptides.

The nature of the α peptide is not well defined since steps 2, 3 and 4 (Fig. 5) can potentially produce the α peptide. The α peptide of the homogeneous TdT is most likely composed of a mixture of peptides averaging about 10,000-dalton.

Experiments on autolysis (utilizing endogeneous proteases) of TdT peptides in the crude calf thymus extracts produced no useful results. No significant proteolysis could be demonstrated by self-digestion of these extracts, suggesting the presence of inhibitors. Reproducible proteolysis of TdT peptide can be demonstrated when the crude extracts were incubated in the presence of sodium dodecyl sulfate, but the resultant peptides have different molecular weights from those observed during purification. Early steps in the purification of TdT appear to remove these protease inhibitors and result in gradual degradation of the TdT peptide. The degradation of TdT peptides can be easily monitored with the immunoblot technique during initial stages of purification. Examination of other peptides present in the fractions by Coomassie Blue staining of polyacrylamide gels shows that proteolysis is not general since major peptides present in the calf thymus extract are not degraded. These observations suggest a rather specific proteolysis of TdT which may be related to the in vivo function of the enzyme.

REFERENCES

1. L. M. S. Chang and F. J. Bollum, J. Biol. Chem. 246:909-916
       (1971).
2. F. J. Bollum, Proc. Nat'l. Acad. Sci. (USA) 72:4119-4122 (1975).
3. F. J. Bollum and M. Brown, Nature 278:191-192 (1979).
4. M. Deibel and M. S. Coleman, J. Biol. Chem. 254:8634-8640
       (1979).
5. H. Towbin, T. Staehlin and J. Gordon, Proc. Nat'l Acad. Sci.
       (USA) 76:4350-4354 (1979).
6. F. J. Bollum, and L. M. S. Chang, J. Biol. Chem. (in press)
       (1981).
7. L. M. S. Chang, Biochem. Biophys. Res. Comm. 44:124-131 (1971).
8. U. K. Laemmli, and M. Favre, J. Mol. Biol. 80:575-599 (1973).

ACKNOWLEDGEMENT

    This investigation was supported in part by a grant, CA 23262,
from the National Cancer Institute, DHHS.

PURIFICATION AND CHARACTERIZATION OF MULTIPLE FORMS OF TERMINAL TRANSFERASE FROM HUMAN LEUKEMIC CELLS

Martin R. Deibel, Jr., Mary Sue Coleman and John J. Hutton

Department of Biochemistry, University of Kentucky, Lexington, Kentucky and Department of Medicine, University of Texas, San Antonio, Texas

Terminal deoxynucleotidyl transferase catalyzes the polymerization of deoxynucleoside triphosphates onto the 3'-OH terminus of a single stranded oligo($\geq$3 residues) or polydeoxynucleotide initiator. This non-template directed DNA polymerase was first detected in calf thymus tissue. Terminal transferase carries out, in the presence of divalent cation (Mg^{+2}), a linear condensation polymerization reaction which is dependent solely on the concentration of 3'-OH groups and deoxynucleotide monomer. As a biochemical reagent, terminal transferase has enjoyed widespread popularity in the synthesis of defined oligomers and polymers, in recombinant DNA technology and in sequence analysis of oligodeoxynucleotides (reviewed by Bollum, 1981).[1] The enzyme was purified to apparent homogeneity from calf thymus tissue by Chang and Bollum[2] and was found to have an average native molecular weight of 32,000 and subunit molecular weights in SDS of 24,000 (β) and 8000 (α).

Sustained terminal transferase activity is restricted to cortical cells of thymus[3-6] and lymphoid precursors in bone marrow.[7-9] This exclusive association of terminal transferase activity within subpopulations of marrow or thymus of healthy individuals led to the important clinical observation in 1973 of the existence of terminal transferase activity in peripheral blood of a patient with acute lymphoblastic leukemia.[10] Since that initial observation, many laboratories contributed significantly to the establishment of tests for terminal transferase as a valuable tool in the differential diagnosis of leukemia.

In our laboratory, access to clinical materials afforded us the opportunity to carry out detailed analysis of the biochemical properties of human terminal transferase. We chose for enzyme

purification studies, human lymphoblasts derived from therapeutic
leukapheresis. Terminal transferase activity in lymphoblasts ranges
from 25 to 1000 units/10^8 cells as compared with 10-25 units/10^8
cells in calf thymus tissue. From individual leukapheresis procedures we routinely obtained 100-1000g. of packed cells.

Our initial experiments were aimed at devising a purification
scheme which would be suitable for the human TdT from lymphoblasts.
Three classes of potential affinity ligands were analyzed; (1) a dye
(Cibacron Blue F3GA) known to bind to proteins possessing unique
nucleotide binding sites; (2) substrate affinity columns which
possessed immobilized ligands mimicking the monomer in the reaction,
deoxynucleoside 5'-triphosphate; (3) substrate affinity columns which
possessed immobilized ligands mimicking the initiator, a homo-polymer
of deoxyadenylic acid, or thymidylic acid.

The first of these reagents, Cibacron Blue F3GA, is a relatively
inexpensive dye which has found wide applicability in enzyme purification. The binding of human terminal transferase to Cibacron Blue
F3GA-Sepharose columns was considered because the dye "recognizes"
oligodeoxynucleotide sites on the enzyme. Kinetic experiments
demonstrated that the free dye is a competitive inhibitor of the
initiator, $p(dA)_{\overline{50}}$ (at fixed, saturating concentrations of dGTP)
(Figure 1A), but a non-competitive inhibitor of the deoxynucleoside
5'-triphosphate, dGTP (at fixed, saturating concentrations of the
initiator, $p(dA)_{\overline{50}}$) (Figure 1B). The immobilized form of the dye
proved to be highly efficient in binding large quantities of human
terminal transferase from crude tissue extracts.

A second class of affinity supports tested was immobilized
deoxynucleoside 5'-triphosphates. dATP was covalently attached to
Sepharose either through direct coupling or through a 6-carbon
spacer. Only the latter analog was shown to bind human terminal
transferase. Attempts to elute bound terminal transferase using
substrate were unsuccessful, and optimal utilization of the column
required elution by increased ionic strength. Nevertheless, substantial purification was afforded by this procedure. The utilization of Cibacron Blue F3GA-Sepharose and dATP-Sepharose columns in
the purification of human terminal transferase is documented in
Table 1. This protocol was effective in four large scale enzyme
purifications from human leukemic lymphoblasts. Specific activities
of greater than 100,000 units/mg were generally obtained using this
protocol.

The third type of affinity chromatography, representing
immobilization of the substrate initiator, was subsequently introduced into the procedure. Oligo(dT)cellulose (6-15 residues) was
extremely valuable in yielding homogeneous human enzyme. This
column procedure was included at the end of the purification scheme
since it not only was highly specific for human terminal transferase,

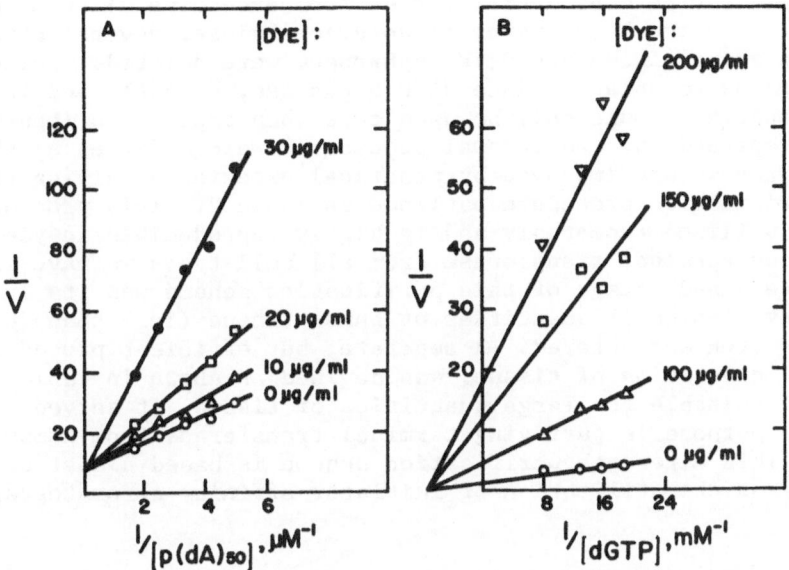

Figure 1.   Inhibition of human terminal transferase by Cibacron Blue F3GA (dye).   (A) Effect of varying p(dA)$_{50}$ concentration at fixed, saturating concentrations of dGTP on the reaction rate (B) Effect of varying dGTP concentration at fixed, saturating concentrations of p(dA)$_{50}$ on the reaction rate.

Table 1.   PURIFICATION OF TERMINAL TRANSFERASE FROM 165 GRAMS OF HUMAN BLASTS

	FRACTION	PROTEIN	SPECIFIC ACTIVITY	TOTAL ACTIVITY	YIELD
		mg	units/mg	units	%
I.	Crude Extract	31,625	16.7	1,055,950	100.0
II.	Cibacron Blue F3GA–Sepharose	3,200	117.1	749,555	71.0
III.	Hydroxylapatite–I	675	1,168	788,555	74.6
IV.	CM–Sephadex–I	187	2,743	514,350	48.7
V.	dATP–Sepharose	84	4,571	384,000	36.3
VI.	Hydroxylapatite–II	14.6	24,483	357,445	33.8
VII.	CM–Sephadex–II	3.7	121,725	450,380	42.6

but also possessed a very high enzyme binding capacity. This factor
allowed for an elution of purified enzyme in concentrated form,
thereby preventing substantial losses from excessive dilution of the
protein. As a result of using oligo(dT)cellulose, several chromatographic steps, including dATP-Sepharose, were deleted. This was
considered as advantage, since dATP-Sepharose, a costly and labile
column material, could only be used once when exposed to impure
enzyme preparations. Additional procedures were modified as the
overall process was improved by practical experience, giving the
final purification procedure outlined in Table 2. This procedure
has been utilized repeatedly and is highly reproducible in yielding
homogeneous terminal transferase from all cell types we have examined. One disadvantage of this purification scheme was its in-
efficiency with small quantities of human tissue (less than 50 g)
which we often encountered. A separate, but efficient procedure
for small quantities of tissue, was devised as shown in Table 3.
While not suitable for large quantities of tissue, it served a
necessary purpose in purifying terminal transferase from human
thymus. This alternate purification scheme is based almost exclusively on the utilization of initiator affinity chromatography.

Table 2. PURIFICATION OF TERMINAL TRANSFERASE FROM 132 GRAMS OF
          HUMAN BLASTS

	FRACTION	PROTEIN mg	SPECIFIC ACTIVITY units/mg	TOTAL ACTIVITY units	YIELD %
I.	Crude Extract *	9,000	38.9	350,525	100
II.	Phosphocellulose	510	433	220,850	63
III.	Hydroxylapatite-I	67.5	1,015	68,515	19.5
IV.	Cibacron Blue F3GA-Sepharose	28	3,950	110,590	31.5
V.	Hydroxylapatite-II	16.8	7,728	129,830	37.0
VI.	Oligo(dT)cellulose	0.8	131,760	101,270	28.9

*Initial stages of the purification were conducted in the presence
of 1 mM PMSF

Table 3. PURIFICATION OF TERMINAL TRANSFERASE FROM 15 GRAMS OF
          HUMAN THYMUS

	FRACTION	PROTEIN mg	SPECIFIC ACTIVITY units/mg	TOTAL ACTIVITY units	YIELD %
I.	Crude Extract	161.2	55.8	9,000	100
II.	Oligo(dT)cellulose	1.75	3,900	6,825	75
III.	Oligo(dA)cellulose*	0.24	29,315	7,035	78
IV.	Hydroxylapatite	0.11	55,180	6,070	67

*Oligo(dA)cellulose was prepared in our laboratory by polymerization
of 50 to 100 residues of deoxyadenylic acid onto commercially
available oligo(dT)cellulose using terminal transferase.

All of the enzyme preparations which were undertaken by the
three procedures outlined above, were shown to conform to our criteria for terminal transferase homogeneity: (a) the presence of a
single polypeptide on SDS polyacrylamide gels or the presence of
the previously reported two subunit composition observed for the
calf thymus enzyme;[2] (b) a single band on non-denaturing isoelectric
focusing gels; (c) a specific activity that corresponds to protein
preparations that were originally[2] defined as homogeneous, and (d)
a single species of terminal transferase activity on non-denaturing
polyacrylamide gels.

Substantial variation in the molecular structure of terminal
transferase was apparent on SDS polyacrylamide gels (Figure 2).
The following species were observed: (a) a protein with a subunit
molecular weight of 62,000 was isolated from two patients with
acute lymphoblastic leukemia and from MOLT 4 cells[11]; (b) a protein
with a subunit molecular weight of 42,500 (average molecular weight)
was isolated from two patients with acute lymphoblastic leukemia;
(c) a protein that migrated as a single polypeptide with M_r = 42,500
was isolated from two patients with chronic myelogenous leukemia in
blast crisis; (d) a protein comprised of two non-identical subunits
of 27,000 and 10,000, respectively, was isolated from two additional
patients with chronic myelogenous leukemia in blast crisis. The
subunit structure of (d) is characteristic of the homogeneous enzymes
purified from human and calf thymus. By isolating the purified
enzyme from human thymus, we demonstrate in SDS gels that the
molecular weight differences we have previously seen between terminal
transferases from human lymphoblasts and calf thymus may be tissue
related[12].

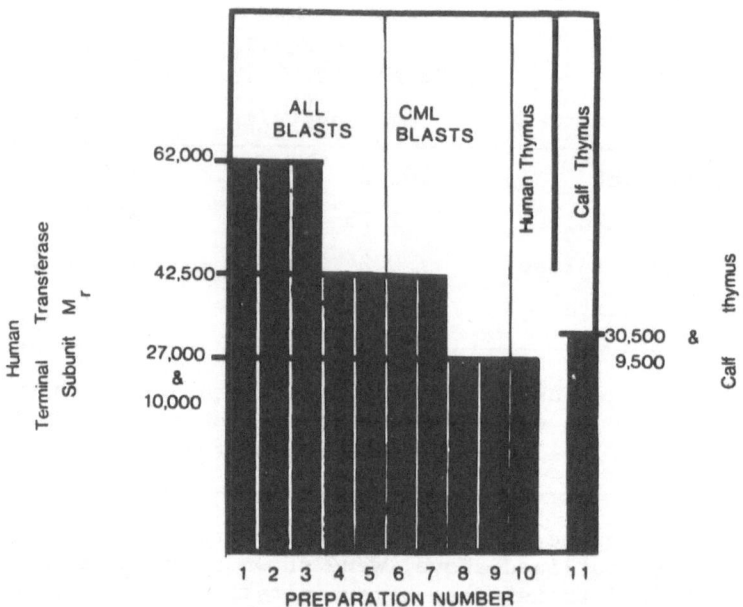

Figure 2. Subunit molecular weights of human terminal transferase
purified to homogeneity from several sources. Preparations 1,2,
and 3 were characterized as acute lymphoblastic leukemia, non-B
and non-T marked; preparation 3 was a MOLT 4 clone grown in our
laboratory, which was not T-marked at the time of enzyme purification; preparations 4 and 5 were not typed for surface markers; ALL
blasts were obtained from patients with adult acute lymphoblastic
leukemia; preparations 6,7,8 and 9 were CML blasts isolated from
patients with chronic myelogenous leukemia in blast crisis, whose
cells were found to contain terminal transferase; of these, preparations 6 and 9 were not typed for surface markers; preparation 7
cells formed fewer than 1% E-rosettes; preparation 8 contained 92%
blasts, 22% of which formed E-rosettes (this group of cells were
classified as T-marked). Preparation 7 cells were extracted and
terminal transferase was initially purified in the presence of
1mM PMSF.

     The subunit structures of the enzymes from thymus of calf and
human origin are comparable. Each has two non-identical subunits
of molecular weight 27,000 and 10,000 (human) and 30,500 and 9,500
(calf). An example of each type of subunit structure on SDS gels
is summarized in Figures 3,4,5 and 6.

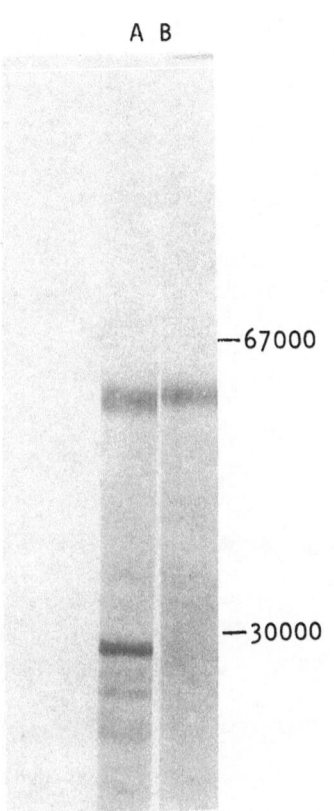

Figure 3. SDS gel electrophoresis of human terminal transferase from blasts of some patients with acute lymphoblastic leukemia; (A) Molecular weight markers, bovine serum albumin (M_r = 67000); carbonic anhydrase (M_r = 30000) (B) Human terminal transferase.

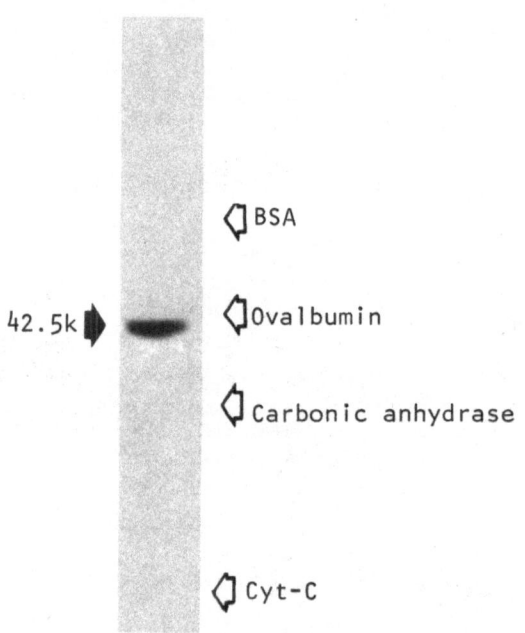

Figure 4.   SDS gel electrophoresis of human terminal transferase
from blasts of some patients with acute lymphoblastic leukemia
and chronic myelogenous leukemia in blast crisis.

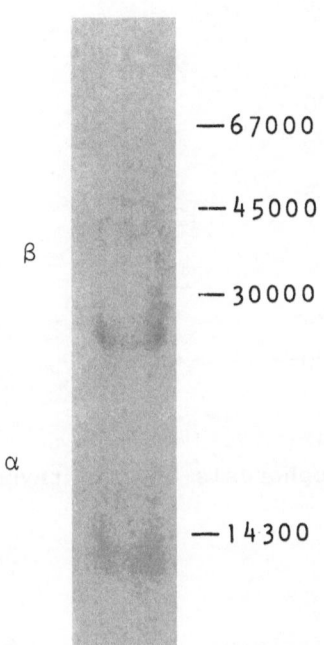

Figure 5. SDS gel electrophoresis of human terminal transferase from blasts of some patients with chronic myelogenous leukemia in blast crisis and from all samples of normal human thymus.

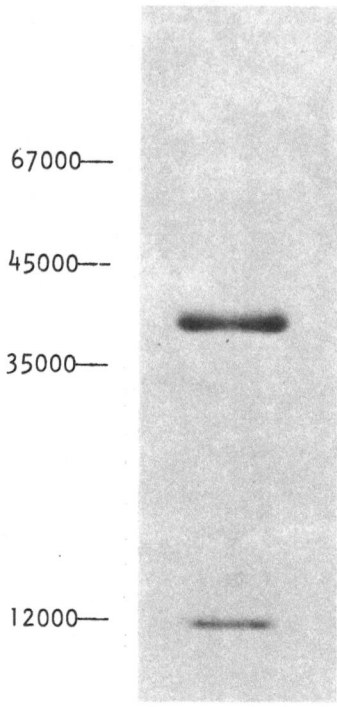

Figure 6. SDS gel electrophoresis of calf thymus terminal transferase.

        Because the SDS gel patterns of the M_r = 62,000 form of the
human terminal transferase were so different from the traditional
two subunit structure characteristic of calf thymus terminal transferase[2], properties of the native form of the enzyme from human
blasts were investigated. The molecular weight of human terminal
transferase (SDS subunit molecular weight of 62,000) was estimated
by non-denaturing polyacrylamide gel electrophoresis utilizing the
procedure of Hedrick and Smith[13]. By this method, the migration
of human terminal transferase as a function of acrylamide concentration can be utilized in conjunction with the corresponding migration
of standard proteins in estimating native molecular weight. A
typical example, ascertained by using the quantitative enzyme assay
for terminal transferase in gel slices, is shown in Figure 7A. By
construction of a standard curve against other proteins of known
molecular weight, a value for terminal transferase of M_r = 60,000
was determined (Figure 7B). An independent determination was made
by gel filtration on Sephadex G-100 and on Biogel A 0.5m (Figure 8).
By these two methods, molecular weights of 60,000 and 67,000 were
obtained respectively. Homogeneity of all preparations of human

terminal transferase was ascertained by the presence of a single protein band coincident with terminal transferase activity on non-denaturing isoelectric focusing gels in polyacrylamide. A typical example is shown in Figure 9. All terminal transferase preparations examined exhibited a pI of 8.2 ± 0.3, regardless of the tissue source or purification protocol.

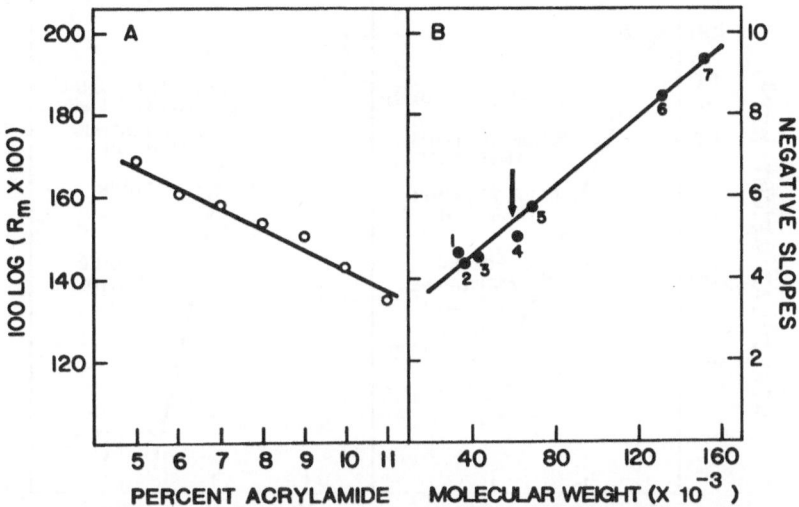

Figure 7. Molecular weight estimation of native human terminal transferase isolated from some patients with acute lymphoblastic leukemia. (A) Dependence of the migration of terminal transferase activity on non-denaturing polyacrylamide gels as a function of acrylamide concentration. The location of terminal transferase activity on gels was determined by using the quantitative enzyme assays in gel slices. (B) Estimation of native molecular weight of human terminal transferase using the method of Hedrick and Smith[13] on non-denaturing polyacrylamide gels. Standards included carbonic anhydrase (1), carboxypeptidase A (2), ovalbumin (3), inorganic pyrophosphatase (4), bovine serum albumin (5), bovine serum albumin dimer (6), and lactate dehydrogenase (7). Each protein was subjected to electrophoresis at several gel concentrations between 5 and 12% acrylamide, and R_m values were determined. The negative slopes of each standard protein as well as terminal transferase were calculated by linear regression analysis of the line generated by plotting the following variables: 100 log (R_m x 100) versus the acrylamide concentration. Subsequently, a standard curve of negative slopes versus molecular weight for the marker proteins was calculated by linear regression. For terminal transferase activity, a negative slope of 5.31 (see A) was obtained, which corresponds to a M_r = 60,000 (see arrow in B).

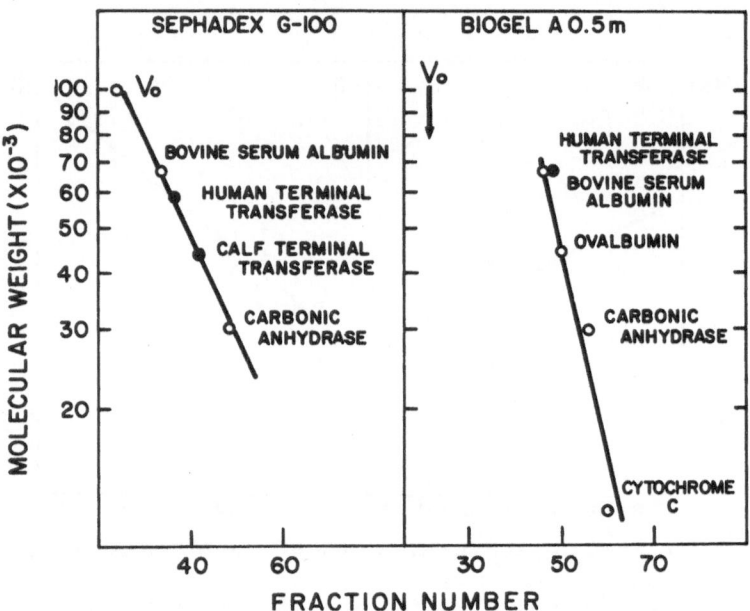

Figure 8. Estimation of the native molecular weight of human
terminal transferase by chromatography on Sephadex G-100 (A), and
on Biogel A 0.5m (B).

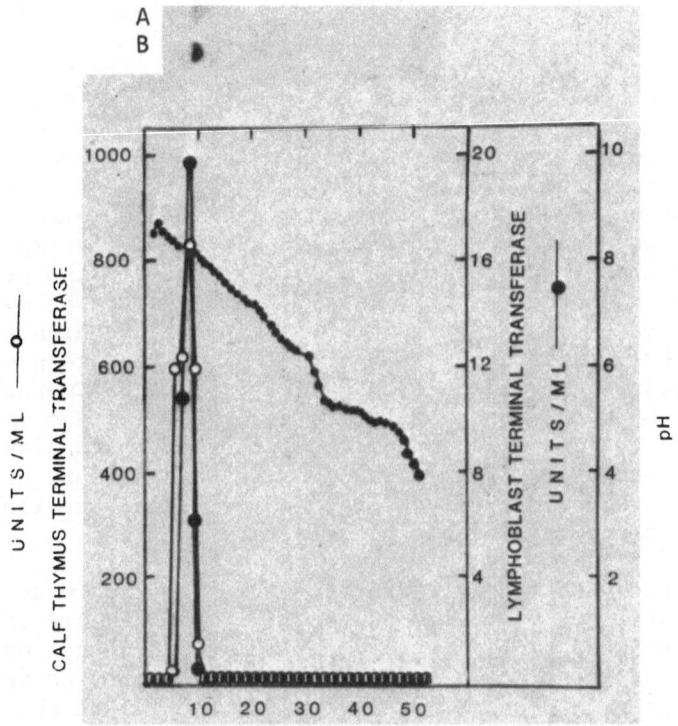

Figure 9. Typical analytical non-denaturing isoelectric focusing polyacrylamide gel electrophoresis profiles of human (A) and calf thymus (B) terminal transferase. The activities corresponding to each protein band are shown in the accompanying figure.

Because of the variation in molecular weights of terminal transferase from human tissues, it was expected that the kinetic properties would differ as well. Unexpectedly, no differences in kinetic properties for any species of human terminal transferase could be discovered. A series of experiments aimed at determining the physical properties of the enzyme were undertaken to ascertain whether the extra protein in the human enzyme (large molecular weight form) had a catalytic function independent of the function generally attributable to the calf thymus enzyme which was studied in great detail by Bollum[2]. In the process, the mechanism of the reaction catalyzed by human terminal transferase was determined, and many physical characteristics attributed to the human enzyme were elucidated.

There are three substrates for human terminal transferase to be considered: (1) the monomer, dATP; (2) the initiator, $p(dA)_{\overline{n}}$ (where $\overline{n}$ is an oligomer of polydeoxyadenylate with an average chain length of n); (3) the metal ion involved in catalysis, existing primarily in a complex with the two substrates, dATP and $p(dA)_{\overline{n}}$. Because of the complexity of analyzing all possible substrates in the reaction catalyzed by terminal transferase, all kinetic determinations were conducted exclusively with nucleotides of adenine. Thus, dATP, but not dGTP polymerization was examined, since dGTP polymerization was designed specifically for relatively impure enzyme preparations (aggregation of poly(dG) protects the product polymer against exonuclease activities). This feature of dGTP polymerization makes it undesirable for assays of the highly purified enzyme. Polymerization with Mg^{2+}:dATP is linear to greater than 80% of substrate utilization[14]. The simplest parameter to elucidate was the metal ion specificity for the enzymatic reaction. Optimization of the metal ion under conditions in which both monomer and initiator were at saturation was considered necessary in order to avoid complex kinetic interactions, and to reduce the kinetic determination to one involving a bi-substrate reaction.

Two initiators of widely different chain lengths, $p(dA)_6$ and $p(dA)_{\overline{50}}$, were utilized with dATP as monomer. The results of this analysis are depicted in Figure 10. Maximum rates of polymerization for both long and short chain initiators were achieved as follows: $Mg^{2+} > Zn^{2+} > Co^{2+} >> Mn^+$. The rather large differential in the optimum concentrations and maximum activities of the two most commonly utilized metal ions, Mg^{2+} and Mn^{2+}, suggested that the use of Mg^{2+} as divalent cation in the human terminal transferase reaction system was warranted. Furthermore, these results cast some doubt on the applicability of Mn^{2+} in the overall analysis of human terminal transferase kinetics.

The kinetic mechanism of purified human terminal transferase was examined using two substrates, dATP as monomer, and $p(dA)_n$ as initiator (n from 4 through 14, and 50). In order to analyze the reaction mechanism in the simplest manner, a long chain initiator $(p(dA)_{\overline{50}})$ was utilized initially for the following reasons: (a) the K_m and V_{max} for the polymerization of $p(dA)_{\overline{50}}$ and its immediate products (n>50) are unchanged during the time of the enzymatic reaction. Therefore, the determination of average rates of polymerization are actually indicative of initial velocities for this substrate; and (b) oligodeoxyadenylates of chain length n>50 are readily precipitated by acid on glass fiber papers, facilitating rapid and reliable kinetic analyses.

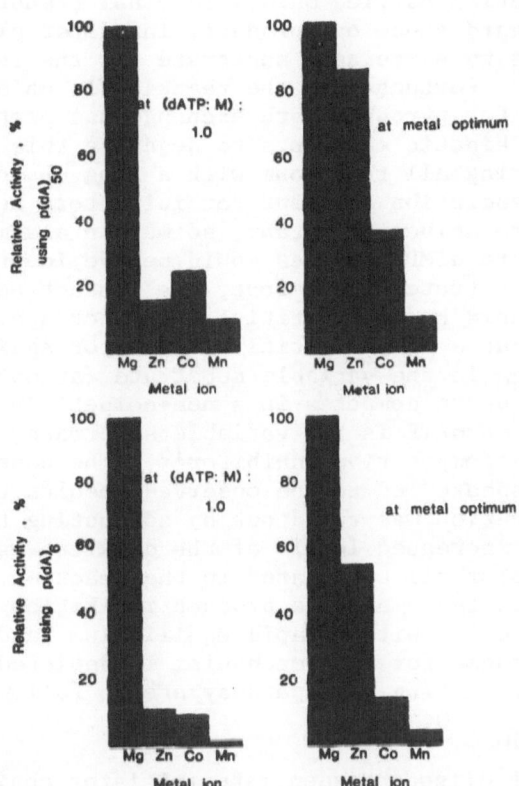

Figure 10. Metal ion specificity of human terminal transferase
using two different chain length polydeoxyadenylate initiators,
$p(dA)_6$ and $p(dA)_{50}$. Left, effect of maintaining dATP: metal ratio
of 1.0; Right, effect on catalysis at the respective optimal concentration of each metal ion.

We confirmed that the reaction mechanism for human terminal transferase was random sequential by initial velocity experiments in which both substrates were varied independently at sub-saturating concentrations[15]. The family of lines generated intersect at a common abscissa, indicating that the addition of one substrate does not facilitate the addition of the second substrate. Further elucidation of the mechanism required product inhibition studies. The enzymatic reaction carried out by terminal transferase is unique in this regard since one product, initiator plus one dAMP residue, continues to serve as a substrate for the reaction in the forward direction. Fortunately, the reaction is unidirectional since no evidence for pyrophosphate exchange has been observed[14]. The complexity of kinetic constants to describe this phenomenon was reduced by performing all reactions with a long chain initiator, $p(dA)_{\overline{50}}$. The dissociation constant for initiators having 50-75 residues remains relatively constant, so we can assume that initiator plus one or more dAMP residues would behave identical kinetically to the original substrate. Therefore, the product as $p(dA)_{\overline{50}}$ plus n dAMP residues would be a competitive inhibitor against the substrate, $p(dA)_{\overline{50}}$, but a non-competitive inhibitor against the monomer, dATP. When $p(dA)_{\overline{50}}$ is the variable substrate (at saturating dATP), inorganic pyrophosphate competes in a non-competitive fashion. On the other hand, when dATP is the variable substrate, inorganic pyrophosphate is a competitive inhibitor[15]. The possibility that inorganic pyrophosphate led to the observed results through chelation of divalent cation was ruled out by conducting the same experiments using increased levels of the divalent cation, $MgCl_2$, sufficient to complex all substrates in the reaction. Thus, the initial velocity patterns and the product inhibition patterns are qualitatively in accord with a rapid equilibrium random mechanism (Table 4). The scheme for this mechanism is depicted in Figure 11. The constants representing each pathway are as follows: K_{dATP} = 110 µM and $K_{p(dA)_{\overline{50}}}$ = 0.4 µM.

The effect of oligodeoxyadenylate initiator chain length on terminal transferase activity was determined under conditions in which the monomer dATP was present at saturating concentrations. The resulting dissociation constants and maximum velocities of the enzyme as a function of initiator chain length are shown in Figure 12. The K_m and V_{max} values for oligodeoxyadenylates in the range n = 6 through 14, and 50, indicate that affinity of the enzyme for the growing chain and the corresponding rate of polymerization are directly proportional. The apparent deviation in substrate utilization as compared with the calf thymus enzyme[15] occurs at the tetranucleotide level, in which the human enzyme binding constant differs by more than 2 orders of magnitude over the K_m for initiators which have more than 5 nucleotide residues. The possible significance of this discrepancy in the utilization of short chain initiators by the human and calf thymus terminal transferases is not known.

Table 4. KINETICS OF HUMAN TERMINAL TRANSFERASE

I.   INITIAL VELOCITY STUDIES
      dATP versus $p(dA)_{\overline{50}}$ Non-Competitive

                Intersection on the abscissa

          The two substrates are bound at independent sites

II.   PRODUCT INHIBITION STUDIES

	Varied S_O	Product	Pattern
1.	dATP	PP_i	Competitive
2.	$p(dA)_{\overline{50}}$	PP_i	Non-Competitive
3.	dATP	$p(dA)_{\overline{51}}$	Non-Competitive*
4.	$p(dA)_{\overline{50}}$	$p(dA)_{\overline{51}}$	Competitive*

*Expected patterns of inhibition
Dissociation constants and binding sites for $p(dA)_{\overline{50}}$ to $p(dA)_{\overline{75}}$
are statistically identical

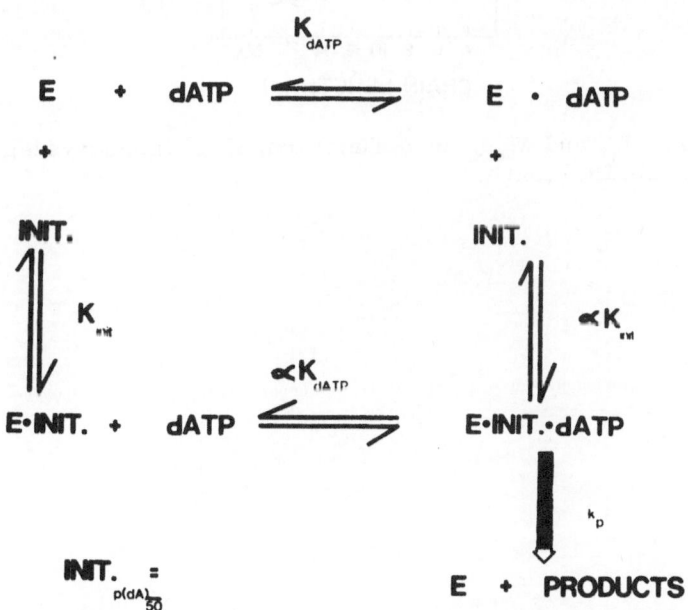

Figure 11. Kinetic mechanism for human terminal transferase as
determined by initial velocity and product inhibition experiments;
$K_{dATP} = 110\mu M$; $K_{p(dA)_{\overline{50}}} = 0.4\mu M$ and $\alpha = 1.0$.

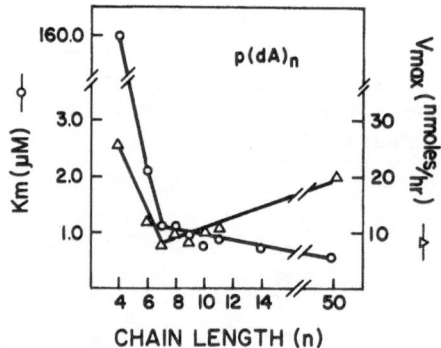

Figure 12. K_m and V_{max} as a function of oligodeoxyadenylate initiator chain length.

The effect of other adenine nucleotides on the binding of dATP to human terminal transferase was also examined. For these experiments, the initiator, $p(dA)_{\overline{50}}$, was maintained at saturation, while dATP and the potential inhibitors were varied at sub-saturation. ATP is a competitive inhibitor of dATP addition to $p(dA)_{\overline{50}}$ by purified human terminal transferase. Apparent K_i values of 0.3 mM were obtained for this inhibition, which is comparable or less efficient than the binding of dATP itself. This seems to rule out the possibility that ATP exerts a special regulatory effect on terminal transferase activity, when utilizing our assay conditions. Hydrolysis products of ATP, ADP and AMP, as well as 3'5'-cyclic AMP possessed no inhibitory action of their own on the enzymatic reaction using dATP as substrate. 2'3'-Dideoxyadenosine 5'-triphosphate (ddATP), a known inhibitor of most DNA polymerases, was also tested as an inhibitor of human terminal transferase. It was shown to be a competitive inhibitor of dATP polymerization, with an apparent $K_i = 0.062$ mM (Figure 13). Since this nucleotide is a chain terminator by virtue of lacking a 3'-OH group, this result is not at all surprising. 1,N[6]-etheno-dATP (ε-dATP) and dITP were examined as inhibitors of the reaction. Both compounds are competitive inhibitors of dATP polymerization, with K_i(app) of 5 μM and 350 μM for the ethenodATP and dITP, respectively. The reason for the potent inhibition of terminal transferase activity by etheno-dATP (Figure 14) is not known, but could be due to rigid alignment of the adenine moiety in the proper binding orientation.

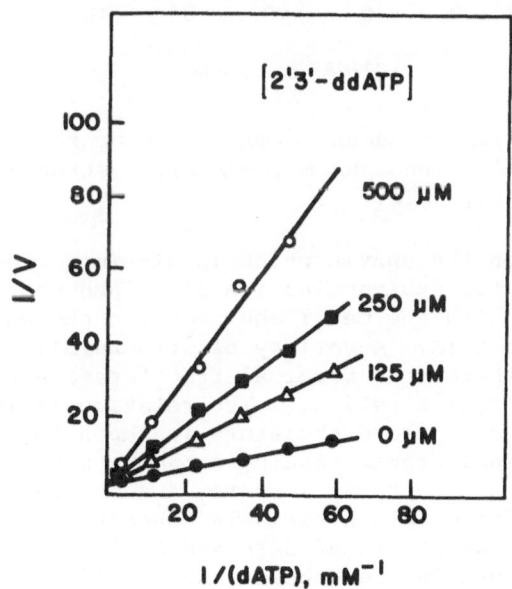

Figure 13.   Inhibition of human terminal transferase by the chain terminator, 2',3'-dideoxy ATP.

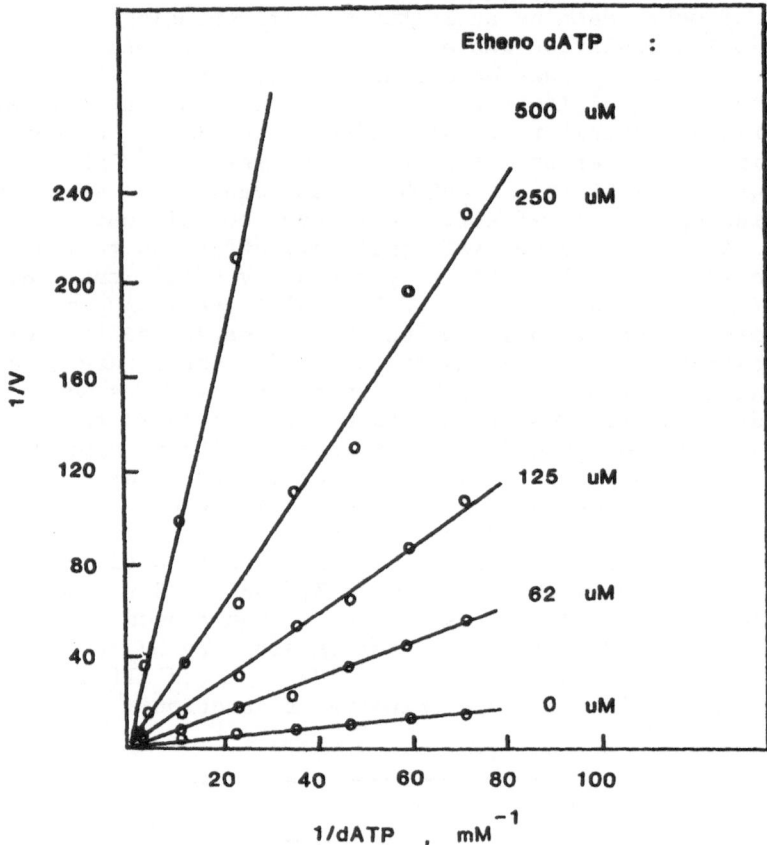

Figure 14.   Inhibition of human terminal transferase by $1,N^6$-etheno-
2'-deoxyadenosine 5'-triphosphate (ε-dATP), a structural analog of
the natural substrate, dATP.

A metal site on the enzyme which is distinct from the divalent
cation requirement for deoxynucleoside 5'-triphosphates, is implicated from studies with the metal chelators, o-phenanthroline,
α,α'-bipyridyl, and EDTA. A variety of structurally different metal
chelators were inhibitors of terminal transferase activity at milli-
molar to nanomolar concentrations. The relative inhibition of
terminal transferase by these chelators is shown in Figure 15.   The
inhibition of terminal transferase by orthophenanthroline, but not
by the meta- or para-structural isomers, and by α,α'-bipyridyl was
shown to be competitive versus variable concentrations of $p(dA)_{\overline{50}}$
(at saturating concentrations of dATP and Mg^{2+}), but noncompetitive
versus dATP (at saturating concentrations of $p(dA)_{\overline{50}}$).   The effect
of variable concentrations of o-phenanthroline on terminal transferase activity was also examined at saturating concentrations of

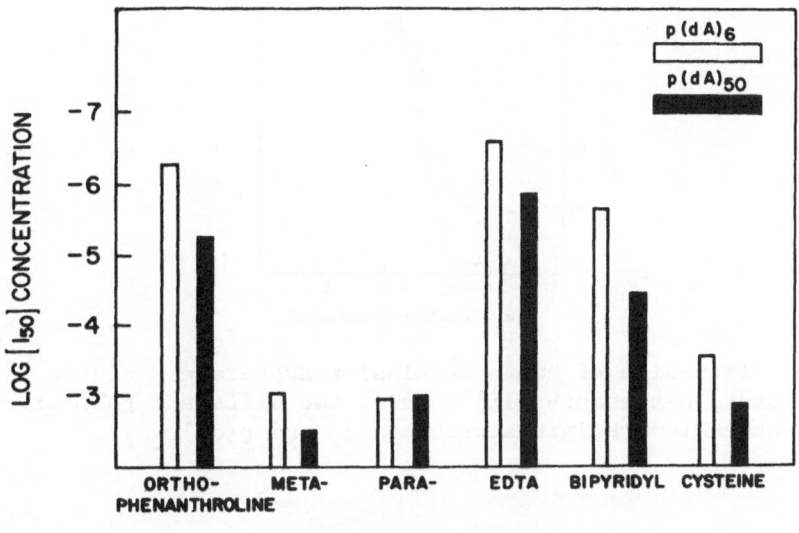

Figure 15.   Inhibition of human terminal transferase by several
metal chelators, using two different chain length polydeoxyadenylate
initiators, p(dA)$_6$ and p(dA)$_{50}$.   The I$_{50}$ value represents the
metal chelator concentration at which 50% of the human terminal
transferase activity is inhibited.

all substrates, utilizing either p(dA)$_{\overline{50}}$ or p(dA)$_6$ as initiators.
The specific titration is documented in Figure 16.   Because of the
excess Mg^{2+} included in each determination (3 mM), removal of
divalent cation from complexation with the substrates was not
anticipated.

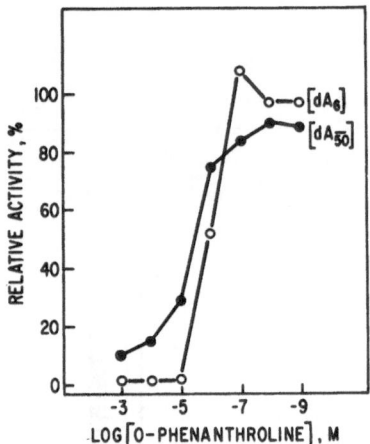

Figure 16. Titration of human terminal transferase activity by the metal chelator, o-phenanthroline, using two different polydeoxyadenylate chain length initiators, p(dA)$_6$ and p(dA)$\overline{_{50}}$.

Our purpose in preparing homogeneous terminal transferase from human lymphoblasts was to determine the structure of the human enzyme in the two major types of leukemia in which it is commonly found (ALL and CML-BC). Isolation of the purified proteins allowed us to demonstrate that, regardless of the source or molecular weight, the protein is catalytically active and appears to be homogeneous in structure. A determination of the kinetic mechanism was made using human TdT, and we obtained as a consequence a model system for the comparable, yet far more complex eucaryotic DNA polymerases. The reaction catalyzed by human TdT follows a rapid equilibrium random mechanism as determined by initial velocity and product inhibition studies with oligodeoxyadenylate initiators. A metal site (possible Zn) on the enzyme, which is distinct from the divalent cation requirement for deoxynucleoside 5'-triphosphates was implicated from studies with several metal chelators. The putative metal site was localized in the vicinity of the initiator binding site by kinetic analyses. The identification of this metal site awaits further spectroscopic determinations

The existence of three distinct molecular weight forms of TdT has led us to speculate that terminal transferase may be synthesized as a large molecule and cleaved to smaller forms as hemopoietic cells differentiate. While it is tempting to postulate a physiological processing event for this enzyme, we recognize that an investigation of the origin of multiple forms of TdT should include artifactual degradation and multiple isozymes coded by different genes for TdT.

REFERENCES

1. F. J. Bollum, Terminal transferase: experienced biochemical reagent seeks biochemical assignment, Trends in Biochemical Sciences 6:41 (1981).

2. L. M. S. Chang, and F. J. Bollum, Deoxynucleotidyl polymerizing enzymes of calf thymus gland. V. Homogeneous terminal deoxynucleotidyl transferase, J. Biol. Chem. 246: 909 (1971).

3. M. S. Coleman, J. J. Hutton, and F. J. Bollum, Terminal deoxynucleotidyl transferase and DNA polymerase in classes of cells from rat thymus, Biochem. Biophys. Res. Commun. 58: 1104 (1974).

4. I. Goldschneider, K. E. Gregoire, R. W. Barton, and F. J. Bollum, Demonstration of terminal deoxynucleotidyl transferase in thymocytes by immunofluorescence, Proc. Natl. Acad. Sci. U.S.A. 74:734 (1977).

5. G. Janossy, J. A. Thomas, F. J. Bollum, S. Granger, G. Pizzolo, K. F. Bradstock, L. Wong, A. McMichael, K. Ganeshaguru, and A. V. Hoffbrand, The human thymic microenvironment: an immunohistologic study, J. Immunol. 125:202 (1980).

6. L. M. S. Chang, Development of terminal deoxynucleotidyl transferase activity in embryonic calf thymus gland, Biochem. Biophys. Res. Commun. 44:124 (1971).

7. M. S. Coleman, J. J. Hutton, P. DeSimone, and F. J. Bollum, Terminal deoxyribonucleotidyl transferase in human leukemia, Proc. Natl. Acad. Sci. U.S.A. 71:4404 (1974).

8. K. E. Gregoire, I. Goldschneider, R. W. Barton, and F. J. Bollum, Intracellular distribution of terminal deoxynucleotidyl transferase in rat bone marrow and thymus, Proc. Natl. Acad. Sci. U.S.A. 74:3993 (1977).

9. K. E. Gregoire, I. Goldschneider, R. W. Barton, and F. J. Bollum, Ontogeny of terminal deoxynucleotidyl transferase-positive cells in lymphohemopoietic tissues of rat and mouse, J. Immunol. 123:1347 (1979).

10. R. McCaffrey, D. F. Smoler, and D. Baltimore, Terminal deoxy-
     nucleotidyl transferase in a case of childhood acute
     lymphoblastic leukemia, Proc. Natl. Acad. Sci. U.S.A. 70:
     521 (1973).

11. M. R. Deibel, Jr., and M. S. Coleman, Purification of a high
     molecular weight human terminal deoxynucleotidyl transferase,
     J. Biol. Chem. 254:8634 (1979).

12. M. R. Deibel, Jr., M. S. Coleman, K. Acree, and J. J. Hutton,
     Biochemical and Immunological properties of human terminal
     deoxynucleotidyl transferase purified from blasts of acute
     lymphoblastic and chronic myelogenous leukemia, J. Clin.
     Invest. 67:725 (1981).

13. J. L. Hedrick, and A. J. Smith, Size and charge isomer separa-
     tion and estimation of molecular weights of proteins by
     disc gel electrophoresis, Arch. Biochem. Biophys. 126:155
     (1968).

14. K. Kato, J. M. Goncalves, G. E. Houts, and F. J. Bollum,
     Deoxynucleotide polymerizing enzymes of calf thymus gland.
     II. Properties of the terminal deoxynucleotidyl transferase,
     J. Biol. Chem. 242:2780 (1967).

15. M. R. Deibel, Jr., and M. S. Coleman, Biochemical properties
     of purified human terminal deoxynucleotidyl transferase,
     J. Biol. Chem. 255:4206 (1980).

PURIFICATION AND PROPERTIES OF CHICK TERMINAL

DEOXYNUCLEOTIDYL TRANSFERASE (TdT)

Claude Pénit, Maria-José Gelabert, Catherine Transy,
and Pierre Rouget

Institut de Recherche en Biologie Moléculaire
C.N.R.S. et Université Paris 7
2, place Jussieu - 75251 PARIS CEDEX 05 - FRANCE

INTRODUCTION

The thymus is made up of two anatomical parts : the stroma,
containing the epithelium, and the thymocytes which belong to the
lymphoid series of the hematopoietic cells. The embryonic origins of
these two populations are different : the thymocytes derive from
stem cells which colonize the thymic rudiment (which is initially
purely epithelial). This facts were suggested by Moore and Owen (1)
and the timing of the lymphoid colonization of the embryonic thymus
was established by Le Douarin and Jotereau (2) using a technique
based on the production of interspecific chimeras. Such chimeras
are obtained by grafting the thymic rudiments of chick embryos in
the somatopleure of quail embryos (or vice-versa). Quail cells are
easily distinguished from chick cells using the Feulgen-Rossenbeck
staining of chromatin : the quail chromatin is condensed in a large
mass in the center of the nucleus whereas the chick chromatin is
finely dispersed (3). By an appropriate choice of the respective
ages of the grafted rudiments and of the host embryos, it could be
shown that the epithelial thymus is colonized by stem cells during
two receptive periods : from 6 1/2 days to 8 days and from 12 to
13 days.

TdT activity can be detected in the thymus of the chick embryo
as of the 12th day of incubation. This activity increases until 3
weeks after hatching more rapidly than does the weight of the thymus.
Characteristic modifications of the phosphocellulose profile of TdT
are observable during this period (4). To determine whether these
modifications were due either to a synchronous evolution of TdT
activity in the majority of thymocytes or to a progressive increase
of the percentage of TdT-positive cells, it was necessary to adopt

61

a detection assay of TdT at the single-cell level. For this purpose,
mg amounts of TdT were purified from thymuses of 3-month-old chicks
using a method derived from that of Chang et al. (5) with several
modifications described here. An antibody against chicken TdT was
produced in rabbits, and purified by affinity chromatography on
TdT-sepharose. This antibody was used for the detection of TdT by
immunofluorescence and by the immunoperoxydase method (6). This has
allowed us to confirm the observations of Sugimoto et al. (7), made
with an antibody against calf thymus TdT : a very low percentage of
TdT-positive cells appears in the thymus of chick embryos between
11 and 12 days. This percentage increases almost linearly until
hatching ; when the plateau is reached, 65 to 75 % of the thymocytes
are TdT-positive.

MATERIALS AND METHODS

1. TdT assay. - TdT activity is measured after at least partial
purification of the enzyme by phosphocellulose chromatography (4).
The incubation medium (100 µl) contains : 50 mM Tris HCl pH 7.9,
2 mM DTT, 0.1mM $MnCl_2$, 5 µg/ml $d(pA)_{10-3}$ (Collaborative Research),
0.02 to 0.5 mM dGTP and 1 or 2 µCi of ^{3}H dGTP (Amersham). The
incubation is performed at 37° C during 30 min. In the case of very
diluted TdT fractions, 250 µg/ml BSA is added.

        This assay is similar to that of McCaffrey et al. (8).

2. Purification of chicken TdT. - Chick thymuses were dissected without
thawing from frozen material obtained in Sopravit chick slaughter-
house ; 3 to 5 g of thymus were obtained per chick.

1) Thymus extract : 2 kg of thymus were divided in 500 g fractions.
To each fraction were added 2 l of TEM buffer (50 mM Tris HCl pH 7.9,
0.1 mM EDTA, 1.4 mM 2-mercaptoethanol) containing 0.4 mM phenyl-methyl
sulfonyl-fluorid (PMSF), 750 ml of 3 M KCl, and 250 ml of 10 % poly-
ethyleneimine. After each addition, homogeneization was performed
for 1 min in a Waring blendor at a high setting. The homogenate was
centrifuged at 10 000 x g for 1 h and the precipitate discarded. The
supernatant proteins were precipitated by addition of ammonium sulfate to 65 % saturation, and redissolved in TEMG (TEM + 10 % glycerol). After dialysis the extract was adsorbed batchwise on phosphocellulose (1.5 l), and eluted by TEMG containing 0.6 M KCl.

2) DE52 chromatography : The phosphocellulose fractions containing
TdT activity were concentrated by ammonium sulfate precipitation,
dialysed against TEMG containing 50 mM KCl and passed through a DE52
column (4 x 47 cm). The TdT is excluded in these conditions.

3) Phosphocellulose chromatography : The TdT excluded from DE52, was
adsorbed onto a phosphocellulose column (5 x 35 cm) and eluted by
a 4 l gradient of TEMG containing 0.1 to 0.8 M KCl.

4) <u>DNA-cellulose chromatography</u> : DNA cellulose was prepared using a modification of the method of Alberts (9). The DNA/cellulose ratio was 1 % (w/w). The phosphocellulose-purified enzyme, dialysed against TEMG + 25 mM KCl was adsorbed onto a DNA cellulose column (2.8 x 7.5 cm) and eluted by a 300 ml linear gradient from 25 to 30 mM KCl. The fractions containing TdT were concentrated by dialysis against TEM containing 60 % glycerol (v/v). This enzyme (nearly 60 % pure) was used for the immunization of rabbits.

5) <u>Hydroxylapatite chromatography</u> : Potassium citrate-washed hydroxyl-apatite (Biorad HTP) was equilibrated in 0.1 M potassium phosphate buffer (pH 7.5) containing 0.1 mM EDTA, 1.4 mM β-mercaptoethanol and 20 % glycerol. The DNA cellulose enzyme was adsorbed on the hydroxyl-apatite column (1 x 10 cm) and eluted by a 500 ml gradient from 0.1 to 0.3 M potassium phosphate buffer (pH 7.5). Since this step leads to a good purification of the enzyme, with a low yield, it was per-formed only when highly purified TdT was needed.

6) <u>Polyacrylamide gel electrophoresis (PAGE)</u> : Samples of purified TdT were subjected to gel electrophoresis on 15 % polyacrylamide gels containing 0.1 % SDS, using the method of Laemmli (10). The gel was staining with Coomassie blue, fixed with 30 % methanol and 7.5 % acetic acid, and dried under vacuum.

3. <u>Production and purification of anti-chicken TdT.</u> - The techniques used for rabbit immunization and antibody purification have already been published (6). Briefly, 100 to 300 µg of DNA-cellulose-purified TdT were adsorbed onto alum and injected into rabbits intramuscularly. Booster injections of 100 µg of TdT were done at 3 week intervals. Blood was taken after the third injection from the central ear artery, and IgG were purified by DEAE cellulose chromatography and ammonium sulfate fractionation. Anti-TdT was purified by affinity chromatography on TdT-sepharose (prepared using the hydroxylapatite-purified enzyme).

4. <u>Immunohistological methods.</u> - TdT was detected in cell suspensions using the immunofluorescence and immunoperoxydase methods. The tech-niques used have been published elsewhere (6).

RESULTS

1. <u>Purification of TdT.</u> - The behaviour of chick TdT along the successive purification steps was very similar to that of the calf enzyme, which we have also purified several times using the same method.

Figures 1 and 2 show, as illustrations, the phosphocellulose and DNA cellulose chromatography that we obtained.

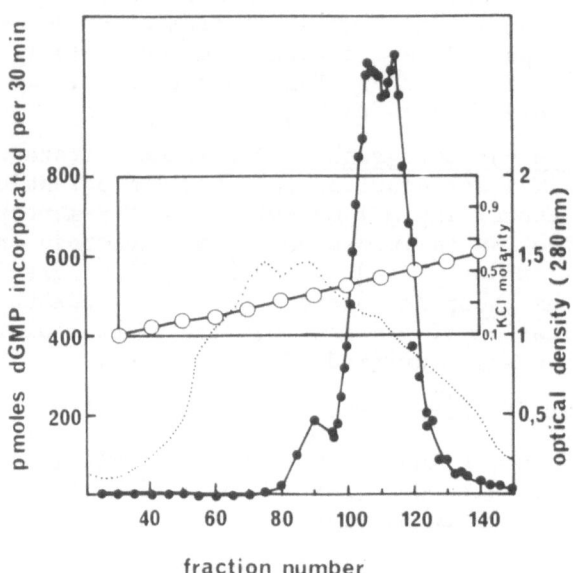

Fig. 1.   TdT chromatography on phosphocellulose. Chromatagraphy
and assay conditions are described in "Materials and
Methods". Enzymatic activity (●−●) with 5 µl of each
fraction and 0.05 mM dGTP. Optical density at 280 nm
(.....) KCl molarity (O−O).

The DNA cellulose chromatography is very useful, by its good
yield and high purification efficiency. It can be replaced by oligo-
dT-cellulose chromatography.

In some cases, a G 150 or S 200 gel filtration step was added
between the phosphocellulose and DNA-cellulose steps.

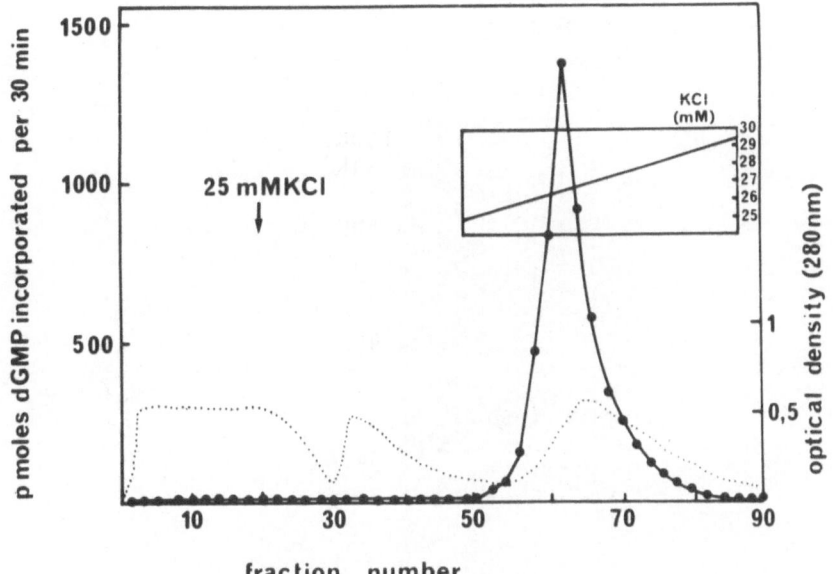

Fig. 2. TdT chromatography on DNA-cellulose. Conditions identical
         to Figure 2, with the exception that aliquotes of each
         fraction were 10 µl. (●—●) : TdT activity. (.....) :
         optical density at 280 nm.

        Figure 3 shows the result of polyacrylamide gel electrophore-
sis of hydroxylapatite-purified TdT : only one band of protein is
visible.

2. <u>Molecular weight</u>. - From its migration properties in SDS PAGE
compared to several proteins as references, the molecular weight of
chicken TdT is estimated to be nearly 40 000. G 150 gel filtration
gave a slightly higher value (45 000). No multiplicity of protein
bands was seen in the presence of SDS. Similar results were obtained
for several preparations.

        The same estimations practiced with the calf enzyme purified
using the same method gives a molecular weight of 55 000 to 60 000.

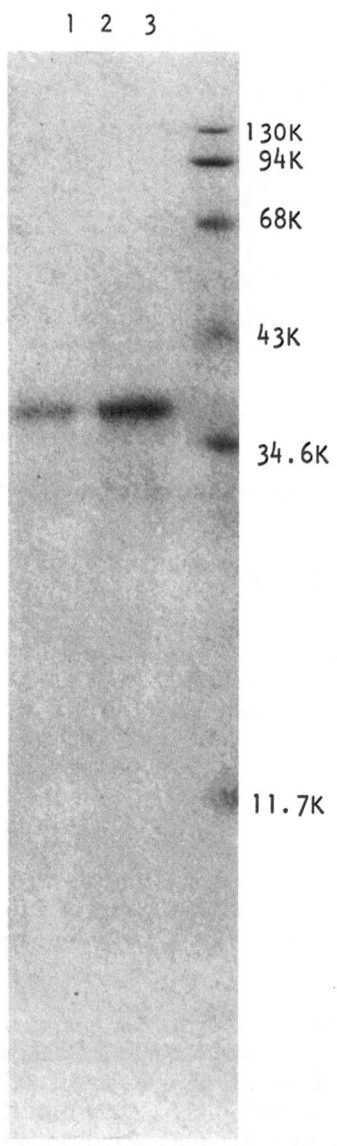

Fig. 3.   SDS-Gel electrophoresis of hydroxylapatite-purified chicken
           TdT. Migration from top to bottom. See Materials and Methods
           for details. 1 : 5 µg of TdT. 2 : 10 µg of TdT. 3 : 1 µg of
           each of the following markers : β galactosidase (130 000),
           phosphorylase A (94 000), Bovine serum albumine (68 000),
           ovalbumine (43 000), carboxypeptidase A (34 600), and cyto-
           chrome C (11 700).

3. Enzymatic properties. - The enzymatic properties of chicken TdT are very similar to those of the calf enzyme (12).

The effect of ionic strength (KCl optimum at 20 to 50 mM) of divalent ion molarity (narrow optimum at 0.1 mM for Mn^{2+}, wide optimum from 0.5 to 2 mM for Co^{2+}) have been studied.

The Table 1 shows the relative efficiencies of the different combinations of initiators and substrates in two assay system; the Tris-Mn^{2+} system, which we routinely use, and the cacodylate/Co^{2+} system. The d(pA)$_{10}$-dGTP combination appears to be the better one in both systems, at least in the conditions used.

The apparent K_M of chicken TdT is 2.2×10^{-5} M for dGTP and 4.10^{-7} M for d(pA)$_{10}$.

Table 1

Substrate / Initiator	dATP		dCTP		dGTP		dTTP	
	Tris/Mn^{++}	Caco/Co^{++}	Tris/Mn^{++}	Caco/Co^{++}	Tris/Mn^{++}	Caco/Co^{++}	Tris/Mn^{++}	Caco/Co^{++}
d(pA)$_{10}$	1.3 (2.54)	13.6 (19)	8.3 (15.5)	48.2 (67)	100 (186)	100 (139)	0.8 (1.54)	15.5 (21.54)
d(pC)$_{10}$	0.5 (0.94)	1.8 (2.54)	0.2 (0.54)	1.8 (2.64)	23 (43)	14.6 (20.3)	0	2.4 (3.4)
d(pG)$_{10}$	0.9 (1.76)	2.5 (3.58)	0.9 (1.70)	3.5 (5)	23 (42.5)	8.4 (11.74)	0	3.3 (4.6)
d(pT)$_{10}$	0.4 (0.4)	0.5 (0.7)	0	0.9 (1.36)	71.5 (71.5)	10. (15)	0.5 (1.06)	2.05 (2.86)
none	0.4 (0.4)	0.2 (0.3)	0.1 (0.2)	0.47 (0.66)	0.5 (1.04)	0.5 (0.78)	0.2 (0.5)	0

Activity of chick TdT with various initiator substrate combinations. Incubations were performed for 1 h at 37° C with 0.4 µg of hydroxyl-apatite-purified TdT in a reaction volume of 100 µl. The results are expressed in pmoles of dXMP (numbers in parentheses) and as percentages of the combination giving the better incorporation (i.e. d(pA)$_{10}$ and dGTP) taken as 100 %. The Tris-Mn^{++} system is described in Materials and Methods. The Cacodylate-Co^{++} system contains 0.2 M Na Cacodylate (pH 7.2), 1 mM 2-mercaptoethanol, 1 mM CoCl , other conditions are identical to those of the Tris assay. All initiators were from Collaborative Research and the radioactive dXTP from Amersham.

4. <u>Immunological properties of chicken TdT</u>. - As already published
(6), antibodies against chicken TdT were obtained in rabbit and
purified by affinity chromatography on TdT-sepharose.

The antibody was detected by neutralization of the enzyme
activity : an inhibition of at least 80 % of 0.04 units of TdT was
obtained in presence of 3 to 4 µg of IgGs isolated from the serum of
the immunized rabbits. As shown in Figure 4, the inhibition depends
of the amount of TdT present. The purified antibody was also tested
in a precipitation assay (6).

The same IgGs did not inhibit α and β DNA polymerases from
chicken embryos. Antibodies against calf TdT, obtained using similar
methods, cross-reacts with chicken TdT, and, inversely, the anti-
chicken TdT inhibits human and calf TdT.

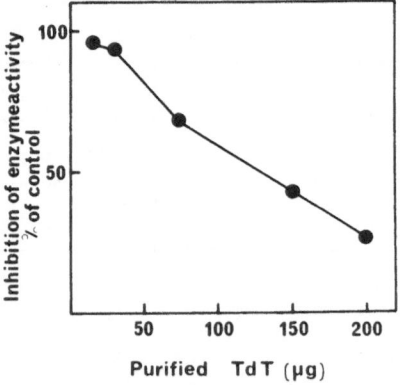

Fig. 4.   Titration of anti-Tdt IgG with increasing amounts of
          purified enzyme. The concentration of IgG is 50 µg/100 µl.
          The neutralization conditions have been described (6).

Antibodies directed against α and β DNA polymerases from chick embryo (13) do not inhibit chicken TdT.

After affinity chromatography, the neutralizing activity of anti TdT was increased at least 10 fold.

5. TdT positive cells during ontogeny of avian thymus. - As already published (6), TdT appears in few thymus cells at the 12th day of incubation. Figure 5 shows the evolution of the percentage of TdT positive cells and figure 6 a typical picture of immunoperoxydase detection of TdT in 18 day embryo thymus cells.

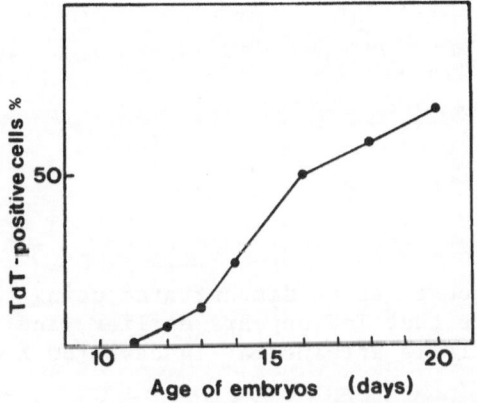

Fig. 5. Evolution of the percentage of TdT-positive thymocytes dur-
         ing the development of the chicken thymus. The detection
         was performed using the immunoperoxydase technique.

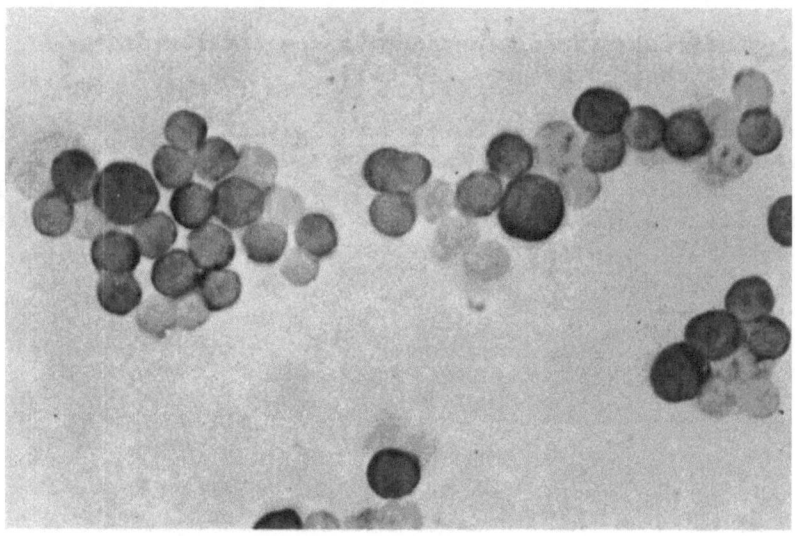

Fig. 6.   Immunoperoxydase detection of TdT in a cell suspension of
          thymus from 18 day chick embryo. Magnification: x 1 200.

        A similar process can be demonstrated using the quail thymus,
with that difference that TdT appears earlier, and the plateau value
of TdT positive cells is attained at 14 days (60 % of TdT positive
cells).

DISCUSSION

        Chick TdT appears to be very similar to the mammalian enzymes.
Compared with the calf TdT, no significant differences could be found
in its chromatographic and catalytic properties. The only apparent
difference concerns the molecular weight of the enzyme. The values
obtained by SDS-PAGE (40 000) and by gel filtration (45 000) appeared
higher than the determinations performed with the pure calf enzyme
(12), and no subunit structure could be found. Recently, new datas
have been reported by several authors, showing a more complex situa-

tion : in fact, the low molecular weight forms of TdT derive from a 60 000 daltons protein (14, 15). Until now, no correlation between these multiple molecular weights of the catalytically active enzyme has been shown with the multiple chromatographic forms described by several authors (4 , 8) . More sophisticated methods could be used to study this problem, and to determine whether the existence of multiple forms of TdT is due to a degradative process, which could be related to the disapearance in the enzyme at the late steps of T cell differenciation. Chick TdT is also similar to mammalian TdT by its immunological properties, as shown by the wide cross-reactivity of anti chick TdT with the enzymes from different mammalian species. This close antigenic relationship between TdTs from different origins was also shown by others (16-17) and reflects a remarquable stability of the antigenic determinants.

The purification of TdT and the production of a specific antibody, presented in the results, were undertaken mainly to obtain a tool for the study of the expression of the enzyme during the differentiation of T cells. This process takes place mainly in the thymus; in the young adult animal, thymocyte differentiation and cell traffic are difficult to delineate. In the embryo, it is possible to study these phenomena at their initial steps i.e. the first inflow of stem cells and the first acquisition of differentiation markers. TdT appears in the thymus later than the lymphoid stem cells, in the chick (4, 6, 7) but also in mammals (18). Moreover, TdT-positive cells represent an increasing percentage of thymocytes, suggesting an intrathymic differentiative process. We want to present here the general method that we have recently began to use to study the relations existing between the changes in TdT expression during the thymic ontogeny and the successive waves of stem cells in the thymic rudiment.

Several hypothesis can be proposed :

1. TdT expression in thymocytes is related to the age of the lymphoid cells which populate the thymus, according to a genetic program of these cells.

2. TdT expression is related to the origin of the stem cells, and predetermined by the organ where they have been produced (bone marrow, other hematopoietic areas in the embryo).

3. TdT expression is related to the effect of an inductive signal produced by the thymus (and more precisely to the stationnary part of the thymus: reticulo-epithelial cells). The production of this signal is age dependent.

4. TdT expression is a very complex phenomenon and is related to the maturation of stem cells together with that of the thymic reticulum.

The method used to make a choice between these hypotheses is mainly based on the detection of TdT-positive cells in grafted thymuses.

Such grafted thymuses are produced by F. Jotereau, following the techniques already described (2, 3). By an appropriate choice of the ages and species (quail or chick) of grafted thymuses and hosts, we obtain thymuses in which the age of the reticulum is different of that of the thymocytes, and in other cases, in which the different waves of stem cells come from a different species.

These studies are not achieved, but the method has already shown, by its preliminary results, that the third hypothesis (major role of the maturation of the thymic reticulum in TdT induction) is the most probable. The inductive signal could belong to the group of thymic hormones : in fact, some of these hormones have been shown to induce TdT (19, 20).

ACKNOWLEDGEMENTS

We wish to thank P.C. Kung and F. Chapeville for helpfull discussion.

This work was supported by INSERM (ATP 59-78-91, grant 005).

BIBLIOGRAPHY

1. M.A.S. Moore, and J.J.T. Owen, Experimental studies on the development of the thymus, J. Exp. Med., 126:715 (1967).
2. N. Le Douarin, and F. Jotereau, Tracing of cells of the avian thymus through embryonic life in interspecific chimeras, J. Exp. Med., 142:17 (1975).
3. N. Le Douarin, A biological cell labelling technique and its use in experimental embryology, Dev. Biol.,30: 217 (1973).
4. C. Pénit, and F. Chapeville, Developmental changes in terminal deoxynucleotidyl transferase of the chicken thymus, Biochem. Biophys. Res. Commun., 74:1096 (1977).
5. L.M.S. Chang, and F.J. Bollum, Deoxyribonucleotide-polymerizing enzymes of calf thymus gland. V. Homogenous terminal deoxynucleotidyl transferase. J. Biol. Chem.,246:909 (1971).
6. P. Rouget, and C. Pénit, Terminal deoxynucleotidyl transferase during the development of chicken thymus, Cell. Diff. 9:329 (1980).
7. M. Sugimoto, and F.J. Bollum, Terminal deoxynucleotidyl transferase in chick embryo lymphoid tissues. J. Immunol., 122:392 (1979).
8. R. McCaffrey, T.A. Harrison, R. Parkamn, and D. Baltimore, Terminal deoxynucleotidyl transferase activity in human leukemic cells and in normal human thymocytes, New-Engl. J. Med., 292:775 (1975).

9. B. Alberts, and G. Herrick, DNA-cellulose chromatography, Meth. Enzymol. 21:198 (1971)

10. U.K. Laemmli, Cleavage of structural proteins during the assembly of the head of bacteriophage T4, Nature 241:88 (1970).

11. N. Le Douarin, Ontogeny of hematopoietic organs studied in avian embryo interspecific chimeras. in : "Differentiation of normal and neoplastic hematopoietic cells" (Clarkson B., Till J. and P. Marks, eds.) pp. 5-31. Cold Spring Harbor Laboratory, New-York (1978).

12. F.J. Bollum, Terminal deoxynucleotidyl transferase. in : "The Enzymes" (P.D. Boyer, ed.) Vol. 10 pp. 145-171. Academic Press, New-York (1974).

13. G. Brun, L. Assairi, and F. Chapeville, Immunological relationships between chick embryo polymerases, J.Biol. Chem. 250:7320 (1975).

14. F.J. Bollum, and M. Brown, A high molecular weight of terminal deoxynucleotidyl transferase, Nature 278:191 (1979).

15. A. Silverstone, L. Sun, O.N. Witte, and D. Baltimore, Biosynthesis of murine terminal doexynucleotidyl transferase. J. Biol. Chem. 255:791 (1980).

16. P.C. Kung, P.D. Gottlieb, and D. Baltimore, Terminal deoxynucleotidyl transferase. Serological studies and radioimmunoassay, J. Biol. Chem. 251:2399 (1976).

17. F.J. Bollum, Antibody to terminal deoxynucleotidyl transferase, Proc. Natl. Acad. Sci. USA 72:4119 (1975).

18. K.E. Gregoire, I. Goldshneider, R.W. Barton, and F.J. Bollum, Ontogeny of terminal deoxynucleotidyl transferase-positive cells in lymphohemopoietic tissues of rat and mouse. J. Immunol. 123:1347 (1979).

19. N.M. Pazmino, J.N. Ihle, and A.L. Goldstein, Induction in vivo and in vitro of terminal deoxynucleotidyl transferase by thymosin in bone marrow from athymic mice, J. Exp. Med. 147:708 (1978).

20. Y. Cayre, A. De Sostoa, and A.L. Silverstone, Induction of a subset of thymocytes inducible for terminal transferase biosynthesis, J. Immunol. 126:553 (1981).

21. I. Goldshneider, A. Ahmed, F.J. Bollum, and A.L. Goldstein, Induction of terminal deoxynucleotidyl transferase and Lyt antigen by thymosin : identification of multiple subsets of prothymocytes in mouse bone marrow and spleen, Proc. Natl. Acad. Sci. USA 78:2469 (1981).

RESTRICTION ANALYSIS OF CLONED HETEROPOLYMERIC DNA SYNTHESIZED
IN VITRO BY TdT

G.Damiani, E.Palla, V.Sgaramella, A.I.Scovassi
and U.Bertazzoni

Istituto CNR Genetica Biochimica ed Evoluzionistica
Via S.Epifanio 14, 27100 Pavia, Italy

INTRODUCTION

It is known that the enzyme terminal deoxynucleotidyl transferase (TdT) catalyses in vitro the polymerization of the four
deoxynucleosides triphosphates in a sequence which can be considered largely random. The resulting copolymers can be varied in
their composition e.g. by changes in the kind of divalent cation
used in the reaction (1). The only experimental data available on
the sequence derive from the incorporation of the four precursors
and from the enzymatic degradation and nearest neighbor analysis
of the product (2,3). None of these approaches yields direct
insights on the actual sequence of the polymeric chains synthesized
in vitro by TdT.

In order to gain some evidence related to this point, and
thus possibly to the unknown function of TdT in vivo, we have
synthesized in vitro single-stranded DNA heteropolymeric chains,
converted them into double-stranded structures, and cloned via an
interspecific E.coli-B.subtilis plasmid. Clones characterized by
the phenotype expected following insertion of the heteropolymers
in the single Pst I site present in the pBR322 moiety of our
vector,have been isolated. The plasmids have been extracted and
subjected to restriction and preliminary sequence analysis.

75

RESULTS

<u>Synthesis and cloning of TdT Heteropolymers</u>

Figure 1 gives an outline of the scheme followed in the synthesis of the heteropolymers. For the cloning step these have been "tailed" by the addition of heteropolymeric sequences (oligo-dC) catalyzed again by TdT. In this way such sequences can be annealed to complementary homopolymeric (oligo-dG) sequences tailed in a similar way to the Pst I linearized plasmid pHV 33 (4): addition of dG residues to the protruding 3'-ends of the plasmid and intracellular replication of the recombinant plasmids restores an intact Pst I recognition site (5'-CTGCAG-3'), which is cut at the A-G bond.

Transformation of $CaCl_2$ treated E.Coli cells (strain HB101) with the annealed mixture (5) gave six transformants with the correct Tetracycline and Chloramphenicol resistant (Tc^R, Cm^R) and Ampicilline sensitive (Ap^S) phenotype. Plasmids were extracted from four of the six transformants and DNA size determined by gel electrophoresis: they appeared larger than the vector, and following restriction with Pst I, two of them (pPV505 and pPV506) showed the presence of inserts of ca.450 and 200 BP respectively (see Fig.2). We have decided to analyse in some detail only the largest heteropolymer.

<u>Restriction and Sequence Analysis</u>

The inserts were recovered from agarose gels using the perchlorate (6) of the DEAE-81 paper strips modified methods (7). Aliquots of the pPV505 insert were subjected to a variety of restriction endonucleases: Bam HI, Sal I, Bgl I, Eco RI, Hind III, Alu I, Bsp I, Cfo I, Hind II, Hinf I, Hpa II, Mnl I, Sau 3A. The first five which recognize a target of six nucleotides, did not cut the insert, which seems not to contain any site for Sau 3A either. All the other restriction endonucleases used recognized at least one site on the cloned heteropolymer. The relevant 8% polyacrylamide gel is shown in Fig.3.

For the sequencing of this insert, we have labelled either the dephosphorylated 5'-OH ends with $\gamma^{32}P$ ATP and polynucleotide kinase (8) or the 3'-OH ends with $\alpha^{32}P$ 3' dATP (cordycepin) in the presence of TdT (9). The labelling were followed by cutting of the insert with either Cfo I or Hind II, separation of the two

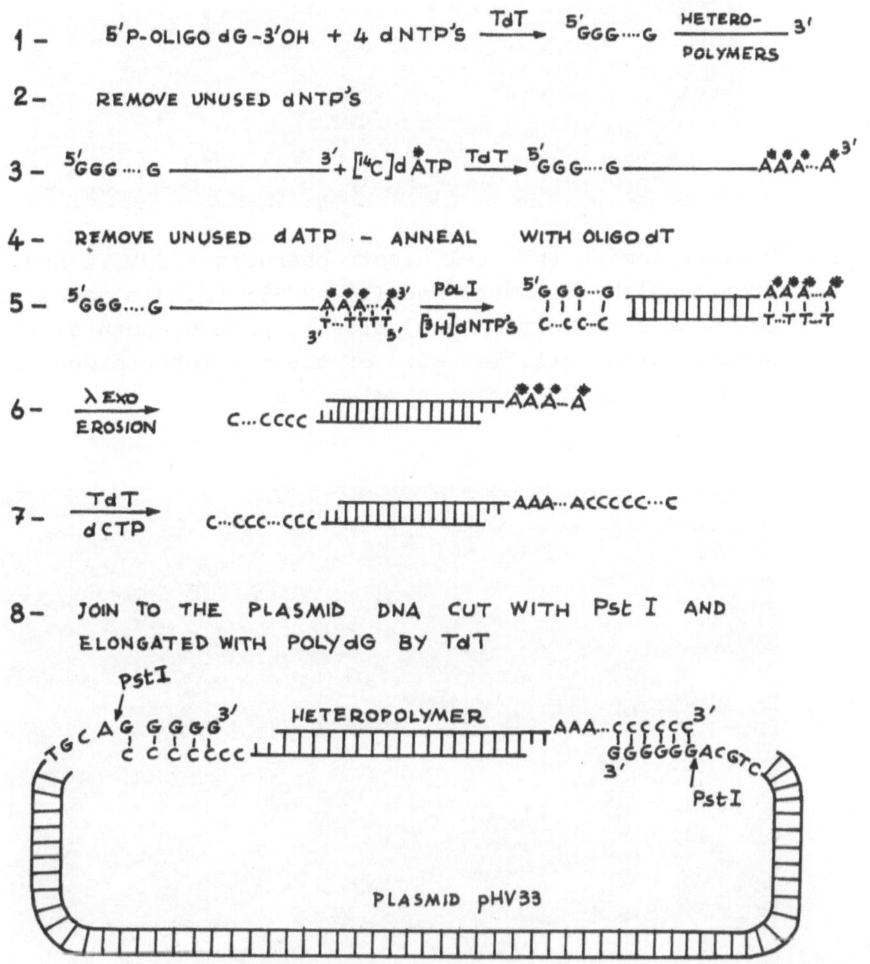

Fig.1. Synthesis and cloning of DNA heteropolymers. The synthesis is started by addition of the four deoxynucleotides to an oligo(dG)$_{15}$-initiation. The 3'OH ends are further elongated with dA residues by TdT action and the resulting single stranded heteropolymer copied by DNA polymerase I by using ^{3}H dNTPs. After erosion of 5' ends with λ exo and elongation of 3'termini with dC residues, the double stranded product is annealed to the dG-tailed vector.

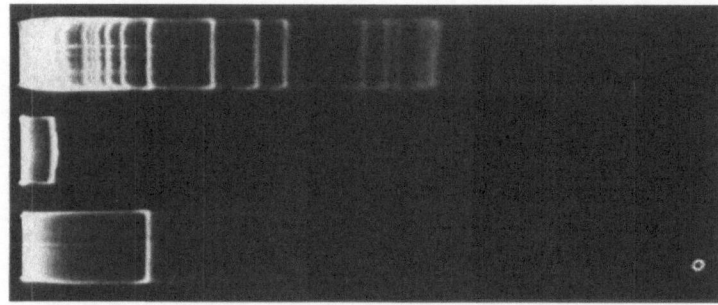

Fig.2. Polyacrylamide (8%) gel electrophoresis (200V, 2 hrs) of:
(top lane) MW standard Hae III-pHV33; (middle lane) plasmid
pPV 505: and (bottom lane) plasmid pPV506, both after digestion with PstI. The size of the two inserts were close
to 450 and 200BP, respectively.

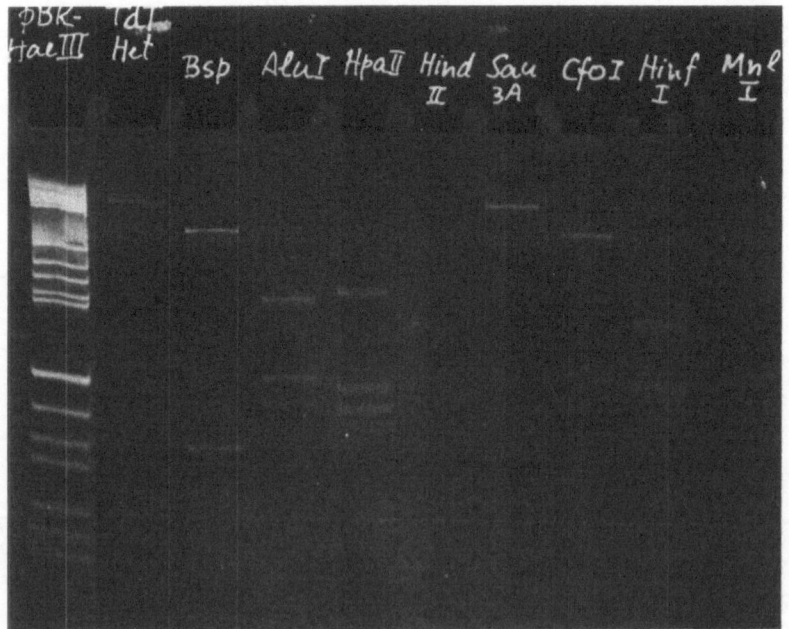

Fig.3. Polyacrylamide (8%) gel electrophoresis of the insert of
pPV505, recovered from the gel as described in the text,
and digested with the restriction endonucleases listed in
the figure;conditions were those suggested by the enzymes
vendors; gel electrophoresis as in Fig.2.

resulting labelled fragments, and sequencing according to Maxam
and Gilbert (8). Preliminary results indicate the presence, in
addition to the homopolymers introduced during the synthesis of
the product, of the heteropolymeric sequences.

DISCUSSION

    The strategy used for the synthesis of the TdT heteropolymers,
as shown in Fig.1, involves the production of homopolymeric sequences alien to the scope of our project, but useful for the cloning step. In particular, the addition of poly-dA sequence instead
of poly dG at the 3'-end of the single-stranded heteropolymer was
decided in order to avoid any intramolecular annealing of the two
tails. Our sequencing data have indeed revealed the presence of a
long homopolymeric stretch of dA residues (as.80). Alternative
approaches are being investigated in order to reduce the proportion
of such homopolymeric sequences, as well as of those at the tails.

    The cloning of the resulting heteropolymers has been per-
formed using an interspecific vector, in view of the possibility
that they could unfavourably affect the viability of either the
E.coli or the B.subtilis host. In fact we have obtained, after
E.coli transformation, only few clones. This result is consistent
with a negative interference of the TdT heteropolymers in this
host. Unpublished observations indicate that, even in the absence
of ampicillin, the growth rate of the six E.coli clones harboring
plasmids with the TdT products are lower than those of control
cells containing only the vector. It is possible that some of our
heteropolymeric sequences are expressed in a product fused to the
β -lactamase (5). Their effects on the growth of B.subtilis have
yet to be studies.

    As for the analysis of the sequence inserted in pPV505, which
is the largest fragment we have isolated and studied so far, it is
noteworthy to stress that it contains a variety of restriction
sites (see Fig.3): in particular, the recognition target for MnI I
(CCTC) seems to be present between 12 and 15 times. The preliminary
data for the sequencing have confirmed the presence of the sites
for Bsp .I and Hpa II. It is clear that only the knowledge of the
complete sequence of all the available inserts could allow us to
propose any hypothesis on the nature of the in vitro products of
TdT.

AKNOWLEDGEMENTS

    G.D.is a predoctoral student, E.P. (presently at Assoreni, ENI,
S. Donato) and A.I. S. are postdoctoral fellows (Accademia Naz. Agri-
coltura and Scuola Perfezionamento Genetica, respectively) at the Dept.
of Genetics, Pavia. V. S. acknowledges the support of a grant of the
Finalized Project "Control of Tumor Growth", C.N.R. and of a grant
from Montedison Spa.
    This publication is contribution n. 1810 of the Radiation Prot-
ection Programme (DGXII) of the European Community Commission.

REFERENCES

1. K. Kato, J. M. Goncalves, G. E. Houts and F.J. Bollum,Deoxynucleo-
    tide-polymerizing enzymes of calf thymus gland. II Properties of
    the terminal deoxynucleotidyl transferase, J. Biol. Chem. 242:
    2780 (1967).
2. F. J. Bollum, Preparation of oligodeoxynucleotides, in "Procedu-
    res in Nucleic Acid Research" (G. Cantoni and D. Davies,eds)
    pp. 592-599, Harper, New York (1966).
3. R. L. Ratliff, A. W. Schwartz, V. N. Kerr, D. L. Williams, D.G.
    Ott and F.N. Hayes, Heteropolydeoxynucleotides synthesized with
    Terminal Deoxynucleotidyl Transferase. II Nearest neighbor fre-
    quencies and extent of digestion by micrococcal Deoxyribonucle-
    ase. Biochemistry, 7:412 (1968).
4. S. D. Ehrlich, DNA cloning in Bacillus subtilis. Proc. Natl.
    Acad. Sci. USA, 75: 1433(1978)
5. L. Villa-Komaroff, A. Esfradiatis, S. Broome, P. Lomedico, R.
    Tizard, S. Naber, W. L. Chick and W. Gilbert, A bacterial clone
    synthetizing proinsulin, Proc. Natl. Acad. Sci. USA 75:3727(1978).
6. R. C. A. Yang, J. Lis and R. Wu, Elution of DNA from agarose gels
    after electrophoresis, in "Methods in Enzymology", R. Wu, ed.,
    vol. 68, p. 176, Academic Press,(1979).
7. C.Winberg and M. L. Hammarskjold, Isolation of DNA from agarose
    gels using DEAE-paper. Application to restriction site mapping
    of adenovirus type 16 DNA, Nucleic Ac. Res., 8:253 (1980).
8. A. M. Maxam and W. Gilbert, Sequencing end-labeled DNA with base-
    specific chemical cleavage, in "Methods in Enzymology",L.Grossman
    and K. Moldave,eds, vol. 65, p. 499, Academic press, (1980).
9. C. P. D. Tu and S. N. Cohen, 3'-end labeling of DNA with ^{32}P
    cordycepin-S'-triphosphate, Gene 10: 177 (1980).

TERMINAL DEOXYNUCLEOTIDYL TRANSFERASE HELP DNA POLYMERASE α BYPASS THYMINE DIMER

Shonen Yoshida[a], Sigeo Masaki[a], Hiromu Nakamura[b] and Toshiteru Morita[b]

[a]Department of Biochemistry, Institute for Developmental Research, Aichi Prefecture Colony, and [b]Department of Experimental Radiololy, Aichi Cancer Center Research Institute
Kasugai, Aichi, 480–03, and Chikusa–ku, Nagoya, 464 Japan

Introduction

The physiological function of terminal deoxynucleotidyl transferase (EC 2.7.7.31) is not known. In 1974, Baltimore (1) proposed a model in which terminal transferase acts as a somatic mutator at a special region of immunoglobulin gene during the maturation of immunocompetent cells. On the other hand, a speculative mechanism for induced mutagenesis in the eukaryotic cell was also discussed in connection with the possible induction of terminal transferase in the cell which has damaged DNA (2). In order to verify these hypothesis, further biochemical analysis may be required with respect to a possible modulation of the DNA polymerase reaction by terminal transferase. Recently, we have shown (3) that the replication by DNA polymerase α stops at the site of thymine dimer induced by ultraviolet–irradiation on the template poly(dT). The addition of terminal deoxynucleotidyl transferase to this system enhanced the DNA synthesis to the control level and it concomittantly increased the incorporation of the mismatched deoxynucleotides into the newly synthesized poly(dA) strand. The sizes of newly synthesized DNA were smaller with ultraviolet irradiated template but they increased to the control level with the addition of terminal deoxynucleotidyl transferase to the system. These results suggest that terminal deoxynucleotidyl transferase can help DNA polymerase α "bypass" thymine dimers in vitro by the formation of mismatched regions at the positions opposite to pyrimidine dimers on the template.

RESULTS AND DISCUSSION

        In order to separate the newly synthesized poly(dA) from the
template poly(dT), we used an alkaline Cs_2SO_4 density gradient
centrifugation.   Poly(dA) banded at a lower density (1.36) than
poly(dT)(1.46).   As initiator strand, we used oligo(rA) because
terminal transferase hardly use oligoribopolymers.   We used 10S
DNA polymerase α (4) purified from calf thymus for the present
study.
<u>Reaction on the ultraviolet-irradiated poly(dT) template</u>.   Poly(dT)
irradiated with UV at 750 J/m , at which every poly(dT)strand (1000
nucleotides long) would have 14-16 thymine dimers showed the template
activity of 1/5 of the control.   However, the fidelity of the
reaction was not influenced by UV-irradiation.   When terminal
transferase was added to this system, the amount of poly(dA)
synthesis increased up to 85-100 % of the unirradiated control and
the amount of the incorporation of incorrect deoxynucleotide
([^3H]dGMP) into the newly synthesized [^{14}C]poly(dA) strand increased
strikingly (Fig. 1, B and C).   Rate of the misincorporation (G/A)
calculated from the re-centrifuged peak fraction was 1/100, 60 times
more than that of the control (1/6000).   These results indicate
that terminal deoxynucleotidyl transferase can act at 3'-OH ends of
the growing chains which have been extended by DNA polymerase α and
then blocked at or near the sites of thymine dimers.   The amount
of misincorporation into the newly synthesized strands was
proportionally increased with the dose of terminal transferase
(Fig. 1B and C) but the amount of the incorporation of dAMP
(correct deoxynucleotide) stayed constant near the control level
(Fig. 1).
<u>Size of the newly synthesized poly(dA) on the UV-irradiated poly(dT)</u>
The reaction products, poly(dA), were separated by the Cs_2SO_4
density gradient centrifugation as in Fig. 1 and their sizes were
determined by alkaline sucrose gradient centrifugation.   As shown
in Fig. 2, the size of poly(dA) synthesized on the irradiated
poly(dT) (2.4S on average, approx. 50 bases long) is considerably
smaller than the control (4.0S on average, approx. 200 bases long).
In the presence of terminal transferase, the product showed a size
similar to that of the control.
<u>Reaction on the unirradiated poly(dT) template</u>   The effect of
terminal transferase on the DNA synthesis by 10S polymerase α was
also measured with the unirradiated poly(dT) template.   The rate
of misincorporation into the poly(dA) peak (G/A) slightly increased
(1.5-fold) by the addition of terminal transferase which is 5-times
as much as 10S DNA polymerase α.   The amount of poly(dA) synthesis
was not affected(3).
<u>Conclusion</u>   In the presence of terminal deoxynucleotidyl transferase,
DNA polymerase α (10S DNA polymerase α, Ref. 4) could replicate the
ultraviolet-irradiated poly(dT), as well as the unirradiated one,
with respect to both the amount and the size of reaction product,
poly(dA).   But the poly(dA) newly synthesized by this dual enzyme

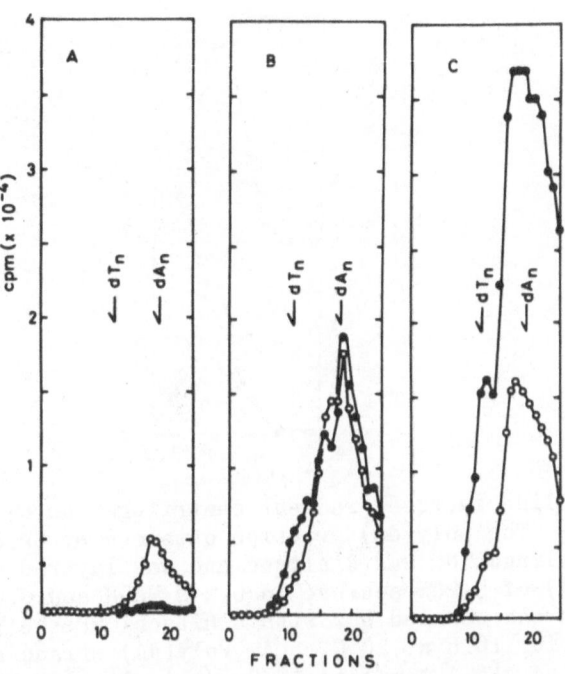

Fig. 1. Alkaline Cs_2SO_4 density gradient centrifugation of the reaction products between 10S DNA polymerase α and ultraviolet-irradiated poly(dT) (1). The reaction mixture (0.125 ml) contained 80 mM Tris·HCl (pH 7.5), 100 μg/ml bovine serum albumin, 5 μg UV-irradiated poly(dT) (750 J/m), 0.25 μg oligo(rA) , 0.5 mM $MnCl_2$, 4 mM 2-mercaptoethanol, 0.1 mM [^{14}C]dATP (9.5 cpm/pmol), 4.32 μM[^{3}H] dGTP (2930 cpm/pmol), and enzyme(s) as indicated. A. The reaction product by 20 units of 10S DNA polymerase α only. B. The reaction product by 20 units of 10S DNA polymerase α and 50 units of terminal transferase. C. The reaction product by 20 units of 10S DNA polymerase α and 100 units of terminal transferase. Incubation was carried out at 37°C for 60 min. O , correct deoxynucleotide (dAMP) incorporated; ● , incorrect deoxynucleotide (dGMP) incorporated. 6 ml of the solution of Cs_2SO_4 containing 0.2 M NaOH and reaction product (refractive index 1.3688) was centrifuged at 4200 rev./min for 62 h at 23°C. Positions of poly(dT) and poly(dA) was indicated.

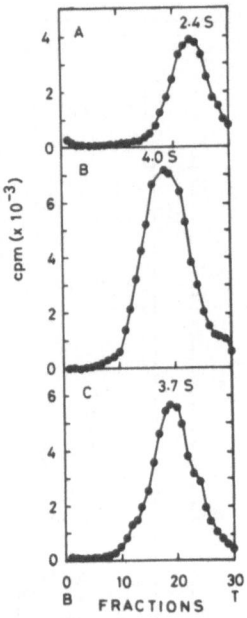

Fig. 2.   Alkaline sucrose gradient centrifugation of poly(dA)
products (2).   The poly(dA) fraction obtained by Cs_2SO_4 density
gradient centrifugation was dialysed and was layered on top of the
gradient (5 ml) of 5–20% sucrose in 0.2 M NaOH and 0.8 M NaCl.
Centrifugation was carried out with a Hitachi RPS 65 Ti rotor at
45000rev./min for 16 h at 20°C.   A. Poly(dA) strand synthesized on
UV–irradiated poly(dT) template (750 J/m) with 10S DNA polymerase α
B. Poly(dA) strand synthesized on control poly(dT) template(1).

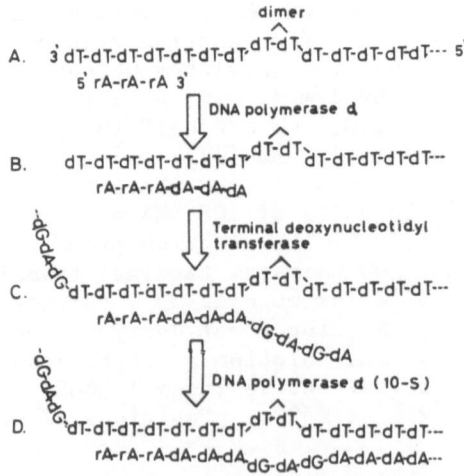

Fig. 3.   Scheme of a possible cooperation of DNA polymerase α (10S)
and terminal deoxynucleotidyl transferase at or near the sites of
thymine dimers on template strands.   For details, see text.

system contained a much higher rate of misincorporation than that with the unirradiated template. These results may be explained as follows. Terminal transferase can use the 3'-OH ends of growing chains as initiators, when the reaction of DNA polymerase α is bolcked at or near the thimine dimers on the template strands (Fig. 3A and B). The single-stranded strechs formed by terminal transferase might reassociate to template strands beyond the thymine dimers and will be further extended by 10S DNA polymerase α (Fig. 3C and D). These results might indicate that terminal transferase can act as a mutator polymerase in the cell which has DNA lesions such as pyrimidine dimers.

REFERNCES

1. Baltimore, D. (1974) Nature 248, 409-411
2. Villani, G., Boiteux, s. and Radman, M. (1978) Proc. Natl. Acad. Sci. U.S.A. 75, 3o37-3041
3. Yoshida, S., Masaki, S., Nakamura, H. and Morita, T. (1981) Biochim. Biophys. Acta, 652, 324-333
4. Masaki, S., and Yoshida, S. (1978) Biochim. Biophys. Acta 521, 74-88

INHIBITORS OF DNA POLYMERASES:

THEIR SELECTIVITY AND MODE OF ACTION

Prakash Chandra, Ilhan Demirhan and Uwe Ebener

Center of Biological Chemistry, Laboratory of Molecular
Biology, University Medical School, Theodor-Stern-Kai 7
Frankfurt(Main), W. Germany

1. INTRODUCTION

The first DNA polymerase in mammalian cells was discovered by
Bollum (1958), but only in the last years the existence of multiple forms of DNA polymerases was recognized in these cells. In the
recent years, there has been a great interest in the purification
and characterization of different eukaryotic DNA polymerases, and
to study the functional role(s) of each of these enzymes. The field
of DNA polymerases gained an unexpected importance as in 1970, a
novel DNA polymerase was discovered as a constituent of retrovirus
(Temin and Mizutani, 1970; Baltimore 1970), now known as reverse
transcriptase (RT). In addition, several new DNA-polymerizing activities are induced in eukaryotic cells infected with DNA viruses.
Another unique enzyme which also exhibits a deoxyribonucleotide
polymerizing activity is the terminal deoxyribonucleotidyl transferase (TdT). This enzyme, though a cellular enzyme, is unique in
the sense that, so far, only thymus tissue has been shown to contain it. In the strictest sense TdT is not a DNA polymerase as it
does not require a template to direct the DNA synthesis; nevertheless, it does catalyze the incorporation of deoxyribonucleotides
into DNA and thus we consider it here as one of the DNA synthetic
enzymes. Like RT the TdT also merits considerable attention as a
biochemical marker'for certain types of leukemias, as will be discussed by Dr. Sarin and Dr. Mertelsmann is this proceeding.

All these additional DNA polymerase activities introduced into
the cell or induced by viruses have become important tools in detecting viral "footprints" in neoplastic cells, or as in the case
of TdT establishing the type of leukemic disease. Particularly in
human malignancies where the retrovirus replication is strictly

restricted, the biochemical and immunological characterization of RT
has played an important role to implicate the association of retro-
viruses to human malignancies (Gallo et al., 1970, 1975; Chandra et
al., 1978, 1980; Chandra & Steel, 1977, 1980; Chandra & Vogel, 1981;
Ebener et al., 1979; Welte et al., 1979; Ohno et al., 1977; Witkin
et al., 1975).

Apart from the basic interest in understanding the functional
role of these enzymes, these additional DNA polymerizing activities
may serve as a useful tool in designing antiviral and anticancer
drugs (Chandra et al., 1975, 1977, 1980a).One of the major concerns
of our laboratory has been to develop selective inhibitors of these
polymerases, in particular the inhibitors of reverse transcriptase.
A large part of these studies has appeared in several reviews
(Chandra et al., 1977, 1977a, 1980a). In this communication we will
describe our recent studies on the inhibition of cellular DNA-polymerase, RT and TdT by some substrate analogs and template-primer
analogs.

2. EFFECT OF 5-ALKYL-2´-DEOXYURIDINE-5´-TRIPHOSPHATES ON THE DNA-POLYMERIZING ENZYME ACTIVITIES

The use of chemically modified deoxynucleoside trip-osphates
as analogs of nucleic acid bases, is a well-known strategy to block
DNA synthesis catalyzed by the DNA polymerase reaction, or synthesize DNA species with altered functional activity. However, in view
of their importance as antiviral compounds, only such substrate
analogs are of value which are specifically utilized by the viral
induced enzymes. Using this approach, Cheng et al. (1976) described
significant selectivity of 5-alkyl-substituted deoxyuridine analogs
against herpes-virus-induced enzymes; 5-ethyl-deoxyuridine (Et.-dU)
and 5-propyl-deoxyuridine (Pr.-dU) are known to possess antiherpetic
activities (Gauri, 1979). In view of this, we were interested to
look for the substrate replacement activity of 5-ethyl- and 5-
propyl-analogs of deoxyuridine triphosphates (Fig. 1)in the DNA-
polymerase reaction catalyzed by the cellular DNA polymerases, the
reverse transcriptase and terminal transferase.

The studies reported herein were carried out using DNA polymerase-α from calf thymus, DNA polymerase-ß from the Rhabdomyo-
sarcoma tissue of a child, DNA polymerase-γ from the human ovarian
tumor tissue, reverse transcriptase from RLV-infected mouse spleen
and terminal transferase from calf thymus. All the enzymes were
purified by column chromatogrpahy procedures, as described earlier
(Chandra and Steel, 1977). The purity of these enzymes was characterized by SDS-gel electrophoresis and using primer-templatespeci-
fic for their activity. The terminal transferase was a coomercial
purified preparation from BRL, Neu Isenburg, Germany.

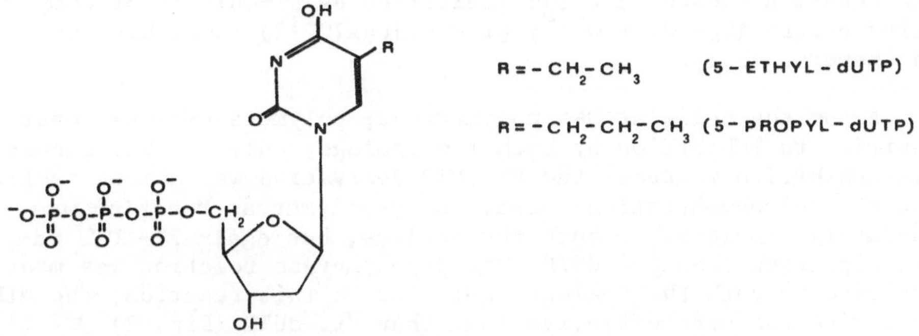

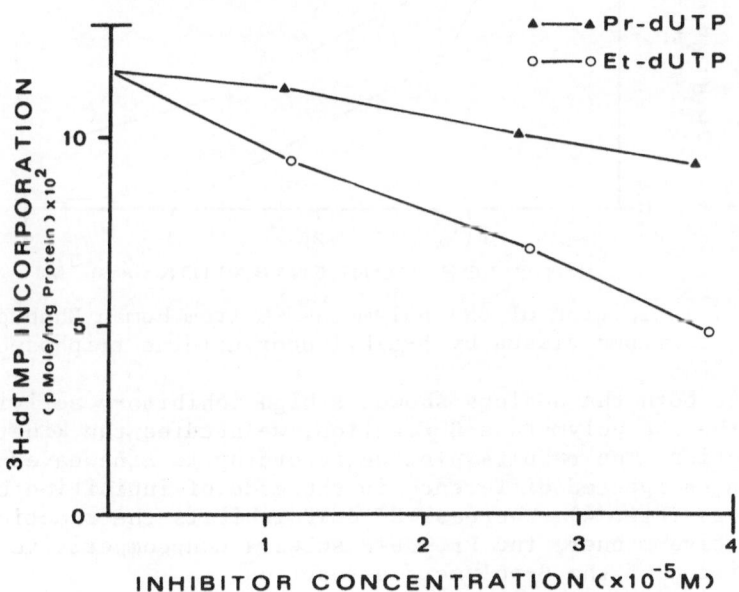

Fig. 1. Chemical Structures of 5-Alkyl-Deoxyuridine Triphosphates.

As follows from Fig. 2, the 5-ethyl-derivative (ET-dUTP) inhibits the reverse transcriptase reaction catalyzed by (rA)n.(dT)12
significantly; whereas, under similar experimental conditions 5-
propyl-derivative (Pr.-dUTP) fails to inhibit the RT-reaction.

Fig. 2. Inhibition of (rA)n.(dT)12-catalyzed reverse transcriptase activity from RLV-infected spleen by 5-
alkyl-deoxyuridine triphosphate.

The kinetics of RT-reaction plotted according to Lineweaver-Burk equation showed that the inhibition by ET-dUTP is of competitive nature (K_m= 0.4×10^{-10}; Ki = 0.166×10^{-15}) (Demirhan, Ebener and Chandra, 1981).

Among the cellular DNA polymerases, polymerase-α was least sensitive to inhibition by both the analogs; only ET-dUTP showed some inhibition whereas, the Pr.dUTP derivative was almost ineffective at the concentrations used. The γ-polymerase reaction was moderately sensitive to both the analogs, but again ET-dUTP was more effective than Pr.-dUTP. The ß-polymerase reaction was most sensitive to both the analogs, but also in this reaction, the ethyl derivative was more effective than than Pr.-dUTP (Fig. 3)

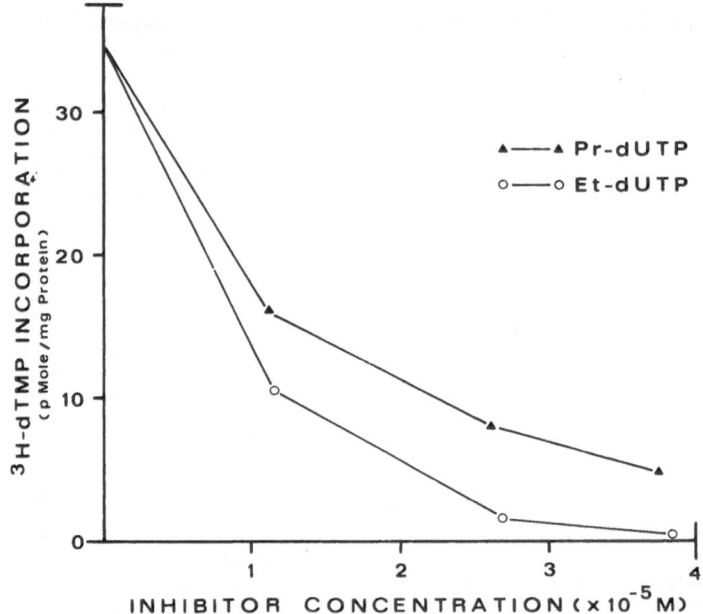

Fig. 3. Inhibition of DNA polymerase-ß from human Rhabdomyo-
        sarcoma tissue by 5-alkyl-deoxyuridine triphosphates.

Since both the analogs showed a high inhibitory activity against the DNA polymerase-ß reaction, we studies the kinetic of this reaction. The results plotted according to Lineweaver-Burk showed an unexpected difference in the mode of inhibition by the two analogs (Fig. 4). Whereas, ET-dUTP inhibits the reaction in a competitive manner, the Pr.-dUTP shows a non-competitive type of inhibition of the ß-polymerase reaction.

Recent studies from our laboratory have shown that the terminal transferase activity from calf thymus is highly sensitive to both these analogs. Using poly dA as starter(primer) the incorporation of ^{3}H-dTMP was inhibited to almost 90% by ET-dUTP and about 50% by Pr.-dUTP (Table 1)at a concentration of 3×10^{-6} M. As shown in Table 1, the inhibitory effect of these analogs at this concentration on other DNA polymerases is not significant.

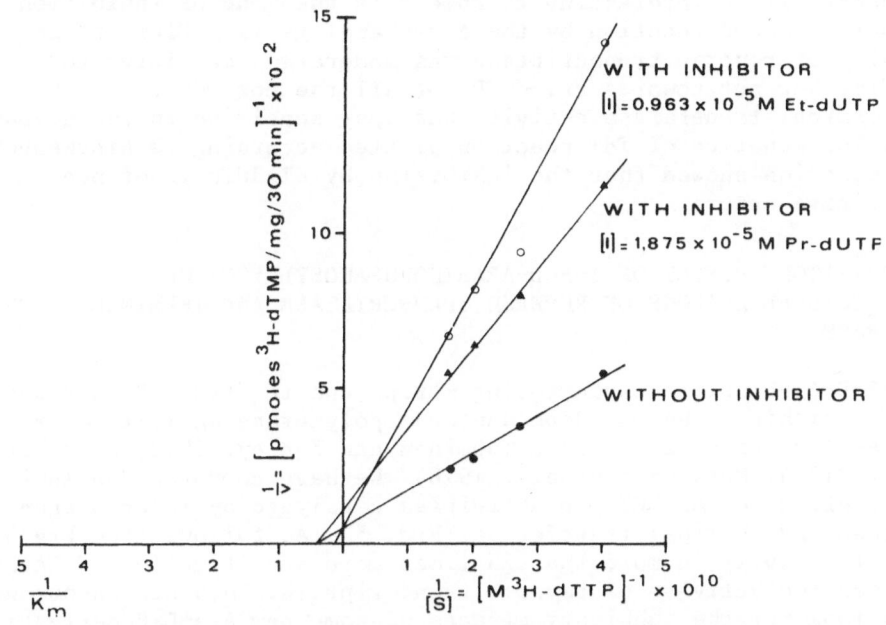

Fig. 4. Lineweaver-Burk plot of the kinetic data of polymerase
ß inhibition by 5-alkyl-deoxyuridine triphosphates.

Table 1.   Inhibitory Effects of 5-Alkyl-Deoxyuridine-Triphos-
phates on the Cellular DNA polymerases,reverse transcriptase and terminal transferase activities.

ENYZME SYSTEM	^{3}H-dTMP INCORPORATION (% OF CONTROL)	
	ET-dUTP[a]	Pr.-dUTP[a]
Cellular Polymerases		
α	98	98
ß	81	87
γ	94	100
Reverse Transcriptase	89	97
Terminal Transferase	14	42

a) Concentration = 3.0×10^{-6}M

Summarizing the studies on 5-alkyl-substituted deoxyuridine
triphosphates, one can say that among the cellular polymerases, the
α-polymerase reaction is least sensitive to these analogs; whereas,
the ß-polymerase reaction is most sensitive to the action of these
compounds. It is interesting to note that the mode of inhibition of
the polymerase-ß reaction by these two analogs is different.The
activity of reverse transcriptase was moderately sensitive to
ET-dUTP, but not towards Pr.-dUTP. Of all the enzymes tested by
us, terminal transferase activity was most sensitive to these analogs. The kinetics of TdT reaction plotted according to Lineweaver-
Burk equation showed that the inhibition by ET-dUTP is of non-
competitive nature.

3. INHIBITORY EFFECT OF 1-ß-D-ARABINOFURANOSYLCYTOSINE
   STRUCTURAL ANALOGS ON REVERSE TRANSCRIPTASE AND TERMINAL TRANS
   FERASE

1-ß-D-Arabinofuranosylcytosine -triphosphate (Ara-CTP) has been
shown to inhibit the RNA-dependent DNA polymerase activity of retro-
viruses (Müller et al., 1971; Touminen and Kenney, 1972; Schreker
et al., 1974; Matsukage et al., 1978). We have compared the inhibitory effect of Ara-CTP on activities catalyzed by reverse transcriptase and terminal transferase (Fig. 5). As follows from Fig.5,
the TdT activity is more than 25 times more sensitive towards Ara-
CTP than the activity of reverse transcriptase. This has encouraged
us to look for the inhbitory effects of some new Ara-CTP derivatives
with substitutions in the sugar moiety, recently synthesized by Dr.
David Shugar, Warsaw. The structures of these derivatives are shown
in Fig. 6.

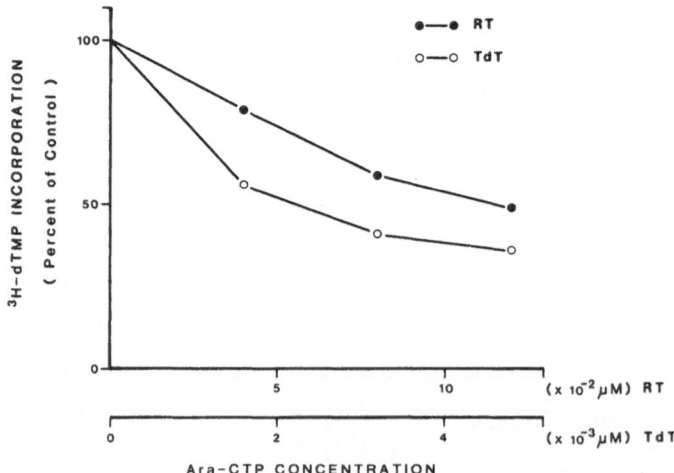

Fig. 5. Inhibitory Effect of Ara-CTP on the Enzyme Activities
        catalyzed by reverse transcriptase and terminal trans-
        ferase.

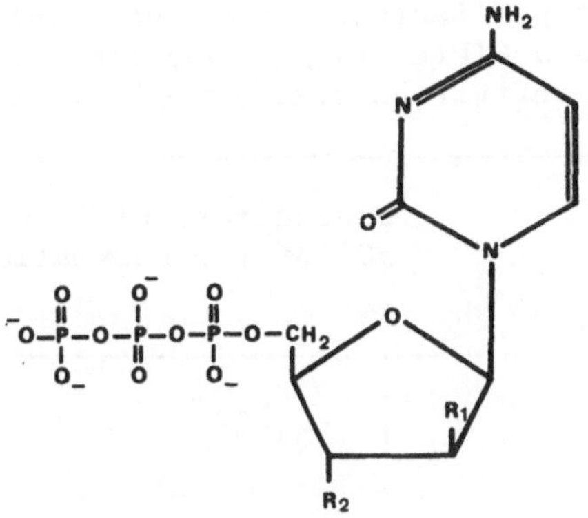

	R_1	R_2
Ara-CTP	-OH	-OH
2'-OMe-Ara CTP	-OCH$_3$	-OH
3'-OMe-Ara CTP	-OH	-OCH$_3$
2',3'-OMe-Ara CTP	-OCH$_3$	-OCH$_3$

Fig. 6. Chemical Structures of Ara-CTP Derivatives.

The inhibitory effects of Ara-CTP and structural analogs on the enzyme actvities catalyzed by reverse transcriptase and terminal transferase are summarized in Table 2. The data depict molar concentrations of different compounds needed to inhibit the 50% of enzymatic activity. The studies with these derivatives lead to two interesting observations: 1) The O-methylation in shugar moiety leads in general to a decrease in the inhibitory response of these compounds; 2) None of the O-methylated derivatives is able to inhibit the reverse transcriptase activity. Therefore, although the O-methylation of Ara-CTP retards its inhibitory potency towards TdT but, it contributes towards its selectivity since the RT activity is not inhibited by these derivatives.Based on this observation , we have tested some derivatives of thymidine triphosphate where the O-methylation was carried out in pyrimidine moiety (Fig. 7) in-stead of in the shugar moiety. The inhibitory effect of these analogs is also shown in Table 2. As follows from results, the O^2-methyl derivative of TTP is almost as potent an inhibitor of TdT reaction as Ara-CTP. At the same time, this derivative does not inhibit the RT reaction. A similar type of behavior is exhibited by the O^4-methyl derivative.Further studies on the inhibitory effects of these derivatives on the eukaryotic cellular DNA polymerases are in progress to determine their selectivity towards terminal transferase.

Table 2.   Inhibitory Effects of O-methyl substituted derivati-
           ves of Ara-CTP(cf Fig.6) and TTP (cf Fig.7) on the
           enzymic activities catalyzed by RT and TdT.

| Compound Used | Concentration(M) needed to Inhibit 50% of the Enzymatic Activity | |
	RT	TdT
Ara-CTP	11.5×10^{-5}	2.2×10^{-6}
2´-OMe-Ara CTP	$-a)$	1.32×10^{-5}
3´-OMe-Ara CTP	$-$	3.70×10^{-5}
2´,3´-OMe-Ara CTP	$-$	4.8×10^{-5}
O^2Me-TTP	$-$	2.2×10^{-6}
O^4Me-TTP	$-$	5.8×10^{-6}

a) No inhibition was seen at the concentrations used.

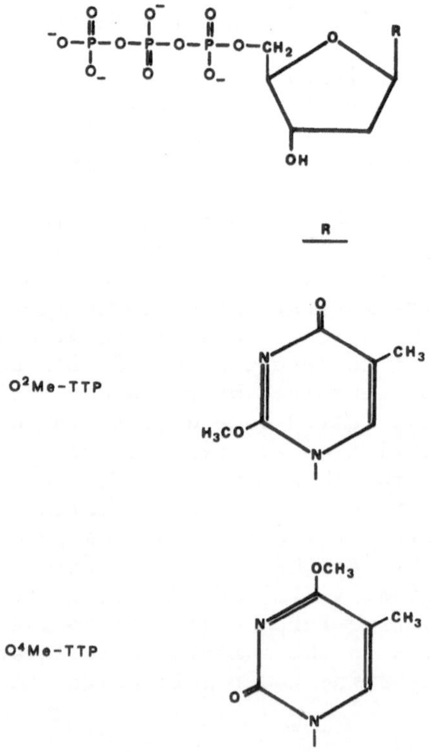

Fig.7. Chemical structures
        of Tyhmidine triphos-
        phate analogs.

4. ANTITEMPLATE APPROACH TO DEVELOP SELECTIVE INHIBITORS OF DNA SYNTHESIS

Single-stranded polyribonucleotides are known to act as efficient templates for the viral DNA polymerases in the presence of a complimentary oligo-deoxyribonucleotide primer. Chemical modification of such templates would be expected to alter the interaction between the template and the viral enzyme (Chandra et al., 1977, 1977a, 1980a). This appears to be a very useful approach for designing inhibitors of viral DNA polymerases which might find application in the chemotherapy of cancer (Chandra, Kornhuber & Ebener,1979).

Our efforts to develop compounds that inhibit viral DNA polymerases by interacting directly with the enzyme led to the discovery of a polycytidylic acid analog, containing 5-mercapto substituted cytosine bases, a partially thiolated polycytidylic acid. This compound, abbreviated as MPC, was found tobe a relatively specific inhibitor of viral reverse transcriptase (Chandra et al., 1980a).In view of the biochemical selectivity of MPC, and its antileukemic effect in animals (Chandra, Kornhuber and Ebener, 1979), MPC has been used clinically to treat childhood leukemia (Kornhuber and Chandra, 1979; Chandra et al., 1977a, 1980a).

In the following pages we would like to describe our recent studies on the inhibition of various DNA polymerases with some poly-A analogs. These analogs are: poly (2-methylthioadenylic acid) or $ms^2A)n$; poly (2-ethylthioadenylic acid) or $(es^2 A)n$; and poly (2'-fluoro-deoxy-adenylic acid) or $(dAfl)n$.

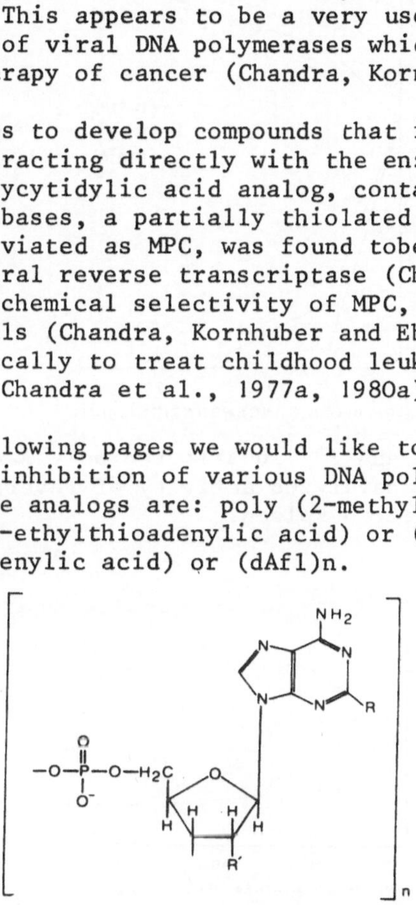

R = −S−CH₃. R'=−OH: poly(2-methylthioadenylic acid). $(ms^2A)_n$

R = −S−CH₂CH₃. R'=−OH: poly (2-ethylthioadenylic acid). $(es^2A)_r$

R = −H. R'=−F: poly (2'-fluoro-2'-deoxyadenylic acid). $(dAfl)_n$

Fig. 8. Chemical Structures of Polyadenylic Acid Analogs.

The effect of these three poly A analogs on (rA)n.(dT)12-primed DNA polymerase activity of RLV is shown in Fig. 9. As follows from results, $(ms^2A)n$ shows a maximum inhibition of the enzyme activity followed by $(es^2A)n$. On the contrary, $(dAfl)n$ showed a concentration dependent stimulation of the RT reaction. However, in the presence of (rC)n.(dG)12 as template-primer, the RT reaction was significantly inhibited by (dAfl)n (Fig. 10). This biphasic effect is understandable since (dAfl)n acts in the latter situation as antitemplate; whereas, this analog does not compete with (rA)n .(dT)12 as antitemplate or, it acts itself as template alike poly A.

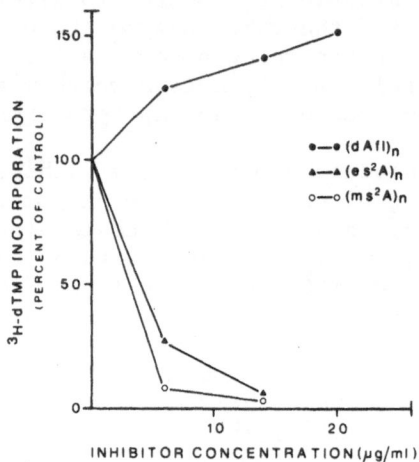

Fig. 9. Effect of 2- and 2′- substituted poly A analogs on the (rA)n.(dT)12-catalyzed activity of reverse transcriptase from RLV-infected spleen.

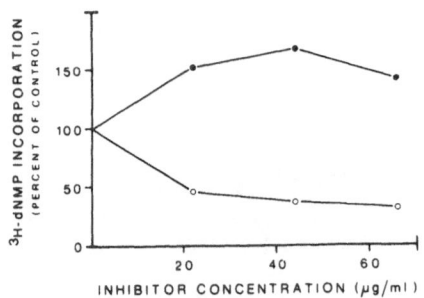

Fig. 10. Effect of poly (2′-fluoro-2′-deoxy-adenylic acid):dAfl on the exogenous activity of reverse transcriptase (RLV),catalyzed by (rA)n.(dT)12.
●————● (rA)n. (dT)12
O————O (rC)n. (dG)12

A very similar spectrum of inhibition was observed in the DNA polymerase -γ reaction, catalyzed by (rA)n.(dT)12. Both the 2-substituted derivatives inhibited the γ-polymerase reaction strongly, whereby the $(ms^2A)n$ analog was a better inhibitor. The 2´-substituted derivative, $(dAfl)n$, showed a concentration dependent stimulation of the DNA polymerase-γreaction.

The main conclusion from these experiments is that, $(es^2A)n$ and $(ms^2A)n$ inhibit the $(rA)n.(dT)12$-catalyzed reactions (RT and DNA polymerase-γreaction), whereas, the $(dAfl)n$ derivative stimulates the $(rA)n.(dT)12$-dependent reaction. It was therefore of interest to look for their relative activity in the DNA polymerase-ß reaction, catalyzed by $(dA)n.(dT)12$. As follows from Fig. 11, the polymerase-ß reaction is inhibited by all the three analogs. The fact that $(dAfl)n$ inhibits the $(dA).(dT)12$-catalyzed reaction indicates that $(dAfl)n$ may have a secondary structure similar to $(rA)n$, than to $(dA)n$.

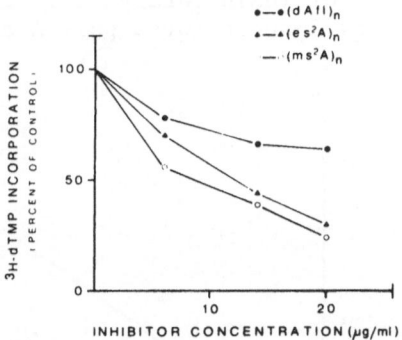

Fig. 11. Inhibition of $(dA)n.(dT)12$-catalyzed polymerase-ß reaction by poly A analogs.

The effect of $(ms^2A)n$ and $(es^2A)n$ on the poly-dA catalyzed reaction of terminal transferase is shown in Fig. 12. Both the compounds inhibit TdT activity, but in this case $(es^2A)n$ is a better inhibitor. Figure 13 shows the Lineweaver-Burk plot of the kinetic data of TdT reaction in the presence of $(es^2A)n$. It is evident that $(es^2A)n$ inhibits the TdT reaction competitively.

The effect of $(dAfl)n$ and other polynucleotides on the poly-dA catalyzed TdT reaction is shown in Fig. 14. Unexpectedly, we found a concentration-dependent stimulation of the TdT reaction when poly U was added into the reaction mixture. This may be due to some conformational effect due to interaction between the two complimentary strands. To our knowledge, this type of effect has not been reported so far. poly rA and $(dAfl)n$ which have no complimenatrity to the starter, remain almost ineffective when added to the reaction mixture.

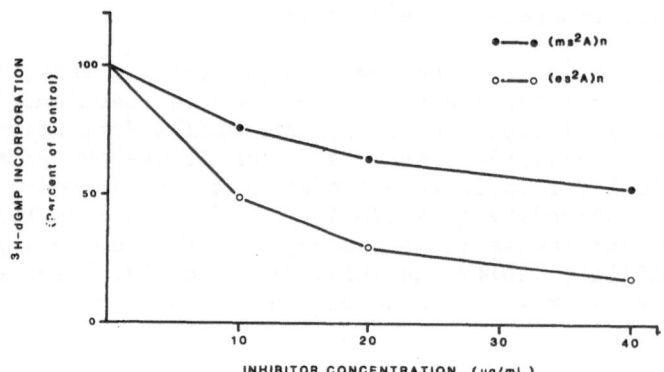

Fig. 12. Inhibition of (dA)n-catalyzed Terminal Transferase
Reaction by 2-substituted poly A-analogs.

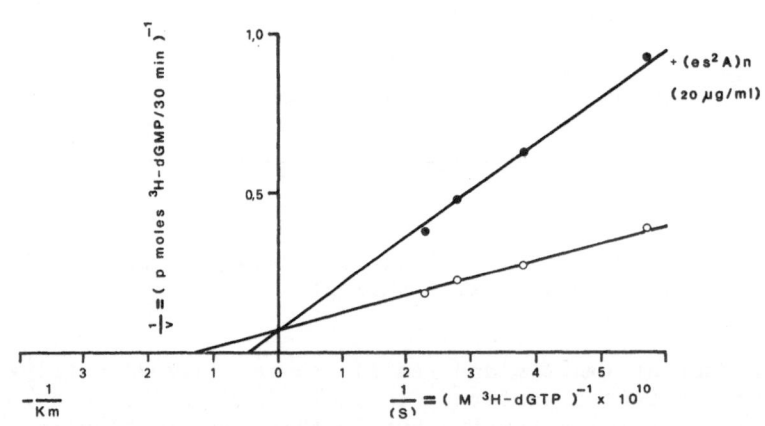

Fig. 13. Lineweaver-Burk Plot of the Kinetic Data of TdT
Inhibition by (es²A)n.

To determine the selectivity of (dAfl)n action on polymerases,
we have also studied its effect on DNA polymerase-α from calf
thymus. As follows from Table 3, (dAfl)n derivative stimulates the
reaction, whereas, (es²A)n shows a slight inhibition of the reaction.
The (ms²A) derivative did not show any significant effect at the
inhibitor concentrations used.Thus, the only polymerase which is
sensitive to (dAfl)n is the polymerase-ß.In this respect, it may
serve as a good indicator for the detection of polymerase-ß activity.

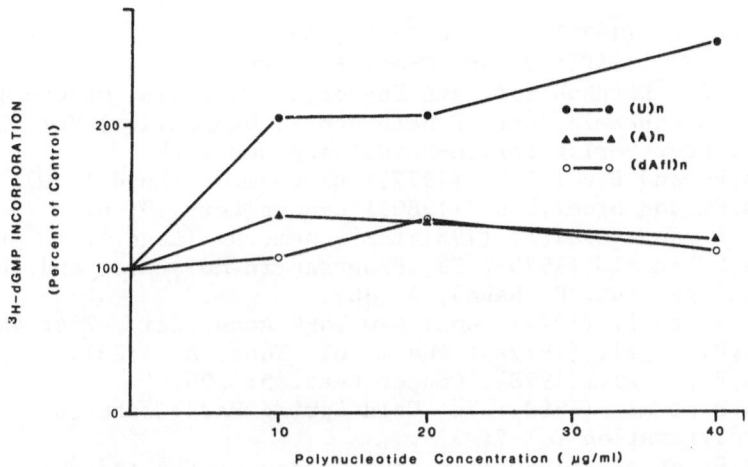

Fig. 14. Effect of (dAfl)n and other Polynucleotides on the
Poly dA-catalyzed Terminal Transferase Reaction.

Table 3.   Influence of 2- and 2'- Substituted Polyadenylic
Acid Analogs on the DNA Polymerase -α Activity from
Calf Thymus.

System	[^{1}H]dTMP incorporation into DNA (cpm/reaction mix)	% of control
Without enzyme	262	
Complete	9469	100 (9207)[a]
Inhibitor[b]		
$(ms^1A)_n$	7544	82
$(es^1A)_n$	6576	71
$(dAfl)_n$	13462	146

[a] Denotes the dTMP incorporation in complete system after the value of the blank probe
(without enzyme) was subtracted. Values for inhibitor experiments were similarly
corrected for blank.
[b] The inhibitor concentration was 20 μg/ml of the reaction mixture.

ACKNOWLEDGEMENTS
This work was supported by grants from Stiftung Volkswagenwerk
(Grant No. 14 0305). We thank Dr. Erik de Clercq (Belgium) for the
gift of poly A analogs, Dr. K.K. Gauri (Hamburg) for 5-alkyl-substi-
tuted derivatives, and Dr. David Shugar (Warsaw) for O-methylated
derivatives of Ara-CTP and TTP.

REFERENCES

Baltimore, D. (1970): Nature 226: 1209.
Bollum, F.J. (1958): J. Am. Chem. Soc. 80: 1766.
Chandra, P., Kornhuber,B. and Ebener,U. (1979)In, Modern Trends in
    Human Leukemia (Edt. R.Neth and K. Mannweiler) Vol. 3: 145,
    Springer-Verlag Berlin-Heidelberg-New York.
Chandra,P. and Steel,L.K. (1977): Biochem.J. (Lond.) 167: 513.
Chandra,P. and Steel,L.K. (1980)! Cancer Lett. 9: 67.
Chandra,P. and Vogel,A. (1981): Biochem. J. (Lond.), in press
Chandra,P. et al. (1975): In, Progress in Molecular and Subcellular
    Biology (Edt. F. Hahn), 4: 167.
Chandra,P et al. (1977): Ann. New York Acad. Sci., 284: 444.
Chandra,P. et al. (1977a): Pharmacol. Ther. A 1:231.
Chandra,P. et al. (1978): Cancer Lett. 5: 299.
Chandra,P et al. (1980): In, Cold Spring Harb.Conferences on Cell
    Proliferation Vol 7: 775.
Chandra,P. et al. (1980a):In, Inhibitors of DNA and RNA polymerases
    (Edts. P.S.Sarin and R.C.Gallo),pp47, Pergamon Press,Oxford.
Cheng,Y.C., Domin,B., Sharma,R. and Bobek,M. (1976): Antimicrobial
    Agents and Chemther. 10: 119.
Demirhan,I., Ebener,U. and Chandra,P. (1981): To be Published
Ebener,U., Welte,K. and Chandra,P. (1979): Cancer Lett.7:179.
Gallo,R.C., Yang,S.S. and Ting,R.C.(1970): Nature 228:927.
Gallo, R.C. et al. (1975): Cold Spring Harb.Symp.Quant.Biol.
    39: 933.
Kornhuber,B. and Chandra,P.(1979): In, Antiviral Mechansims in the
    Control of Neoplasia (Edt. P. Chandra)pp 577, Plenum Press,NY.
Matsukage,A. et al. (1978): Cancer Res. 38: 3076.
Müller,W.E.G., Zahn,R.K. and Seidel,H.J. (1971): Nature New Biol.
    232:143
Ohno,T. et al. (1977):Proc.Natl.Acad.Sci.,USA 74: 764.
Schrecker,A.W., Smith,R.G. and Gallo,R.C.(1974): Cancer Res.34:286
Temin,H.M. and Mizutani,S. (1970): Nature 226:1211.
Touminen,F.W. and Kenney,F.T.(1972):Biochem.Biophys.Res.Commun.
    48:1469.
Welte,K.,Ebener,U. and Chandra,P. (1979):Cancer Lett.7:189.
Witkin,S.S.,Ohno,T. and Spiegelman,S.(1975):Proc.Natl.Acad.Sci.,
    USA 72: 4133.

DEVELOPMENT OF ANTIBODIES AND IMMUNOLOGICAL METHODS FOR TERMINAL TRANSFERASE ASSAY

F. J. Bollum

Biochemistry, USUHS

4301 Jones Bridge Road, Bethesda, MD 20814

INTRODUCTION

Antibody against homogeneous terminal transferase (1) has turned out to be an essential reagent for biological and biochemical investigations on the occurrence and structure of the enzyme. Carefully prepared antibody preparations have allowed characterization of the normal cells that contain TdT (2-4), analysis of leukemic cell populations (5), and, most recently, new insight into the conservation and structure of terminal transferase peptides (6). Although preparation, characterization and use of monospecific immune reagents is a fairly routine procedure, it is of value to record some of the experience we found to be worthwhile.

Preparation of Antigen

Homogeneous calf thymus TdT is used as antigen. It is prepared as previously described (7), and before use it is checked for homogeneity by electrophoresis in 12.5% acrylamide containing 0.1% SDS. Not all preparations are clean, so we select one that shows only the β and α chains when 5 μg is loaded onto the gel. About 2 mg of pure antigen with full enzyme activity is required to immunize four 2 kg New Zealand male rabbits. Each rabbit is given 50 μg antigen per injection and about 10 total injections are needed. For continued production on a routine basis more antigen will be needed. Preparation of the immunoadsorbent column also requires at least 1 mg and preferably 5 mg of pure antigen.

The homogenous antigen is dialyzed or diluted into phosphate buffered saline (PBS) to a concentration of 100 μg TdT per ml. The solution is brought to 0.1% glutaraldehyde by addition of a 5%

solution with good mixing. The protein is allowed to cross-link
at room temperature for 20 minutes, during which time a slight
turbidity may form. The reaction is stopped by adding 1/40th volume
of 1% NaBH$_4$ solution. The solution is mixed carefully and allowed
to degas overnight before use.

Immunization of Rabbits

An injection schedule is set up according to Table 1.
This schedule is based on our successful experiences and works
100% of the time. All of the sera are labeled and stored at -20°
until further work up. The critical point seems to be the 7th week,
at which time all animals have received 4 injections of cross-linked
antigen. Analysis of the test bleedings will regularly be negative
until then, at which time weak precipitating (undiluted serum vs.
5 µg per ml TdT) and neutralizing (usually 1:16 or better) activity can be seen. If antibody cannot be detected at this point,
one must decide whether to continue with cross-linked antigen or
discontinue on non-responders.

The boost and bleed schedule using native antigen in responding
animals is based on the idea that one should be able to increase
the response to native determinants in the antigen once immunity is
established. Antibody production to this booster is maximum at 9
to 11 days after injection and large bleedings are taken from the
ear at this time. We use rabbit restrainers and bleed directly into
beakers from 18 ga syringe needles in the central ear vein. The
11 day serum is as active as the 9 day serum and we usually pool
these two samples to reduce inventory. Titer is defined as the
dilution giving 50% direct inhibition of 1 unit of terminal transferase activity and titers of 1:128 to 1:256 are routine. After
this rather extensive bleeding the rabbit is allowed to recover for
three to four weeks. The boost and bleed schedule can be repeated
for at least six months, at which time each rabbit should have provided about 360 ml of antiserum from the six volume bleedings. The
quality of the antiserum may vary during prolonged immunization and
each preparation must be tested to ensure proper practical specificity and titer.

Crude Immunoglobulin Preparation

Serum samples from two large volume bleedings of a single
rabbit (9th and 11th day after boost) are pooled to obtain about
60 ml of serum. If serum from several rabbits is processed simultaneously all volumes are adjusted to 60 ml for convenience. The
serum is diluted with an equal volume of PBS at room temperature
and immunoglobulins are precipitated by gradual addition of an
equal volume of saturated ammonium sulfate. The precipitate is
collected by centrifugation and washed two times by resuspension
in 50% saturated ammonium sulfate and centrifugation. Each pellet

Table 1. Immunization Schedule

	Bleed	Inject
Week 1	Preimmune (5 ml)	0.5 ml cross-linked antigen in 0.5 ml complete Freund's adjuvant to toe pads.
Week 3	Test bleed (5 ml)	0.5 ml cross-linked antigen in 0.5 ml incomplete Freund's adjuvant 0.25 ml subcutaneously over each limb to the back.
Week 5	Test bleed (5 ml)	0.5 ml cross-linked antigen in 0.5 ml incomplete Freund's adjuvant 0.25 ml subcutaneously over each limb to the back.
Week 7	Test bleed (5 ml)	0.5 ml cross-linked antigen in 0.5 ml incomplete Freund's adjuvant 0.25 ml subcutaneously over each limb to the back.
Week 8	Test all sera for precipitation by microdiffusion. Test all sera for neutralization of TdT activity.	
Week 9		0.5 ml native TdT in 0.5 ml incomplete Freund's adjuvant, to the back as above in all rabbits showing immune response.
+ 9 days	Volume bleed (70 ml)	None
+ 11 days	Volume bleed (70 ml)	None
Week 12 + 13	Rest	
Week 14 —	Repeat week 9 schedule, on a monthly basis as required for antibody production.	

is resuspended to a final volume of 60 ml with PBS and re-precipit-
ated by addition of an equal volume of saturated ammonium sulfate
and washed with 50% saturated ammonium sulfate as before. The
final pellet is dissolved in 60 ml of PBS as before and dialyzed
extensively against PBS at 4°. The dialyzed crude IgG is stored
at −20° until further purification. This procedure should provide
60 ml of crude IgG at 10–15 mg protein per ml. The crude IgG
should not be considered pure on any criteria, since many other
serum proteins still contaminate this fraction.

Immunoadsorbent Column

Controlled pore glass/Glycophase-G, 550 Å pore diameter, is
weighed out, suspended in water and degassed in a vacuum dessicator.
For each mg of pure TdT to be coupled, 1 gm of glass is required.
The glass particles are suspended in 50 ml of 10 mM $NaIO_4$ per gm of
glass. After 15 minutes the $NaIO_4$ is washed out well with 0.1 M
K_2HPO_4. TdT is then mixed in at 1 mg per gm of glass and the solution is allowed to mix for one hour at 4°. The coupling is termi-
nated by addition of 1 ml of 1% $NaBH_4$ per 50 ml of reaction. The
degree of coupling is assessed by measuring A^{280} of the supernatant
solution. Better than 95% coupling is always achieved. The glass-
TdT column is washed extensively with PBS containing .02% NaN_3 and
stored at 4° until used. For longer storage suspend in 50% glycerol
PBS with NaN_3 at −20°. Equilibrate the column with PBS before use.

Processing for TdT-specific IgG

A 3 ml TdT-glass column is equilibrated with 20 ml PBS. Add
12 ml of crude IgG solution to the column and allow to pass through
at 16 drops per minute. The pass-through fraction and a 5 ml PBS
wash is saved for reprocessing. The column is washed further with
20 ml PBS and this wash is discarded. The column is then eluted with
20 ml 0.5 M NaCl containing 50 mM glycine buffered at pH 3.0 and
1 ml fractions are collected into 13 x 100 tubes, each containing
50 µl of 1 M TRIS·Cl, pH 8.4. Absorbancy of the fractions at 280 nm
is read and all fractions with absorbancy greater than 0.1 are
pooled. The column is then washed with 20 ml of PBS before a second
load is applied and processed exactly as above. A third load is
processed as above and then the pass through fractions from the
first three column runs are pooled and run through the column again.
The column is then washed and eluted to recover any IgG not absorbed
in the first pass. The four pools of TdT-specific IgG are then
combined. From 36 ml of crude IgG (A^{280} = 10.8 per ml) 36 ml of
TdT-specific IgG (A^{280} = 0.38) was obtained. This solution may be
concentrated or used directly for immunofluorescence titrations.
We usually dialyze into PBS and store at −20°. Material prepared
and stored in this manner has remained active for at least 6 years.
The TdT-specific IgG is usually used at about a 1:10 dilution into
.01% BSA in PBS. Samples are usually shipped in well-sealed Nunc®

tubes in a cardboard mailing tube by ordinary, first class air mail.

The column purification procedure described is carried out at room temperature, and the column stored in a cold box overnight when not in use. We prefer to use relatively small columns to conserve TdT and reuse continuously to avoid bacterial contamination. The schedule suggested above can be modified according to the volume of IgG to be processed and the volume of each load can be adjusted for major changes in serum titer. With the serum titers we have achieved thus far (1:32 - 1:256) the schedule works well as written.

Second Layer Antibody

We have used a high titer goat antirabbit IgG prepared in our own laboratories. This material is converted to an $F(ab')_2$ dimer and then fluoresceinated. Standard reference works should be consulted for these procedures (8). The increasing availability of immuno-adsorbent purified antibodies, and $(Fab')_2$ derivatives from commercial suppliers should relieve the investigator of this task. But all antibodies, primary and secondary, must be appropriately tested before use.

Immunofluorescence

Cell suspensions are prepared in RPMI 1640 tissue culture medium (+ 10% FCS) and adjusted to about 10^6 cells per ml. Aliquots of 10^4 - 10^5 cells are deposited on microscope slides using a cyto-centrifuge. Cryostat sections are prepared by standard methods. All cell preparations are allowed to air-dry briefly and then processed immediately. Bone marrow and peripheral blood smears (preferably lymphocyte-rich fractions) are also prepared by standard methods on microscope slides for handling convenience. Cover slip preparations can be used but are inconvenient for mass handling. All of these slide preparations are rather unstable for TdT fluorescence. Tissue culture preparations, cytospin smears and cryostat sections are stained the same day for best results. Blood and bone marrow smears may keep for two weeks at room temperature and still be readable, but fresh preparations always give the clearest results. Preservation is improved in all cases by storing unfixed slides at room temperature in a dessicator over Drierite®.

The standard two-layer procedure used is outlined in Table 2.

Troubleshooting TdT Immunofluorescence

Microscope light source - The light filtration system commonly used for visualizing surface immunoglogulin fluorescence does not give good results for TdT fluorescence. Narrow band pass filters producing a royal blue colored light (maximum at 495 nm) provides a much reduced background and clear apple green fluorescence. If

Table 2. Indirect Immunofluorescent Technique (After Coons)

---

1. Slide preparation, preferably freshly made. Circle area
    of interest with diamond pencil.
2. Cold methanol fixation (4°C, 30'), air dry.
3. I° Antibody (immunoadsorbent-purified rabbit anti-TdT).
    Apply 15 μl to circled area and keep in humid chamber 30'.
4. Wash three times with fresh PBS in a Coplin jar. Wipe off
    excess PBS.
5. II° Antibody (FITC-goat anti-rabbit IgG, F(ab')$_2$). Apply
    15 μl to circled area and keep in humid chamber 30'.
6. Wash three times with fresh PBS in a Coplin jar. Wipe off
    excess PBS. Apply 1 drop FA amount to # 1 cover slip.
    Invert circled area of slide over drop.
7. Examine by phase and epifluorescence using 40x dry
    objective. Photograph for permanent record using 63x or
    greater immersion objective. Use ASA 200 or greater color
    film. Expose at ASA 800 and push process.

---

the light coming from the objective has a purplish cast new filters
will be required.

  Bad slides - Slide material that has been subjected to hydration and dehydration by storage below room temperature or shipped long
distances by air will often give poor results. The reason for this
problem is that TdT is a rather soluble antigen and can migrate on
repeated hydration. Unfixed, unstained material should be stored and
shipped dry.

  Wet methanol fixative - During humid times of the year absolute methanol stored in refrigerators, or even at room temperature,
tends to absorb moisture. Use of wet methanol will result in
incomplete fixation. Be sure to use methanol that is absolutely
dry for fixation immediateley before staining.

Other Slide Methods

  The most sophisticated immunological marker detection system
uses a combination of surface staining of viable cells in suspension
followed by cytocentrifugation and fixation for TdT staining. For
double markers this procedure is usually done with TRITC and FITC
second reagents to produce two color detection with separate filters. Three color marker systems could be developed for even greater
power in determining cellular phenotypes. With known cellular marker
locations it is possible to do triple markers with only two colors.
Thus, surface antigen A can be stained and capped. Surface antigen B

can be stained and either "patched" or left distributed. The cells are then cytocentrifuged and fixed. Antigen C can then be detected by nuclear or cytoplasmic localization. These procedures have been described in work by Janossy et al. (9).

Immunoperoxidase detection systems offer the advantage of permanent mounting and use of other cytochemical stains. Several laboratories have published procedures for peroxidase-TdT staining. The time required for these procedures is somewhat greater, due to the increased number of staining steps.

ELISA

Immunoreactive TdT peptides can be detected in bone marrow plasma of ALL patients using an ELISA procedure (10). Microtiter plates coated with rabbit anti TdT are allowed to react with plasma dilutions. After suitable washes antigen antibody complexes are detected by alkaline phosphatase conjugates of anti TdT and appropriate substrates. The ELISA methodology is quite rapid and allows quantitative determination of antigen level.

The ELISA procedure has not yet been evaluated in large scale patient studies. The major potential advantage of this procedure would be in detection of enzymatically inactive (and active) TdT fragments in serum, urine, other body fluids, or tissue extracts. This might allow estimation of tumor burden and response to therapy. Non-invasive testing would simplify routine evaluation and screening of large numbers of patients.

General Comments

The availability of methods for enzyme assay, cytochemical tests, and solution methods for TdT analysis provides great flexibility in the detection of TdT. These methods have application in chemical and immunological studies that should provide continued improvement in our understanding of the biology and pathology of TdT.

ACKNOWLEDGEMENT

This research has been partially supported by USPHS grant CA 23262.

REFERENCES

1. F. J. Bollum, Proc. Natl. Acad. Sci (USA) 72:4119-4122 (1975).
2. K. E. Gregoire, I. Goldschneider, R. W. Barton, and F. J. Bollum, J. Immunol. 123:1347-1352 (1979).

3. R. Sasaki, F. J. Bollum, and I. Goldschneider, J. Immunol.
      125:2501-2503 (1980).
4. I. Goldschneider, D. Metcalf, T. Mandel, and F. J. Bollum,
      J. Exp. Med. 152:438-446 (1980).
5. F. J. Bollum, Blood 54:1203-1215 (1979).
6. F. J. Bollum, S. M. Hassur, and L. M. S. Chang in:
      "Leukemia Markers", W. Knapp, ed., Academic Press,
      New York pp 33-40 (1981).
7. L. M. S. Chang and F. J. Bollum, J. Biol. Chem. 246:
      909-916 (1971).
8. D. M. Weir, ed., Handbook of Experimental Immunology,
      Vol. 3, Immunochemistry, Blackwell Publications,
      Oxford, 3rd Ed. (1973).
9. G. Janossy, F. J. Bollum, K. F. Bradstock, and J. Ashley,
      Blood 56:430-441 (1980).
10. S. A. Stass, S. C. Peiper, and F. J. Bollum, Am. J.
      Hematol. 9:429-433 (1980).

USE OF IMMUNOFLUORESCENCE TESTS FOR TdT IN HUMAN HEMATOLOGIC MALIGNANCY

Mary Sue Coleman and John J. Hutton

Department of Biochemistry, University of Kentucky
Lexington, Kentucky and Department of Medicine
University of Texas, San Antonio, Texas

Terminal deoxynucleotidyl transferase (TdT) has been established as a valuable clinical tool in the differential diagnosis of acute leukemia. In pediatric disease, TdT is almost uniformly elevated in tumor cells for "T" cell and "Common" acute lymphoblastic leukemia (ALL) and in selected patients with "Pre B" ALL. In adult leukemias TdT is highly elevated in chronic myelogenous leukemia (CML in blast crisis)(30% of cases) and in undifferentiated leukemia (AUL) in acute phase (50% of cases). Patients with acute myeloid leukemia (AML) will occasionally exhibit high levels of TdT activity (10% of cases).[1] Detection of TdT in leukemic cells is important clinically because it is indicative of responsiveness of the disease to chemotherapeutic protocols containing vincristine and prednisone.

In clinical samples (bone marrow, peripheral blood, testicular tissue, cerebrospinal fluid, and lymph nodes), terminal transferase has been detected by a quantitative radiometric assay and by an immunofluorescence assay which was developed after antibody to terminal transferase became available.

The quantitative assay which we developed utilizes a homopolymer of average chain length 50 residues [$p(dA)_{50}$] and dGTP. This assay is particularly useful in crude extract because, the initiator can withstand several endonucleolytic cleavages and the poly dG which is synthesized during the reaction is resistant to exonuclease activities because the product aggregates. In the leukemic population of ALL and CML blast crisis, the majority of patients exhibit activities ≥ 20 $U/10^8$ cells in BM, and ≥ 10 $U/10^8$ cells in PB. Patients who do not exhibit these levels of activity are often characterized as pre B cell leukemias. Among acute myeloid leukemias, most are ≤ 20 $U/10^8$ cells in BM and ≤ 10 $U/10^8$ cells in PB.

Control marrow is almost always $\leq$20 U/10^8 and PB $\leq$10 U/10^8 cells.
The quantitative assay is very reliable, but it does require extensive cell preparation (2x10^7 cells) and reaction time.

Interest existed in developing antibodies to terminal transferase for a variety of reasons. Bollum first accomplished this
in 1975 when he was able to produce rabbit anti-calf TdT.[2] He later
developed an immunofluorescence assay for the protein.[3] When we
purified human terminal transferase an antibody to the human antigen
was raised in rabbits and in goats.

The human antigen we used was M_r = 42,500. Calf TdT was the
control immunogen (M_r = 30,500 and M_r = 9,500). In rabbits,
acceptable titers were developed against both human and calf antigens. In goats antibody was present, but the titers were too low
for the antibody to be of practical value in the laboratory. When
we tested the rabbit antisera to calf and human TdT anti-human TdT
cross-reacted with lines of identity against all forms of human
enzyme, but spurs were apparent against calf TdT. Anti-calf
TdT cross-reacted with calf TdT, but spurs were apparent with all
forms of human TdT. These data are quantitated in Table 1.

Table 1.   Inhibition of TdT by Rabbit Antisera

Serum	Protein required to neutralize by 50%, 1U of TdT			
	Human TdT (M_r=62,000)	Human TdT (M_r=42,500)	Human TdT (M_r=27,000 + 10,000)	Calf TdT (M_r=30,500 + 9,500)
	µg	µg	µg	µg
Rabbit anti-human TdT	20.8	14.2	10.2	75.4
Rabbit anti-calf TdT	6.7	6.7	5.6	11.3

These data indicate that antigenic sites on the protein are not
equally shared. The large human TdT (M_r=42,500) probably has an
antigenic site which is not present on the calf TdT.

We purified the anti-human TdT by affinity chromatography using homogeneous TdT specifically linked to controlled pore glass.[3] Immunofluorescence assays for TdT antigen were carried out using an indirect staining method. As part of the study of the development of the immunofluorescence test with anti-human TdT, we felt that it was essential to compare the data from quantitative tests for TdT activity with the test for TdT antigen and to assess the reliability of the immunofluorescence test in a clinical setting.

For testing of human samples we used rabbit anti-calf TdT prepared by F. J. Bollum and rabbit anti-human TdT prepared in our laboratory. Slides were prepared in situ and read by two individuals. A positive result is scored if greater than 10% of nucleated cells show reticulated nuclear fluorescence (normal marrow is less than 1%). The quantitative assay was considered positive when ≥ 10 U/10^8 cells were in peripheral blood and ≥ 20 U/10^8 cells were in bone marrow (normal marrow < 3 U/10^8 cells). Bone marrow and peripheral blood data are combined as shown in Table 2.

Table 2. Comparison of TdT in 267 Human Bone Marrow or Peripheral Blood Specimens by Immunofluorescence (IF) or Quantitative Enzyme Assay (RA).

IF positive RA positive	IF positive RA negative
71	14

IF negative RA positive	IF negative RA negative
6	176

A detailed analysis of the use of rabbit anti-calf serum versus rabbit anti-human serum indicated that both of these gave nearly identical results and the data are combined in this Table 2 (207 tests were performed with rabbit anti-calf TdT and 60 tests were performed with rabbit anti-human TdT).

An analysis of individual samples grouped by disease showed that there is no systematic error in detecting TdT protein. That is the cases, in which TdT activity is detected but TdT antigen is not, are not representative of a particular disease state. The lack of correlation between the two types of tests is random and is not confined to bone marrow or peripheral blood or disease diagnosis.

In the second part of the study, we shipped the air dried slides to F. J. Bollum in Bethesda. The shipping experiment was carried out to determine whether dried smears could be sent to a central laboratory for staining. Duplicate slides were then stained and read. The accuracy of the test dropped significantly after shipment as shown in Table 3.

Table 3. The Effect of Air Shipment on Immunofluorescence Test For TdT in 183 Specimens.

IF positive RA positive	IF positive RA negative
11	2

IF negative RA positive	IF negative RA negative
41	127

As is clearly seen in Table 3, the effect of air shipment is to reduce positive immunofluorescence. For this reason, we strongly recommend that slides be prepared for immunofluorescence at the site where the test will be performed.

Utilizing the combined data from rabbit anti-human and anti-calf TdT, the sensitivity and specificity of this test are 97% and 93%. The predictive value of a positive test in this study is 84%. This value has been confirmed by us in a similar study performed with a large group of pediatric patients.[5] We believe that these data indicate that immunofluorescence tests for TdT antigen can be successfully performed in a routine clinical setting.

The care required in handling specimens is clearly illustrated in the decline in sensitivity from greater than 90% when the test is performed on slides prepared in situ to less than 30% for slides shipped by air to a central location for staining.

The small percentage of cases in which a discrepancy is observed between the quantitative enzymatic TdT assay and the immunofluorescence test may be due to technical problems related to slide preparation. Certainly, the divergent results are not specifically associated with any particular disease state or with peripheral cells versus marrow. We must conclude that if results obtained in an immunofluorescence test seem to contradict disease course, then a quantitative enzyme assay is warranted.

It is important to note that the quantitative enzyme assay and the immunofluorescence test do not reflect equivalent data, since it is clear that the quantity of TdT protein per cell may vary among neoplasms. Therefore a prediction of absolute enzyme activity levels is not possible from the immunofluorescence test.

Overall, we believe that immunofluorescence tests for TdT antigen performed in a clinical setting can offer significant advantages in speed and in reduced amount of the required specimen. Some technical difficulties remain with this immunodetection procedure which may be overcome when a sensitive ELISA assay is developed for TdT.

References

1. Hutton, J. J., Coleman, M. S., Keneklis, T. P., and Bollum, F. J., 1979, Terminal Deoxynucleotidyl Transferase as a Tumor Cell Marker in Leukemia and Lymphoma: Results from 1000 Patients, in "Biological Markers for Cancer Diagnosis," M. Fox, ed., Pergamon Press Ltd., Oxford.
2. Bollum, F. J., 1975, Antibody to Terminal Deoxynucleotidyl Transferase, Proc. Natl. Acad. Sci. U.S.A. , 72:4119.

3. Bollum, F. J., 1979, Terminal Deoxynucleotidyl Transferase as a
       Hematopoietic Cell Marker, Blood, 54:1203.
4. Galen, R. S., and Gambino, S. R., 1975, Beyond Normality, in:
       "The Predictive Value and Efficiency of Medical Diagnosis,"
       John Wiley and Sons, New York.
5. Kalwinsky, D. K., Weathrred, W. H., Dahl, G. V., Bowman, W. P.,
       Melvin, S. L., Coleman, M. S., and Bollum, F. J., Value of
       Terminal Deoxynucleotidyl Transferase Determination in
       Childhood Acute Leukemias, in press, Cancer Res.

ONTOGENY OF TERMINAL DEOXYNUCLEOTIDYL TRANSFERASE
CONTAINING LYMPHOCYTES IN RATS AND MICE*

Irving Goldschneider

Department of Pathology
University of Connecticut Health Center
Farmington, Connecticut 06032, USA

INTRODUCTION

In the classic publication by L. Chang (1), the ontogenetic appearance of terminal deoxynucleotidyl transferase (TdT) was documented in the calf thymus by enzymatic analysis. TdT activity was also found in the thymus of pig, rat, rabbit, and chicken. Subsequently, TdT has been identified in high concentrations in the thymus and, in much lower concentrations, in the bone marrow of mice and human beings (2-4). Recently, minor populations of TdT-bearing cells have been demonstrated in bone marrow, spleen, liver, lung and blood of rats and mice by a sensitive and highly specific immunofluorescence assay (5-8). This technique has not only permitted direct quantification of TdT^+ cells, but it has facilitated the study of their morphology, antigenic properties and anatomical distribution.

In collaboration with Dr. F. J. Bollum and his colleagues, we have used the immunofluorescence assay for TdT to study the ontogeny of TdT^+ cells in the rat and mouse. The results of these studies and of those from other laboratories strongly indicate that TdT is restricted to the lymphoid cell series, in which it is associated primarily with the antigen-independent phase of T cell, and possibly of B cell, development.

RESULTS

---

* Supported in part by Grant AI-14743 from the National Institute of Allergy and Infectious Diseases, USPHS, and by Grant CH-111 from the American Cancer Society.

1. Age-Related Distribution of TdT$^+$ Cells

Subsets of TdT$^+$ cells can be distinguished according to their generative kinetics, their anatomical location, and their developmental status (9). Populations of TdT$^+$ cells may appear before or after birth and may be persistent or transient. They may be present in thymus, bone marrow, spleen, liver, lung or blood. They may be prethymic, thymic or postthymic. These relationships are summarized in Figure 1 and Table I.

TdT$^+$ cells first appear in the subcapsular region of the thymus cortex on or about day 17 of fetal life in rats and mice (7). Their appearance correlates with the onset of proliferation and differentiation of thymocyte precursors, which migrate to fetal thymus from hemopoietic tissues (10,11). Results of experiments in which 14 day fetal mouse thymuses were explanted in vitro show that the precursors of TdT$^+$ thymocytes reside in the thymus cortex for at least 3 days before they express TdT (12). The percentage of TdT$^+$ thymocytes increases progressively during the last week of gestation and reaches maximum and relatively stable values of approximately 70% in the neonatal period. The proportion of TdT$^+$ thymocytes decreases slowly after the onset of physiological involution of the thymus.

TdT$^+$ thymocytes appear to be restricted to the thymus cortex in rats and mice. This has been demonstrated by antigenic analysis, cortisone sensitivity, peanut lectin binding, size and buoyant density distribution, and localization in tissue sections (3,5-7,13,14). Approximately 10% of TdT$^+$ thymocytes are large, proliferating lymphoblasts which reside in the subcapsular region of the thymus cortex (15). They have a characteristically diffuse pattern of intranuclear TdT fluorescence, except during mitosis when the TdT is released into the cytoplasm and the nuclear region becomes TdT$^-$. The majority of TdT$^+$ thymocytes are small, non-proliferating cells which are located in the deep cortex. These have a coarse pattern of TdT fluorescence that tends to be associated with the nuclear membrane. In addition, many small TdT$^+$ thymocytes have detectable cytoplasmic TdT fluorescence.

TdT$^+$ cells first appear in bone marrow one or two days after birth and reach peak levels of about 4% by 3 weeks of age (7). These levels are maintained until about 8 weeks of age, after which they slowly decline to reach baseline levels of between 0.5 and 1.0%. Comparable age-associated decreases in TdT enzymatic activity have also been observed (16,17). TdT$^+$ bone marrow cells have a mean diameter of 10.4 μm on cytocentrifuge smears, a moderate amount of deeply basophilic cytoplasm, a prominant Golgi zone, and an indented leptochromatic nucleus which contains several nucleoli (18). They are indistinguishable cytologically and with respect

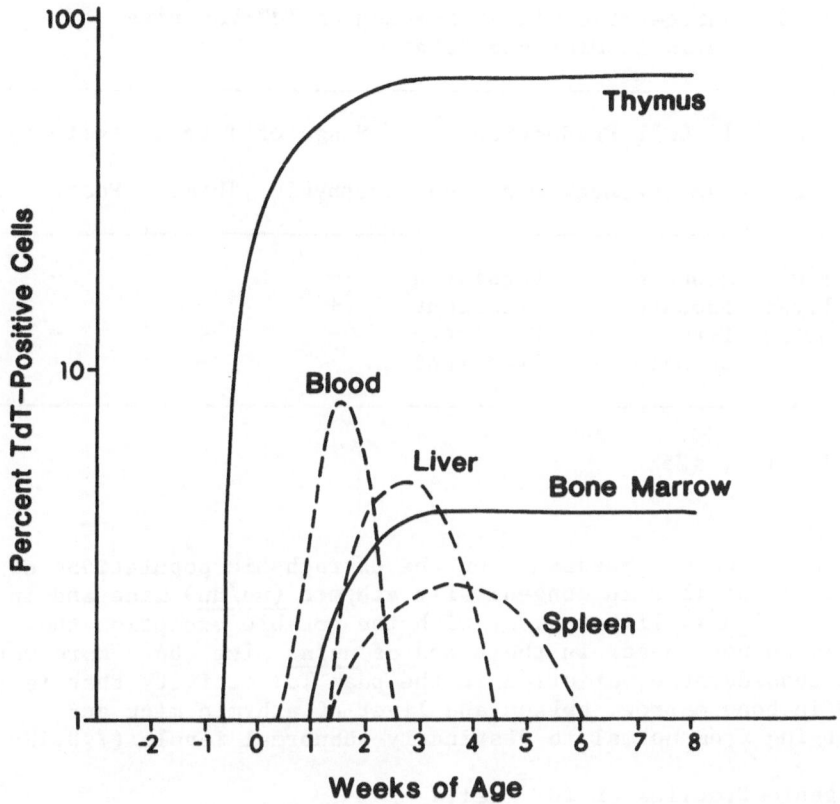

Fig. 1.   Development of TdT-positive cells in the rat as determined
         by indirect immunofluorescence. TdT-positive cells are
         represented as the percentage of total nucleated cells in
         thymus, bone marrow and spleen cell suspensions or as the
         percentage of total lymphoid cells in blood and liver.
         TdT-positive cells in all extrathymic sites were first
         detected (0.1%) within 24 hrs of birth (time 0).

to the pattern of TdT fluorescence from TdT$^+$ cells in the subcapsular
region of the thymus cortex (6).

     Of the transient populations of TdT$^+$ cells shown in Figure 1,
those which appear in spleen and liver most closely resemble TdT$^+$
bone marrow cells morphologically and antigenically, whereas
those which appear in blood most closely resemble TdT$^+$ cortical
thymocytes (see Section 2, below). Each of these populations
appears during the first week of life and disappears by the onset
of puberty (8). However, TdT$^+$ cells reach peak levels at different
times in blood (1-2 weeks), liver (2-3 weeks) and spleen (3-4 weeks).

Table I. Ontogenetic Classification of TdT—Positive
          Cells in Mice and Rats*

Kinetics of TdT$^+$ Cell Production			Stage of T Cell Development		
Location	First appearance	Duration	Prethymic	Thymic	Postthymic
Bone marrow	Neonate	Persistent	+	–	–
Spleen, liver	Neonate	Transient	+	–	–
Thymus cortex	Fetus	Persistent	–	+	–
Blood	Neonate	Transient	–	–	+

*(+), >75%; (–), <25%.

The kinetics of appearance of the extrathymic populations of
TdT$^+$ cells is similar in congenitally athymic (nu/nu) mice and in
their normal (nu/+) littermates, with the notable exception that
TdT$^+$ cells do not appear in the blood of nu/nu mice (8). Moreover,
there is considerable variation in the peak TdT activity that is
attained in bone marrow, spleen and liver of athymic mice and
rats, ranging from normal to distinctly subnormal levels (7,8,19–22).

2. Antigenic Profiles of TdT$^+$ Cells

The antigenic phenotypes of the various populations of TdT$^+$
cells are listed in Table II (23,24). Two major patterns are
evident; a "null" cell pattern and a thymic pattern. The "null"
cell pattern characterizes the TdT$^+$ cells in bone marrow, spleen
and liver. Whereas these cells generally lack markers that are
present on T and B cells, some heterogeneity is evident. Thus in
the rat, approximately 25% of TdT$^+$ bone marrow cells and 65% of
TdT$^+$ spleen cells react with the monoclonal antibody W3/13. This
antibody binds to differentiating and mature hemopoietic cells,
but not with their less differentiated precursors (25; D. Greiner
and I. Goldschneider, unpublished). A similar heterogeneity is
found among TdT$^+$ cells from nu/nu rats, and among TdT$^+$ rat bone
marrow cells generated in vitro (26). Moreover, a majority of
W3/13$^-$ TdT$^+$ bone marrow cells generated in vitro can be induced
to express W3/13 antigen (see ref. 27, these Proceedings). It is
likely therefore that most W3/13$^+$ TdT$^+$ cells in bone marrow and
spleen are derived from W3/13$^-$ TdT$^+$ precursors; although it is
possible that some belong to a separate lineage.

Heterogeneity is also present among TdT$^+$ cells in mouse bone
marrow. As reported by Silverstone et al. (28), approximately

Table II.   Antigenic Properties of TdT-Positive
Cells in Mice and Rats*

	Bone Marrow	Spleen	Thymus	Blood
Mouse				
Thy 1	−	−	+	+
Lyt 1,2/3	−	−	+	+
TL	−	−	+	+
PNA	−	−	+	+
Lyb-2	(±)	N.D.	−	N.D.
Sc-1	(±)	N.D.	−	N.D.
sIg	−	−	−	−
cIg	−	−	−	−
Rat				
Thy 1	+	+	+	+
W3/13	(±)	+	+	+
ALS$_t$	−	−	+	+
A.R.T. 1	−	−	+	+
PNA	−	−	+	+
OX 8	−	−	+	+
W3/25	−	−	+	−
sIg	−	−	−	−
cIg	−	−	−	−

*(+), >75%; (±), 25–50%; (−), <10%, N.D., not done.

50% of TdT$^+$ bone marrow cells from normal mice bear the Lyb-2 alloantigen, and at least 50% of TdT$^+$ bone marrow cells from normal and congenitally athymic mice bear the Sc-1 heteroantigen. The developmental relationship between the various subsets of TdT$^+$ cells is unclear. However, the Lyb-2$^−$ subset could be induced by thymopoietin to express Thy 1 antigen, whereas the Lyb-2$^+$ subset could not (see ref. 27, these Proceedings, for further discussion).

It is evident from Table II that TdT$^+$ bone marrow and spleen cells normally express Thy 1 antigen in rats but not in mice. This reflects a more general difference in the distribution of Thy 1 antigen in these two species. Thus, Thy 1 is present on all lymphohemopoietic precursor cells and on cortical thymocytes in the rat, but is expressed by only about 10% of peripheral T

Table III. Enzymatic Phenotypes of Members of the
            T Cell Series in Rats

Types of Cells	TdT	ADA	PNP	LDH-5'
TdT-positive Bone marrow cells	+	low	high	+
Subcapsular thymocytes	+	low	high	−
Cortical thymocytes	+	high	low	−
Medullary thymocytes	−	low	high	−
Peripheral T cells	−	low	high	−

cells (29-31). Indeed, next to pluripotent hemopoietic stem cells, TdT$^+$ cells are the most strongly Thy 1$^+$ cells in rat bone marrow (18, 31). This observation has enabled us to isolate viable TdT$^+$ bone marrow cells on the fluorescence-activated cell sorter (FACS) at a mean purity of 87% (18).

The presence of TdT in early members of the B cell series of rats and mice is still problematic. Certainly mature B cells and sIg$^+$ bone marrow B cells are TdT$^-$. The detection of low levels of TdT in some Abelson virus-induced murine "pre-B" cell lymphoid cell lines (sIg$^-$cIg$^+$) suggests that TdT may be transiently expressed at an early stage of B cell development (28). However, only one report has claimed to have detected TdT in some normal mouse "pre-B" cells (32). We have not been able to confirm this, but the difference may be due to technical factors in the dual immunofluorescence assay or to extremely small numbers of TdT$^+$ pre-B cells, such as reported for human bone marrow (33). It has not yet been possible to determine whether TdT is expressed by a common T-B lymphoid stem cell, as proposed by Silverstone et al. (28). Indeed, it is not clear that such a stem cell exists in any form (34).

With one exception, the majority of TdT$^+$ cells in blood and thymus cortex have identical antigenic phenotypes. In the mouse it is notable that TdT$^+$ blood cells are TL antigen positive and avidly bind peanut lectin (PNL); in the rat it is notable that they are strongly Thy 1$^+$ and PNL$^+$. The exception is that most rat TdT$^+$ thymocytes react with both the W3/25 and the OX8 monoclonal antibodies, whereas many rat TdT$^+$ blood cells react with only the OX8 antibodies. Inasmuch as rat T cells which have a Thy 1$^+$, OX8$^+$, W3/25$^-$ phenotype are associated with suppressor and/or cytotoxic activities (25,35), this observation may provide some clues as to the potential functions of TdT$^+$ blood cells (see Section 6, below).

3. Enzymatic Profiles of TdT$^+$ Cells

a) Adenosine deaminase (ADA) and Purine nucleoside phosphorylase
(PNP). Inherited deficiencies of either of these nucleotide-
metabolizing enzymes interferes with the normal development of T
lymphocytes in human patients (36,37). Consequently the expression
of these enzymes during normal T cell development is of interest.
We have chosen the rat model for this purpose (38-40).

Our results show that the specific activities of ADA and PNP
vary inversely at defined stages of T cell differentiation (Table
III). ADA activity is 3 to 10-fold higher among TdT$^+$ cortical
thymocytes than among medullary thymocytes, peripheral T cells or
TdT$^-$ bone marrow cells (isolated on the FACS). PNP activity is low
among cortical thymocytes and high among the other cell populations.
Fetal thymocytes have very low ADA levels, which progressively
increase with the onset of thymocyte differentiation; PNP levels
decrease during maturation of the fetal thymus. The same pattern
is seen in the regenerating thymus, in which undifferentiated
subcapsular thymocytes have a low ADA, high PNP profile. That the
inverse relationships between ADA and PNP activities may be obligatory
(or at least tightly associated) was shown by experiments using
selective ADA inhibitors (R. Barton and I. Goldschneider, unpublished).
It was noted that PNP levels increased progressively as ADA levels
decreased. A similar inverse relationship, which seems to be related
to the relative stage of cellular differentiation, has been observed
in a variety of rat T cell leukemia lines (44). Furthermore,
administration of ADA inhibitors in vivo markedly depressed the
regeneration of TdT$^+$ cortical thymocytes but not of TdT$^-$ bone
marrow cells (39).

These results suggest that specific stages of T cell development
may be characterized by the absolute and relative levels of ADA and
PNP; and conversely that selective deficiencies of these enzymes
might affect discrete stages of T cell differentiation. The results
also suggest that there are major changes in purine nucleotide
metabolism during T cell ontogeny which involve several enzymes in
addition to TdT. This inference is strongly supported by the
observations of Hoffbrand et al. (see ref. 42, these Proceedings).

b) 5'-Nucleotidase (5'-NT). This enzyme, which catalyzes the
dephosphorylation of 5'-monophosphates, is associated with the
plasma membrane of mammalian cells. Deficiency of 5'-NT has been
found in some patients with congenital agammaglobulinemia (43).
Using both biochemical and histochemical assays for 5'-NT, we have
found that 5'-NT is preferentially associated with T cells in the
rat, and that the enzyme specific-activity is much higher in peripheral T cells than in thymocytes (44). In addition, our results
suggested that a minor subset of peripheral T cells may have low
5'-NT activity. It is not known whether this subset is related to

the population of TdT$^+$ blood cells, which has the antigenic phenotype of cortical thymocytes and suppressor T cells (see Sections 2
and 6). The point is of interest because low 5'-NT activity is
thought to predispose cells to the accumulation of deoxyguanosine
triphosphates (45) which is selectively toxic for cortical thymocytes,
TdT$^+$ leukemia cell lines, and suppressor T cell precursors (9,46).

c) <u>Lactic dehydrogenase (LDH) isozymes</u>. Analysis of LDH isozyme
profiles in lymphoid cell populations from thymus, lymph node,
spleen and bone marrow of adult rats revealed the presence of an
LDH-5' sub-band only in bone marrow cells (47). The LDH-5' sub-
band was not detected in thymocytes of any age, beginning at day
16 of fetal life; or in lymph node cells from 2 week to 4 month old
animals. In contrast, the LDH-5' sub-band was detected in bone
marrow of 3 day to 4 month old rats, the oldest animals examined.
The sub-band was also detected in spleen and liver from day 20 of
embryonic life to between 3 and 4 weeks postnatal life, after which
it declined rapidly. The kinetics of appearance of the LDH-5' sub-
band closely parallels that of the TdT$^+$ cell populations in these
sites. While not entirely restricted to TdT$^+$ cells, LDH-5' activity
is present in highly enriched TdT$^+$ cell fractions from these sources
and in TdT$^+$ bone marrow cells generated <u>in vitro</u>. Moreover, the
LDH-5' sub-band is present in the majority of TdT$^+$ Gross virus-
induced thymic leukemias in rats. It is of interest that these
leukemias also have the low ADA high PNP phenotype of TdT$^+$ bone
marrow cells, further suggesting that they arise from pre-thymic
target cells.

4.   <u>What is the Nature of TdT$^+$ Cells in Hemopoietic Tissues</u>?

Despite the fact that TdT$^+$ cells first appear in the thymus
during ontogeny, several pieces of evidence strongly indicate that
most TdT$^+$ cells in bone marrow, spleen and liver are prethymic cells:

i)   TdT$^+$ cells develop in bone marrow, spleen and liver of
     both normal and congenitally athymic mice and rats (7,8,
     19-22; and unpublished observations).

ii) TdT$^+$ cells can be generated <u>in vitro</u> in the absence of
     thymic influence in Dexter-type cultures of normal mouse
     bone marrow (48) and of <u>nu/+</u> and <u>nu/nu</u> rat bone marrow
     (26). Some of these newly formed rat TdT$^+$ cells can be
     induced to express T cell markers by incubation with thymo-
     sin α_1 peptide (see ref. 27, these Proceedings).

iii) More than 75% of TdT$^+$ cells from <u>nu/+</u> and <u>nu/nu</u> mouse
     bone marrow and spleen can be induced to express T cell
     antigens by brief incubation with thymosin (19).

iv) TdT$^+$ bone marrow cells can be physically separated from

pluripotent hemopoietic stem cells and myeloid and erythroid progenitor cells on the fluorescence-activated cell
sorter (18,31). Moreover, TdT$^+$ bone marrow cells are virtually eliminated by the administration of cortisone (7,9,
31), and are markedly increased by adrenalectomy (49);
whereas the non-lymphoid elements in bone marrow are not
significantly affected.

v) The great majority of TdT$^+$ bone marrow cells in nu/+ and
    nu/nu rats and mice are sIg$^-$ and cIg$^-$.

It must be cautioned that the above evidence, while suggestive,
does not prove that TdT$^+$ hemopoietic cells normally function as
thymocyte progenitors. We are presently using a quantitative assay
system to test the ability of FACS-purified TdT$^+$ rat bone marrow
cells to repopulate the thymus of irradiated recipients (50).
Preliminary results suggest that many TdT$^+$ bone marrow cells do not
serve efficiently as prothymocytes in this system, but that their
TdT$^-$ precursors may. Thus, for the moment, the fate of prethymic
TdT$^+$ cells is unresolved (see ref. 27, these Proceedings, for
further discussion).

5. What is the Fate of TdT$^+$ Thymocytes

As indicated in Section 1, approximately 75% of cortical
thymocytes are TdT$^+$ and virtually all TdT$^+$ thymocytes are located
in the thymus cortex. Of the TdT$^+$ thymocytes, 10% are large,
proliferating cells which are located in the subcapsular region of
the cortex. The remainder are small, non-proliferating cells which
are located in the deep cortex. Using combined radioautography and
immunofluorescence, we have recently identified 3 subsets of proliferating TdT$^+$ thymocytes in the mouse (15): a minor subset with a
low Thy 1, high H2 phenotype; an intermediate subset with a low Thy 1,
low H2 phenotype; and a major subset with a high Thy 1, low H2 phenotype. These subsets comprise approximately 1.6%, 4.5% and 80% of
total proliferating thymocytes, respectively. We have speculated
that the low (or negative) Thy 1, high H2 subset may be the most
ancestral member of the cortical thymocyte series, inasmuch as it
has an antigenic phenotype similar to that of TdT$^+$ bone marrow cells
and inasmuch as it represents the smallest subset of dividing TdT$^+$
thymocytes. Kinetic labelling studies will be required to establish
precursor-product relationships between these various subsets of
TdT$^+$ thymocytes. However, it seems likely from the work of Fathman
et al. (51) that dividing high Thy 1, low H2 thymocytes are the
immediate precursors of most non-dividing TdT$^+$ cortical thymocytes.
It is also likely that many of these non-dividing cortical thymocytes
die shortly after they are formed (see refs. 51,52 and below).

Additional evidence regarding the generation and fate of TdT$^+$
thymocytes has been provided by an in vitro model of thymocytopoiesis
which restricts cell influx and efflux (12). In this model, explanted
thymus lobes from 14 day mouse fetuses were cultured in vitro for

one to 4 weeks, the lobes were pulsed with ^{3}H-TdR, and the percentages of TdT$^+$ and TdT$^-$ cells were determined at timed intervals. The results suggest that proliferating TdT$^+$ subcapsular thymocytes generate non-dividing TdT$^+$ cortical thymocytes, most of which become TdT$^-$ and die shortly thereafter. This developmental cycle is repeated at 48-72 hour intervals, so that the total number of thymocytes remains relatively constant. This implies that some progenitors of cortical thymocytes persist in the explanted thymus for at least 3 weeks, and that their activation is tightly regulated. Experiments in which cortisone was added to the culture medium showed that most of the proliferative activity of TdT$^+$ cortical thymocytes was unrelated to the generation of medullary thymocytes.

The results of this in vitro model are consistent with in vivo studies of thymocytopoiesis (reviewed in 10,11,52). Two points deserve further emphasis. Firstly, although most cortical thymocytes probably die shortly after they are formed, possibly because they represent self-reactive clones (53,54), large numbers of cortical thymocytes appear to be released to the peripheral lymphoid tissue (55-58). The fate and function of these cells is unknown. Secondly, although there is general agreement that cortical and medullary thymocytes have independent generative kinetics (51,52), it is still not known whether they are derived from a common precursor in the subcapsular region of thymus cortex or whether they arise from separate precursors in cortex and medulla, respectively. We have recently presented evidence supporting the existence of discrete subsets of thymocyte progenitors in bone marrow (19). One subset expresses or can be induced to express TdT, the other does not. We have also described a subset of postthymic TdT$^+$ cells in peripheral blood (23,24). Taken in their aggregate, these results raise the possibility that TdT$^+$ cells may represent a developmentally and functionally discrete lineage of T lymphocytes (see below and ref. 27, these Proceedings).

6. Origin, Function and Fate of TdT$^+$ Blood Cells.

As described in Sections 1 and 2, the transient population of TdT$^+$ cells that appears in the blood of infant mice and rats resembles cortical thymocytes with respect to antigenic phenotype, cell size and pattern of TdT fluorescence. The absence of these cells from the blood of congenitally athymic mice further testifies to their thymic origin.

At their peak, TdT$^+$ cells may account for fully 50% of T cells in the blood of young animals. Yet, despite their high frequency in blood, we have not been able to detect significant numbers of them in lymph nodes, spleen, gut-associated lymphoid tissues or in non-lymphoid tissues. This suggests that TdT$^+$ blood cells either die shortly after they are released from the thymus or that they rapidly acquire a TdT$^-$, TL$^-$, PNL$^-$ phenotype before migrating to peripheral lymphoid tissues. It should ultimately be possible to resolve this question by adoptive transfer experiments.

Another conundrum concerns the meaning of the transitory appearance of TdT$^+$ cells in the blood of young animals, and the failure of these cells to reappear in adult animals even in the face of antigenic challange or thymic regeneration (unpublished results). Do TdT$^+$ blood cells belong to a unique subset of T lymphocytes whose production and function is only required early in postnatal life; or does their appearance simply represent the release in an immature state of a T cell subset which is produced throughout the life of the host? There is support for both possibilities. For example, age-restricted subsets of antigenically, physically and functionally distinct cortical thymocytes and peripheral T cells have been identified in neonatal mice (59,61) and in newly hatched chickens (61). Conversely, there is evidence that thymocytes are released in a less differentiated state from neonatal than from adult thymus (62,63), possibly due to increased vascular permeability. Also, TdT$^-$ T cells with some of the properties of cortical thymocytes have been found in the blood and peripheral lymphoid tissues of adult mice and rats (reviewed in 56; also see 64). Inasmuch as many TdT$^+$ cortical thymocytes can apparently become TdT$^-$ without cell division (12), a scenario could be envisioned in which prematurely released TdT$^+$ cortical thymocytes would rapidly complete their differentiation to TdT$^-$ T cells in the blood of neonatal animals. In the adult host, such differentiation would occur <u>before</u> the T cells were released from the thymus cortex.

Only the most preliminary evidence is available concerning a possible function for TdT$^+$ blood cells. This evidence suggests that at least some of these cells may be precursors of neonatal suppressor cells. Thus in both the rat and mouse, TdT$^+$ blood cells have the antigenic phenotype of neonatal suppressor cells, being high Thy 1$^+$, 0x8$^+$, W3/25$^-$ and high Thy 1$^+$, TL$^+$, Lyt 1$^+$,2$^+$,3$^+$, respectively (25,65). Interestingly, Gershon <u>et al</u>. (66), Folch and Waksman (67) and Mosier <u>et al</u>. (65) have suggested that suppressor T cells in the spleens of <u>mice</u> and rats may be derived from the thymus cortex; and Durkin <u>et al</u>. (58; and unpublished) have demonstrated the emmigration of putative suppressor cells from the subcapsular cortex of thymus of tolerized adult rats. Suppressor cells in lymph node and spleen from these same rats, as well as from neonatal rats, have a high Thy 1$^+$, 0X8$^+$ phenotype ((25); and Goldschneider, unpublished). Inasmuch as fewer than 10% of rat thymocytes are 0X8$^+$, W3/25$^-$ (25), it seems reasonable to infer that they may represent a functionally specialized subset of cortical (possibly subcapsular) thymocytes from which TdT$^+$ rat blood cells are derived. It should be noted that some cytotoxic T cell precursors in the rat also have a high Thy 1, 0x8$^+$, W3/25$^-$ phenotype (35). Although we have not detected the appearance of TdT$^+$ cells in the peripheral lymphoid tissues of tolerized or sensitized rats, Small <u>et al</u>., (68) have reported increased TdT activity in the spleens of tumor-bearing adult mice. They have correlated this increased TdT activity with the appearance of tumor-enhancing T cells, and have suggested that these putative suppressor cells are recent migrants from the thymus cortex.

Thus far we have only tested enriched populations of TdT$^+$ rat blood cells for suppressor cell activity. The results indicate that suspensions of blood cells which are enriched approximately 7-fold for TdT$^+$ cells ($\sim$50% purity) are enriched proportionately in their ability to non-specifically suppress a localized graft-versus-host reaction. However, much greater purification will be required before definitive conclusions can be drawn about the function(s) of TdT$^+$ blood cells.

SUMMARY

Terminal deoxynucleotidyl transferase (TdT) has been postulated to play an important, but as yet undefined, role in the early stages of T lymphocyte (and possibly B lymphocyte) differentiation in rats and mice. The ontogeny of these TdT-containing cells has been traced using a specific immunofluorescent assay for terminal transferase. Persistent populations of TdT$^+$ cells first appear in thymus during late fetal life and constitute the majority of thymocytes in adult animals. TdT$^+$ cells appear in bone marrow shortly after birth and constitute approximately 1-4% of nucleated bone marrow cells in adult animals. Transient populations of TdT$^+$ cells are present in spleen, liver, lung and blood of prepubertal animals.

Terminal transferase seems to be entirely restricted to immature members of the lymphoid cell series during normal emopoiesis. Pluripotent hemopoietic stem cells, and myeloid and erythroid progenitor cells, isolated on the fluorescence-activated cell sorter (FACS), are TdT$^-$; as are differentiating myeloid and erythroid cells and mature T and B cells. The vast majority of TdT$^+$ cells in bone marrow and spleen belong to the T cell lineage, as is indicated by their ability to express T cell differentiation antigens after treatment with appropriate inducing agents. That they (and TdT$^+$ cells in liver) are prethymic cells is shown by their presence in nu/nu mice and rats and by their selective generation in vitro on bone marrow feeder layers. It has been suggested that TdT is also present in some "pre-B" cells and possibly in even less mature members of the B cell lineage in mice and rats. However, this remains to be confirmed.

TdT$^+$ thymocytes are located predominantly, if not exclusively, in the thymus cortex. Those in the subcapsular region are mitotically active and are comprised of low Thy 1$^+$ and high Thy 1$^+$ subsets. In the explanted fetal thymus, TdT$^+$ cortical thymocytes are generated in repeated waves of 24-48 hrs duration for at least 4 wks. Shortly after each generative wave, mitotically inactive TdT$^+$ cells become

TdT – negative, and most then appear to die. The relationship of TdT[+] cortical thymocytes to medullary thymocytes is unclear. However, recent evidence suggests that separate precursor cell pools may exist in thymus cortex and medulla, respectively.

Most of the TdT[+] cells which are present in the blood of juvenile rats and mice appear to be postthymic cells, as is evidenced by their absence in nu/nu mice and by their T cell antigenic phenotypes. It is likely that they are derived directly from cortical thymocytes, inasmuch as they display both TL antigens and receptors for peanut lectin. The fate of these postthymic TdT[+] cells is not yet known. However, recent evidence indicates that they bear differentiation antigens that are associated with neonatal suppressor cells and with precursors of cytotoxic T cells.

REFERENCES

1. Chang, L.M.S. Development of terminal deoxynucleotidyl transferase activity in embryonic calf thymus gland. Biochem. Biophys. Res. Commun. 44: 124 (1971).

2. Coleman, M.S., Hutton, J.J., DeSimone, P., and Bollum, F.J. Terminal deoxynucleotidyl transferase in human leukemia. Proc. Nat. Acad. Sci. USA. 71: 4404 (1974).

3. Kung, P.C., Silverstone, A.E., McCaffrey, R.P. and Baltimore, D. Murine terminal deoxynucleotidyl transferase: Cellular distribution and response to cortisone. J. Exp. Med. 141: 855 (1975).

4. Barr, R.D., Sarin, P.S., and Perry, S.M. Terminal transferase in human bone-marrow lymphocytes. Lancet i: 508 (1976).

5. Goldschneider, I., Gregoire, K., Barton, R.W. and Bollum, F.J. Demonstration of terminal deoxynucleotidyl transferase in rat thymocytes by immunofluorence. Proc. Nat. Acad. Sci. USA. 74: 734 (1977).

6. Gregoire, K.E., Goldschneider, I., Barton, R.W. and Bollum, F.J. Intracellular distribution of terminal deoxynucleotidyl transferase (TdT) in rat bone marrow and thymus. Proc. Nat. Acad. Sci. USA. 74: 3993 (1977).

7. Gregoire, K.E., Goldschneider, I., Barton, R.W. and Bollum, F.J. Ontogeny of terminal deoxynucleotidyl-transferase-positive cells in lymphohemopoietic tissues of rat and mouse. J. Immunol. 123: 1347 (1979).

8. Sasaki, R., Bollum, F.J., and Goldschneider, I. Transient populations of terminal transferase-positive (TdT[+]) cells in juvenile rats and mice. J. Immunol. 125: 2501 (1980).

9. Bollum, F.J. and Goldschneider, I. Terminal deoxynucleotidyl transferase and lymphocyte differentiation. In Membranes, Receptors, and the Immune Response (Kohler, H. and Cohen, E.P., Eds.): Alan R. Liss, Inc., New York. P. 189 (1980).

10. Goldschneider, I. and Barton, R.W. Development and differentia-
    tion of lymphocytes. In The Cell Surface in Animal Embryo-
    genesis and Development (G. Poste and G.L. Nicolson, Eds.):
    Elsevier, Amsterdam. P. 599 (1976).

11. Cantor, H. and Weissman, I. Development and function of subpopu-
    lations of thymocytes and T lymphocytes. Prog. Allergy 20:
    1 (1976).

12. Goldschneider, I., Mandel, T., and Bollum, F.J. Cyclical and
    reciprocal generation of terminal deoxynucleotidyl transferase
    positive and negative cells in explanted fetal mouse thymuses.
    (In preparation).

13. Barton, R., Goldschneider, I. and Bollum, F.J. The distribution
    of terminal deoxynucleotidyl transferase (TdT) among subsets
    of thymocytes in the rat. J. Immunol. 116: 462 (1976).

14. Rothenberg, E. Expression of differentiation antigens in subpop-
    ulations of mouse thymocytes: Regulation at the level of de
    novo synthesis. Cell 20: 1 (1980).

15. Goldschneider, I., Shortman, K., McPhee, D., Linthicum, S.,
    Mitchell, G., Battye, F., and Bollum, F.J. Identification of
    subsets of proliferating low Thy 1 cells in thymus cortex and
    medulla (submitted for publication).

16. Muller, W.E., Zahn, R.K., Geurtsen, W., and Munsch, N. Age-
    dependent alterations of DNA synthesis. Terminal deoxynucle-
    otidyl transferase and DNA polymerase activities in bone marrow
    subpopulations from mice. Mech. Aging Develop. 13: 119 (1980).

17. Pahwa, R.N., Modak, M.J., McMorrow, T., Pahwa, S., Fernandes, G.,
    and Good, R.A. Terminal deoxynucleotidyl transferase (TdT)
    enzyme in the thymus and bone marrow. I. Age-associated
    decline of TdT in humans and mice. Cell. Immunol. 58: 39
    (1981).

18. Goldschneider, I., Metcalf, D., Battye, F., Mandel, T. and Bollum,
    F.J. Analysis of rat hemopoietic cells on the fluorescence-
    activated cell sorter. II. Isolation of terminal deoxynucleo-
    tidyl transferase-positive cells. J. Exp. Med. 152: 438 (1980).

19. Goldschneider, I., Ahmed, A., Bollum, F.J., Goldstein, A.L.
    Induction of terminal deoxynucleotidyl transferase and Lyt
    antigens with thymosin: Identification of multiple subsets of
    prothymocytes in mouse bone marrow and spleen. Proc. Nat.
    Acad. Sci. USA 78: 2469 (1980).

20. Pazmino, N.H., Ihle, J.N., and Goldstein, A.L. Induction in vivo
    and in vitro of terminal deoxynucleotidyl transferase by thymo-
    sin in bone marrow cells from athymic mice. J. Exp. Med. 147:
    708 (1978).

21. Pazmino, N.H., Ihle, J.N., McEwan, R.N., and Goldstein, A.L.
    Control of differentiation of thymocyte precursors in the bone
    marrow by thymic hormones. Cancer Treatment Rep. 62: 1749
    (1978).

22. Hutton, J.J. and Bollum, F.J. Terminal deoxynucleotidyl trans-
    ferase is present in athymic nude mice. Nucleic Acids Res. 4:
    457 (1977).

23. Goldschneider, I. and Bollum, F.J. TdT-positive cells in blood of neonatal rats and mice are derived from cortical thymocytes. 4th Int. Congr. Immunol. Abst. No. 3.1.06 (1980).

24. Goldschneider, I. and Bollum, F.J. Antigenic profiles of terminal deoxynucleotidyl transferase positive (TdT$^+$) cells in blood and lymphohemopoietic tissues of juvenile rats and mice. (submitted for publication).

25. Mason, D.W., Brideau, R.J., McMaster, W.R., Webb, M., White, R.A. and Williams, A.F. Monoclonal antibodies that define T-lymphocyte subsets in the rat. In Monoclonal Antibodies (Kennett, R.H., McKearn,T.J. and Bechtol, K.B., Eds), Plenum Press, New York, P. 251 (1980).

26. Hayashi, J., Goldschneider, I. and Bollum, F.J. In vitro culture of terminal deoxynucleotidyl transferase-positive rat bone marrow cells. Fed. Proc. 39: 1133 (1980).

27. Goldschneider, I. Effects of biological response modifiers on the growth and differentiation of terminal deoxynucleotidyl transferase containing lymphocytes. (in press, these Proceedings).

28. Silverstone, A., Rosenberg, N., Baltimore, et al. Correlating terminal deoxynucleotidyl transferase and cell-surface markers in the pathway of lymphocyte ontogeny. In Differentiation of normal and neoplastic hematopoietic cells (B. Clarkson et al., eds.), Cold Spring Harbor Laboratory, N.Y. P. 433 (1978).

29. Williams, A.F., Barclay, A.N., Letarte-Muirhead, M. and Morris, R.J. Rat Thy-1 antigens from thymus and brain: Their tissue distribution, purification and chemical composition. Cold Spring Harbor Symp. Quant. Biol. XLI: 51 (1976).

30. Ritter, M.A., Gordon, L.K. and Goldschneider, I. Distribution and identity of Thy-1 bearing cells during ontogeny in rat hemopoietic and lymphoid tissues. J. Immunol 121: 2463 (1978).

31. Goldschneider, I., Metcalf, D., Battye, F. and Mandel, T. Analysis of rat hemopoietic cells on the fluorescence-activated cell sorter. I. Isolation of pluripotent hemopoietic stem cells and granulocyte-macrophage progenitor cells. J. Exp. Med. 152: 419 (1980).

32. Landreth, K.S., Rosse, C., and Clagett, J. Evidence for a lineage relationship between cytoplasmic μ^+/surface μ^- and surface μ^+ B lymphocytes in adult murine bone marrow. Fed. Proc. 39: 806 (1980).

33. Janossy, G., Bollum, F.J., Bradstock, K.F., McMichael, A., Rapson, N., and Greaves, M.F. Terminal transferase-positive human bone marrow cells exhibit the antigenic phenotype of common acute lymphoblastic leukemia. J. Immunol 123: 1525 (1979).

34. Abramson, S., Miller, R.G., and Phillips, R.A. The identification in adult bone marrow of pluripotent and restricted stem cells of the myeloid and lymphoid systems. J. Exp. Med. 145: 1567 (1977).

35. Woan, M., McGregor, D.D., and Goldschneider, I. T-cell-mediated cytotoxicity induced by Listeria monocytogenes. III. Phenotypic characteristics of mediator T cells. (Submitted for publication).

36. Parkman, R., Gelfand, E.W., Rosen, F.S., Sanderson, A., and Hirshhorn, R. Severe combined immunodeficiency and adenosine deaminase. N. Eng. J. Med. 292: 714 (1975).

37. Stoop, J.W., Zegers, B.J.M., Hendricks, G.F.M., et al. Purine nucleoside phosphorylase deficiency associated with selective cellular immunodeficiency. N. Eng. J. Med. 296: 651 (1977).

38. Barton, R.W., Martiniuk, F., Hirschhorn, R., and Goldschneider, I. The distribution of adenosine deaminase among lymphocyte populations in the rat. J. Immunol. 122: 216 (1979).

39. Barton, R.W. and Goldschneider, I. Nucleotide-metabolizing enzymes and lymphocyte differentiation. Molec. Cell Biochem. 28: 135 (1979).

40. Barton, R.W., Martinuik, F., Hirschhorn, R., and Goldschneider, I. Inverse relationship between adenosine deaminase and purine nucleoside phosphorylase in rat lymphocyte populations. Cell. Immunol 49: 208 (1980).

41. Greiner, D., Barton, R.W. and Goldschneider, I. Phenotypic characteristics of rat leukemic cell lines. Cancer Res. (in press).

42. Hoffbrand, A.V. Function of terminal deoxynucleotidyl transferase. (in press, these Proceedings).

43. Edwards, N.L., Magilavy, D.B., Cassidy, J.T., and Fox, I.H. Lymphocyte Ecto-5'-nucleotidase deficiency in agammaglobulinemia. Science 201: 628 (1978).

44. Barton, R.W. and Goldschneider, I. 5'-Nucleotidase activity in subpopulations of rat lymphocytes. J. Immunol 121: 2329 (1978).

45. Wortman, R.L., Mitchell, B.S., Edwards, N.L., and Fox, I.H. Biochemical basis for differentiatial deoxyadenosine toxicity to T and B lymphoblasts: Role for 5'-nucleotidase. Proc. Nat. Acad. Sci. USA 76: 2434 (1979).

46. Dosch, H.M., Mansour, A., Cohon, A., Shore, A. and Gelfand, E.W. Inhibition of suppressor T-cell development following deoxyguanosine administration. Nature 285: 494 (1980).

47. Barton, R.W. and Goldschneider, I. Evidence for the cellular origin of Gross virus-induced leukemia in the rat: Description of a unique LDH isozyme sub-band in leukemia lymphoid cells and lymphohemopoietic precursor cells. J. Immunol. 125: 2299 (1980).

48. Schrader, J., Goldschneider, I., Bollum, F.J., and Schrader, S. In vitro studies of lymphocyte differentiation. II. Generation of terminal deoxynucleotidyl transferase positive cells in long term culture of mouse bone marrow. J. Immunol 122: 2337 (1979).

49. Whittum, J.A., Goldschneider, I., Zurier, R.B. Effects of prostaglandin E_1 on terminal deoxynucleotidyl transferase-positive T cell progenitors in NZB/W F_1 and Balb/c mice.

Fed. Proc. 39: 1130 (1980).

50. Greiner, D.L., Goldschneider, I., Barton, R.W. and Lubaroff, D.M. A quantitative assay system for thymocyte regeneration in the rat. Transplant. Proc. 13: 1457 (1981).

51. Fathman, C.G., Small, M., Herzenberg, L.A., and Weissman, I.L. Thymus cell maturation. II. Differentiation of three "mature" subclasses in vivo. Cell Immunol. 15: 109 (1975).

52. Shortman, K. The pathway of T-cell development within the thymus. Prog. Immunol. 3: 197 (1977).

53. Burnet, F.M. The Clonal Selection Theory of Acquired Immunity. Cambridge Univ. Press, London (1959).

54. Jerne, N.K. The somatic generation of immune recognition. Eur. J. Immunol. 1: 1 (1971).

55. Chanana, A.D., Cronkite E.P., Joel, D.D., Williams, R.M., and Waksman, B.H. Migration of thymic lymphocytes: immunofluorescence and ^{3}HTdR labeling studies. Adv. Exp. Med. Biol. 12: 113 (1971).

56. Goldschneider, I. Antigenic relationship between bone marrow lymphocytes, cortical thymocytes and a subpopulation of peripheral T cells in the rat. Description of a bone marrow lymphocyte antigen. Cell Immunol. 24: 289 (1976).

57. Weissman, I.L., Papaioannov, V.E., and Gardner, R.L. Normal and neoplastic maturation of T lineage lymphocytes. Cold Spring Harbor Symp. Quant. Biol. 41: 9 (1977).

58. Durkin, H.G., Carboni, J.M. and Waksman, B.H. Antigen-induced increase in migration of large cortical thymocyte (regulatory cells) to the marginal zone and red pulp of the spleen. J. Immunol. 121: 1075 (1978).

59. Potworowski, E.F. and Nairn, R.C. Origin and fate of a thymocyte-specific antigen. Immunol. 13: 597.

60. Mosier, D.E. and Johnson, B.M. Ontogeny of mouse lymphocyte function. II. Development of the ability to produce antibody is modulated by T lymphocytes. J. Exp. Med. 141: 216 (1974).

61. Droege, W. and Zucker, R. Lymphocyte subpopulations in the thymus. Transplant. Rev. 25: 3 (1975).

62. Colley, D.G., Malakian, A. and Waksman, B.H. Cellular differentiation in the thymus. II. Thymus-specific antigens in rat thymus and peripheral lymphoid cells. J. Immunol 104: 585 (1970).

63. Weissman, I.L. Thymus cell migration. J. Exp. Med. 126: 291 (1967).

64. Piguiet, P.F., Irle, C., and Vassalli, P. "Early post-thymic" T cells: Peripheral T cells with the characteristics of immature thymocytes. 4th Int. Cong. Immunol. Abstr. 3: 3.1.23 (1980).

65. Mosier, D.E., Mathieson, B.S., Campbell, P.S. Ly phenotype and mechanism of action of mouse neonatal suppressor T cells. J. Exp. Med. 146: 59 (1977).

66. Gershon, R.K., Kondo, K., and Lance, E.M. Immuno-regulatory role of spleen localizing thymocytes. J. Immunol. 112: 546 (1974).

67.   Folch, H. and Waksman, B.H.   The splenic suppressor cell.   I.
        Activity of thymus-dependent adherent cells: Changes with age
        and stress.   J. Immunol. 113: 127 (1974).
68.   Small, M., Lasser-Weiss, M., and Daniel, V.   Release of immature
        cells from the thymus during solid tumor growth: Identification
        by assay of TdT activity.   J. Immunol. 123: 259 (1979).

EFFECTS OF BIOLOGICAL RESPONSE MODIFIERS ON THE GROWTH AND DIFFERENTIA-
TION OF TERMINAL DEOXYNUCLEOTIDYL TRANSFERASE CONTAINING LYMPHOCYTES

Irving Goldschneider

Department of Pathology
University of Connecticut Health Center
Farmington, Connecticut 06032 USA

INTRODUCTION

The ability of certain biological response modifiers to induce
the differentiation of lymphoid cells has been utilized to study the
origins and developmental potentials of TdT^+ cells. Most efforts
have concerned the induction of T cell markers on TdT^+ cells or the
induction of TdT itself. The first successful experiments were
conducted with bone marrow cells, but this approach has since been
extended to cells from thymus and peripheral lymphoid tissues. In
addition, the role of growth regulating factors in the generation of
TdT^+ cells has been studied in vitro in cell and organ culture sytems.
Relatively little work has been done on the induction of B cell
markers on TdT^+ cells, and vice versa. Therefore, the present report
will be limited to a description of some of the effects of inducing
agents on TdT^+ members of the T cell lineage. For convenience, the
discussion will be divided into considerations of prethymic, thymic
and postthymic cells. A tentative synthesis of these results with
the data from ontogenetic studies (see ref. 1, these Proceedings)
will be attempted in the concluding remarks.

RESULTS

1. Prethymic Cells

    a. Induction of T cell markers on TdT^+ cells. Several purified
thymic factors are able to induce the expression of T cell-specific
antigens on the surface of presumptive prothymocytes in murine bone
marrow and spleen (2,3). Of these factors, thymopoietin and thymosin
have been shown to induce a majority of TdT^+ bone marrow cells to
express T cell differentiation antigens (4-7). However, significant

133

Table I: Induction of TdT and T Cell Antigens by Thymus Factors*

| | | Source of Cells | | | | |
| | | nu/+ mice | | | nu/nu mice | |
Factor	Induction of	BM	SPL	LN	BM	SPL
Thymosin	TdT	Yes	Yes	No	Yes	Yes
Thymopoietin		No	n.d.	No	No	n.d.
Thymosin	T cell antigens on	Yes	Yes	n.d.	Yes	Yes
Thymopoietin	TdT-positive cells	Yes	n.d.	n.d.	No	n.d.

* n.d., not determined

differences appear to exist between these factors with respect to the subsets of TdT$^+$ cells that are affected.

As indicated in Table I, thymopoietin induces approximately 50% of TdT$^+$ bone marrow cells from normal mice to express Thy 1 antigen after brief incubation in vitro (4). The non-responding TdT$^+$ cells are reported to have an Lyb-2$^+$ phenotype, which characterizes B cell precursors and possibly lymphopoietic stem cells (5). Thymopoietin was not able to induce Thy 1 on TdT$^+$ cells from congenitally athymic (nu/nu) mice. Many of these TdT$^+$ cells display the Sc-1 antigen, which is present on pluripotent hemopoietic stem cells and on primitive lymphopoietic cells (5).

In contrast, thymosin (fraction 5 and α_1 peptide) not only induces more than 75% of TdT$^+$ bone marrow (and spleen) cells from normal mice to express T cell markers, but affects most TdT$^+$ cells from nu/nu mice as well (7). More than 95% of the responding cells express an Lyt 1$^+$ 2$^+$ 3$^+$ phenotype.

Recently, we have demonstrated that thymosin and thymopoietin can induce T cell alloantigens on TdT$^+$ rat bone marrow cells that have been generated in vitro in a modified Dexter culture system (see Section 1f, below).

b. Induction of TdT in prothymocytes. Another difference between the activities of thymosin and thymopoietin concerns their ability to induce TdT$^-$ prethymic cells to express TdT. Thymosin has

been shown to do so; thymopoietin has not (at least not under the
conditions used thus far). Pazmino and colleagues (6,8) have demonstrated significant increases in TdT specific activity in bone marrow
of normal mice treated by in vitro and/or in vivo administration of
thymosin (fraction 5, β_3 and β_4 peptide). Even greater relative
increases in TdT activity were observed in thymosin treated bone
marrow from congenitally athymic mice and neonatally thymectomized
mice. The latter experiments were especially instructive, inasmuch
as thymectomy was followed by a progressive decrease in TdT activity,
which was returned towards normal levels by in vivo and in vitro
treatment with thymosin.

No increase in TdT specific activity has been observed in bone
marrow from normal or nu/nu mice treated with thymopoietin (4,5).

We have used an immunofluorescence assay to study the induction
of TdT by thymosin (7). Our results show that the increase in TdT
specific activity which follows in vitro treatment of bone marrow
with thymosin can be explained in large part by an increase in the
number of TdT^+ cells. Of course it is possible that the absolute
amount of TdT per positive cell is increased as well. As in Pazmino's
studies (6,8), the inducible cell types were almost exclusively
present in the low buoyant density fractions, thereby suggesting that
they are large, actively cycling cells. Some TdT^- spleen cells which
had similar physical properties could also be induced to express TdT.

The kinetics of induction of TdT by thymosin is linear during
the first 4 hours, then reaches a plateau and remains relatively
constant for at least 24 hours. The induction can be inhibited by
actinomycin D, suggesting that RNA synthesis is required (6,8).
There are several apparent discrepancies between the results obtained
by Pazmino et al., (6,8) and by us (7). Pazmino found that high
doses of thymosin fraction 5 inhibited synthesis of TdT, and that the
thymosin α_1 peptide did not induce synthesis of TdT at all. We did
not detect high dose inhibition with thymosin fraction 5, and we
found that the thymosin α_1 peptide is highly effective at inducing
TdT. It is likely that these differences in results are due in large
part to differences in methodologies (dose, lot and purity of thymosin
preparations, conditions of incubation, nature and sensitivity of the
TdT assay, etc.). One other possibility is that the immunofluorescence assay may detect enzymatically inactive as well as enzymatically
active forms of TdT (9).

Double immunofluorescence studies have shown that almost all of
the bone marrow and spleen cells that can be induced to express TdT
by thymosin can also be induced to express Lyt antigens (7). This is
true for nu/nu mice as well as for their nu/+ littermates. However,
not all newly-induced Lyt^+ cells express TdT, and the proportion
which does varies both between bone marrow and spleen and between
nu/+ and nu/nu mice. The results showed that although the percent-

ages of newly induced Lyt$^+$ cells are roughly equivalent ($\sim$8%) in
bone marrow from normal and athymic mice, 80% of these Lyt$^+$ cells
expressed TdT in normal mice as compared to only 20% in athymic mice.
Similarly, the percentage of newly induced Lyt$^+$ spleen cells which
expressed TdT was approximately 3 times higher in normal than in
athymic mice. However, even in normal mice the percentage of newly
induced Lyt$^+$ cells which could be induced to express TdT was substan-
tially lower in spleen than in bone marrow (33% versus 80%).

The preceeding observations are consistent with the following,
experimentally testable, conclusions:

i)   The great majority of TdT$^+$ bone marrow and spleen cells in
     both normal and congenitally athymic mice are prethymic
     cells or, possibly, common precursors of T and B cells.

ii) Two subsets of prothymocytes exist in hemopoietic tissues,
     one of which is normally induced to express TdT (pre-TdT$^+$
     cells), the other of which is not (pre-TdT$^-$ cells). The
     bone marrow may serve as a repository for TdT$^+$ prothymocytes
     and the spleen for TdT$^-$ prothymocytes.

iii) Thymosin can stimulate all subsets of TdT$^+$ bone marrow
     cells to undergo T cell differentiation; including pre-
     TdT$^+$ cells, "early" TdT$^+$ cells and "late" TdT$^+$ cells.
     Thymopoietin only appears to affect "late" TdT$^+$ cells.

iv) A feedback loop exists by which the thymus helps to regulate
     the proportion of prothymocytes that expresses or can be
     induced to express TdT. Thymosin or a similar humoral
     thymic factor might be active in such a loop.

v)   Although thymosin can concurrently induce pre-TdT$^+$ cells to
     express TdT and T cell differentiation antigens in vitro,
     these normally are independently regulated events which may
     be sequentially initiated by several thymus-derived factors
     in vivo.

There have been several other reports of the induction of TdT
in presumptive prothymocytes. Rubenfeld et al. (10) detected the
appearance of TdT in some "null" cells from human and/or mouse bone
marrow and blood when these cells were cultured for 24 to 48 hours
on monolayers of human epidermal cells. These authors have pointed
out the similarities between thymus epithelial cells and skin epith-
lial cells and have postulated that cultured epidermal cells may
release a soluble factor which promotes T cell differentiation (also
see (11)). Jäger and Lau (12) have observed a selective generation
of TdT$^+$ cells from the "null" cell fraction of human blood placed in
diffusion chambers in normal mice. The appearance of TdT$^+$ cells at
about day 6 of culture was preceded by the appearance of cALL$^+$

cells. The cALL antigen is characteristic of neoplastic TdT^+ cells in cases of common acute lymphoblastic leukemia (cALL) and has also been found on the majority of normal TdT^- cells in human bone marrow (13). The stimulus for the selective generation of TdT^+ cells in these cultures is unknown, but it is intriguing that TdT^+ cells are selectively generated in another heterologous system, namely rat bone marrow cells cultured in vitro on mouse bone marrow cell feeder layers (see Section lf below).

c. Effects of adrenal corticosteroid hormones. The sensitivity of TdT^+ bone marrow cells to adrenal glucocorticosteroid hormones has been well documented (14-16). Administration of cortisone results in almost complete depletion of TdT^+ cells by 48 hours. This is followed by a rebound to supranormal levels by day 9 and a return to normal levels by day 14. The reappearance of TdT^+ cells can be accelerated by the in vivo administration of thymosin (17).

In experiments designed to isolate pluripotent stem cells on the fluorescence-activated cell sorter (18,19), we have shown that the depletion of TdT^+ cells is due to their physical disappearance (presumably due to lysis or migration) rather than to their inability to synthesize TdT. Adrenalectomy has the opposite effect, causing hyperplasia of TdT^+ bone marrow cells (20). However, adenalectomy does not cause the reappearance of TdT^+ cells in spleen or liver of adult mice or rats. Thus, endogenous adrenal glucocorticoids are probably important in regulating the rate of production and/or the half life of TdT^+ prethymic cells, but they do not influence the ontogeny of these cells significantly.

Although most TdT^+ bone marrow cells are highly sensitive to cortisone, it is perhaps surprising that many prothymocytes that are capable of generating thymocytes in irradiated recipient mice (21) and rats (D. Greiner and I. Goldschneider, unpublished) are not. Rather, approximately 50% of the thymocyte regenerating activity is associated with the cortisone-resistant, "null" cell-enriched fraction of rat bone marrow as isolated on the fluorescence-activated cell sorter (FACS) (18,22). We do not yet know whether the active cell type in these preparations is a pluripotent hemopoietic stem cell, a lymphoid stem cell, or a pre-TdT^+ thymocyte progenitor. However, it can not be concluded that cortisone-sensitive TdT^+ bone marrow cells do not normally function as thymocyte progenitors under steady state conditions; only that more primitive cell types may do so under circumstances in which massive regeneration of thymocytes is required.

d. Effects of prostaglandin E_1. There is abundant evidence that prostaglandins may function as important immunoregulatory molecules (23-26). It is also likely that prostaglandins help to regulate normal hemopoiesis (27). However, little is known about the effects of prostaglandins on the early stages of lymphopoiesis.

Scheid et al. (28) and Bach and Bach (29) have reported that PGE$_1$ induces the expression of T cell markers on "null" bone marrow cells. We have found that prostaglandin E$_1$ (PGE$_1$), when given in pharmacological doses ($\sim$10 mg/kg body wt), causes a marked decrease in murine TdT$^+$ bone marrow cells (20). This effect is seen in both normal and adrenalectomized mice. We have also found that NZB/W F$_1$ hybrid mice have abnormally high percentages of TdT$^+$ cells in bone marrow, and that this can be returned to normal levels by therapeutic administration of PGE$_1$. Other investigators have noted that NZB/W F$_1$ mice have abnormal regulation of endogenous PGE$_1$ levels (30,31).

It has been suggested that the immunoregulatory activities of PGE$_1$ in vivo involves a series of feedback loops which may include lymphokines, thymic hormones, colony-stimulating factors and interferon (23,27,32). Thus, it is not clear whether the effects of PGE$_1$ on TdT$^+$ bone marrow cells are direct or indirect. It should be noted in this regard that prostaglandin F$_2\alpha$, which may also have immunoregulatory functions, is inactive against TdT$^+$ bone marrow cells (J. Whittum, doctoral thesis); and that PGE$_1$ is at least 100 times more potent than PGF$_{2\alpha}$ at increasing intracellular cAMP levels (26).

e. Effects of Thiabendazole. Thiabendazole (TBZ) (33) a 2-aryl substituted benzimidazole with antihelminthic properties, significantly potentiates cell-mediated immunological responses in mice (34,36). On a weight basis, solubilized thiabendazole is approximately 100 times more potent than the closely related compound, levamisole (S. Elgebaly, et al., unpublished). Hence it was of interest to determine which cell populations are affected, directly or indirectly, by TBZ.

In previous studies we have found that TBZ when given alone causes hyperplasia of cortical thymocytes and germinal center cells; and when given with the thymus-dependent neoantigen, dinitrofluorobenzene (DNFB), TBZ causes marked extramedullary hemopoiesis and hyperplasia of T and B cells (36). More recently, we have found that TBZ, when administered with DNFB, causes a 2- to 4-fold increase in the percentages of pluripotent hemopoietic stem cells (CFU-S) and of TdT$^+$ cells in mouse bone marrow (37). The peak of CFU-S activity (day 3) precedes that of TdT$^+$ cell activity (day 9), suggesting that they may be causally related events. The role of DNFB is unknown, but it may act indirectly to stimulate those members of the T cell series which can modulate the proliferation and differentiation of CFU-S (38). This notion is supported by our observation that bacterial lipopolysaccharide, which stimulates the proliferation of CFU-S but not TdT$^+$ cells, can be substituted for DNFB. When LPS is given with TBZ, both CFU-S and TdT$^+$ cells are increased. This suggests that TBZ may influence activated CFU-S to generate TdT$^+$ cells.

Some insight into the possible mechanism by which TBZ stimulates the production of TdT$^+$ bone marrow cells has recently been provided by our description of a humoral factor that is generated within 3 hours of injection of TBZ + DNFB (S. Elgebaly et al., unpublished). This factor, when transferred to normal mice, causes a sequential increase in number of CFU-S and TdT$^+$ cells in bone marrow. Preliminary mixing experiments suggest that 2 factors may be involved, one stimulated by TBZ and one by DNFB. This is consistent with the observation that LPS, which can be substituted for DNFB in the above system, causes the release of a factor that stimulates CFU-S proliferation (39). The factor that is generated by TBZ might therefore be expected to act at a somewhat later stage in lymphohemopoiesis than the factor generatd by LPS.

f. Growth requirements for the generation of TdT$^+$ bone marrow cells in vitro. The feasibility of generating TdT$^+$ cells in vitro was shown using a modified Dexter culture system (40) in which normal mouse bone marrow cells were co-cultured on a mouse bone marrow feeder layer (41). A maximum of 2% of the nucleated cells in such cultures were found to be TdT$^+$ by 7 days; and relatively stable levels of TdT$^+$ cells could be maintained for as long as 6 weeks. Approximately 40% of the TdT$^+$ cells in these cultures incorporated ^{3}H-TdR after a 30 minute pulse.

We have modified this system further so that TdT$^+$ cells can now be selectively generated in vitro (42). In the present system, rat bone marrow cells are cocultured on mouse bone marrow cell feeder layers. After a transient decrease of both total and TdT$^+$ cells, generation of TdT$^+$ cells occurs. By day 14, the cultures contain approximately 50% TdT$^+$ cells, all of which exhibit rat histocompatibility antigens. Enhanced growth of TdT$^+$ cells is obtained when activated charcoal-extracted fetal calf serum is used to supplement the culture medium. In such cultures the absolute number of TdT$^+$ cells increase 10-to 15-fold during the incubation period. The charcoal removes factors from the serum, including more than 90% of steroid hormones, which inhibit the growth of TdT$^+$ cells (43). As an example, addition of as little as 10^{-7} M dexamethasone to the charcoal extracted serum inhibits the generation of TdT$^+$ cells in these cultures.

The culture of normal rat TdT$^+$ cells in vitro can be further enhanced with various growth-promoting factors. Addition of 10^{-9} M selenium to cultures of normal bone marrow cells increases the number of TdT$^+$ cells above that obtained in control cultures and permits the serum supplement to be reduced from 30% to 5%. Saline extracts of bone marrow are highly effective in enhancing the growth of TdT$^+$ cells, as is purified fibroblast growth factor (FGF) (44). However, extracts of thymus, spleen and brain appear to be ineffective; as do thymosin (fraction 5 and α_1 peptide), thymopoietin II and T cell growth factor (TCGF, Interleukin II) (45). Attempts to substitute thymus or spleen for bone marrow feeder layers have also been unsuccessful.

TdT$^+$ cells can be generated in vitro with equal facility from
bone marrow of normal or congenitally athymic (nu/nu) rats. In both
cases the TdT$^+$ cells lack markers that are normally present on rat T
cells (e.g. A.R.T.-1, W3/25, OX8) or B cells (sIg, cIg). However,
some can be induced to express T cell antigens by incubation with
thymosin and thymopoietin, and all can be induced to express the
W3/13 antigen, which is normally present on differentiating bone
marrow cells (46). It is important to note that the TdT$^+$ cells that
are generated in these cultures do not cause leukemia when injected
into normal or irradiated syngeneic recipients.

The observation that TdT$^+$ bone marrow cells can be selectively
generated in modified Dexter cultures raises the possibility that a
lymphocyte colony-stimulating factor (L-CSF) may be produced in this
system. This possibility is further supported by the stimulatory
effect of bone marrow extracts. Such a factor, if present, would
provide important insights into the regulation of lymphopoiesis at
the progenitor cell level.

2. Thymocytes

a. Induction (and loss) of TdT in thymocytes. Ontogenetic studies have revealed that TdT$^+$ cells first appear in rat and mouse
thymus on or about day 17 of gestation (15). This coincides with the
time at which the thymus epithelium displays evidence of secretory
activity (47) and at which undifferentiated thymocytes undergo rapid
proliferation and differentiation (48-51). That the appearance of
TdT$^+$ cells results from in situ microenvironmental induction of
undifferentiated thymocytes, rather than from migration of TdT$^+$ cells
from extrathymic sources, is shown by the normal development of
TdT$^+$ thymocytes in fetal thymus lobes explanted in vitro at 14 days
gestation (52). As reviewed elsewhere in these Proceedings (1),
kinetic labelling studies with ^{3}H-TdR suggest that proliferating
TdT$^+$ cortical thymocytes generate non-proliferating TdT$^+$ cortical
thymocytes, most of which die shortly thereafter. The generation of
TdT$^+$ cells from precursors in the subcapsular cortex occurs in proliferative waves which appear to be tightly regulated by an intrathymic
feedback mechanism. The ability of some TdT$^+$ thymocytes to serve as
precursors of TdT$^-$ medullary thymocytes is currently a topic of
active debate (reviewed in (1) and (53)), but it is clear that the
vast majority of cortical and medullary thymocytes have independent
generative kinetics (54,55).

Several authors have used inducing agents to attempt to determine whether the precursors of TdT$^+$ cells in adult mouse thymus are
TdT$^-$ cells; and whether immunologically incompetent TdT$^+$ thymocytes
can differentiate into immunologically competent TdT$^-$ thymocytes.
Cayre et al. (56) identified a minor population of PNA$^-$, TL$^-$ thymocytes that could be induced by in vitro incubation with thymopoietin
to express TdT. However it was unclear whether these cells represent-

ed precursors of TdT$^+$ cells or whether they were differentiating
thymocytes that were induced to reexpress TdT. Conversely, Astaldi
et al. (57) have described the loss of TdT and the acquisition of
functional properties by some cortical thymocytes that had been
treated with a thymus dependent serum factor (SF). However, it is
not certain that the cells which became functionally competent were
the same as those which became TdT$^-$. Similarly, thymosin, which
increases TdT activity in mouse bone marrow, was found to decrease
TdT activity in mouse thymocytes (58). Thymosin is also known to
induce functional differentiation of thymocytes (2).

   The problems that are inherent in establishing precursor-product
relationships in the above studies are illustrated by our study of
the effects of dexamethasome and PGE$_1$ on TdT$^+$ thymocytes (J. Whittum
et al., unpublished). The percentage of TdT$^+$ thymocytes is markedly
decreased by treatment with either reagent (see Sections 2b and 2c,
below). The remaining cells have markedly increased mitogenic
reactivity to Con-A, PHA and histocompatibility alloantigens. Somewhat surprisingly, however, the most functionally active of these
cells are strongly agglutinated by peanut lectin suggesting that they
are cortical rather than medullary thymocytes (59). It is possible,
therefore, that these thymocytes are the products of the same PNA$^+$
subset which responds to T cell growth factor (60).

   In another study, Nagasawa and Mak (61) used phorbol esters to
promote differentiation in the human TdT$^+$ lymphoblastic cell line,
MOLT-3. They found a linear increase in the percentage of E-rosette-
positive cells and a reciprocal decrease in the percentage of
TdT$^+$ cells. This induction was accompanied by an arrest in the
proliferative activity of the treated cells. These observations
are reminiscent of those which we have made with an in vitro organ
culture model of thymocytopoiesis in the mouse, in which the "spontaneous" transformation of TdT$^+$ to TdT$^-$ thymocytes was associated with
cessation of proliferative activity by the TdT$^+$ cells (1).

   b. Effects of adrenal corticosteroid hormones. The effects
of adrenal glucocorticosteroid hormones on thymocytes are well known,
although the mechanisms are not. High doses of cortisone cause
massive lysis of cortical thymocytes in vivo and in vitro (62,63), but
do not affect immunologically competent medullary thymocytes (64).
Adrenalectomy causes marked hyperplasia of cortical thymocytes and
inhibits the age-dependent involution of the thymus (65). Mitotic
indices increase markedly in adrenalectomized animals, and pyknotic
indices decrease (54,66). Most of the affected thymocytes in both
instances are TdT$^+$ (20,67). Hence, it is likely that adrenal corticosteroid hormones help to regulate the production and half-life of
TdT$^+$ thymocytes under physiological conditions as well as under conditions of stress. Cortisone may also affect the migratory activities
of thymocytes and their progenitors (68). In addition, it has been
suggested that cortisone may help to regulate the differentiation of

cortical thymocytes (69). The marked increase in functional activity that we have observed in a subset of cortisone-treated PNA$^+$ cells (see Section 2a, above) lends support to this latter hypothesis.

c. Effects of prostaglandin E$_1$. PGE$_1$ is the predominant prostaglandin in thymus, where it is present at a higher concentration than in any other tissue (26,70). When given to normal mice, PGE$_1$ decreases thymic weight and thymocyte numbers in a dose-dependent fashion (20). Maximum effects are seen by 24 hours and regeneration begins by 36 hours. TdT$^+$ cortical thymocytes are preferentially affected. Experiments in adrenalectomized mice showed that these effects probably do not depend on stimulation of adrenal steroidogenesis. Inasmuch as PGE$_1$ does not seem to cause lysis of thymocytes in vivo or in vitro, it is likely that the atrophy of the thymus cortex is caused by interference with cell proliferation (71) or migration.

The decrease in TdT$^+$ cortical thymocytes that follows the administration of PGE$_1$ is associated with a 5- to 10-fold increase in mitogenic responsiveness to Con-A and PHA. Experiments with peanut agglutinin showed that this enhanced responsiveness is due to a concurrent increase in the functional activity of both PNA$^-$ and PNA$^+$ cells. Indeed, the PNA$^+$ cells from PGE$_1$-treated mice were 50- to 100-fold more responsive to stimulation with Con-A than were the PNA$^-$ cells from the same animals. These PNA$^+$ thymocytes also responded well to PHA and to allogeneic histocompatibility antigens. The enhanced responsiveness of these PNA$^+$ cells could only be partially accounted for as an enrichment phenomenon and did not seem to be due to selective removal of PGE$_1$-sensitive suppressor cells. The results, therefore, are consistent with other observations that PGE$_1$, presumably by increasing intracellular cAMP levels, can stimulate T cell differentiation (23-29).

Other evidence linking PGE$_1$ to regulation of thymocytopoiesis is seen in autoimmune NZB/W F$_1$ mice, in which abnormal regulation of PGE$_1$ metabolism (30,31) is associated with abnormal thymus morphology (72,73), abnormal thymocyte regenerative kinetics (74), and abnormal T cell function (75). Examination of thymus sections from NZB/W F$_1$ mice revealed marked hyperplasia of large, pyroninophilic, TdT$^+$ cortical thymocytes (20). This abnormal developmental pattern closely resembles that seen in thymuses of normal mice treated with the immunopotentiator, thiabendazole (see Section 2d, below). In contrast, NZB/W F$_1$ mice have a relatively normal appearing pattern of thymocytopoiesis after the administration of pharmacological doses of PGE$_1$, but not PGF$_{2\alpha}$. Similar treatment also retards the progression of the autoimmune phenomena in these animals (76).

The striking resemblance between the effects of PGE$_1$ and cortisone on TdT$^+$ bone marrow cells and thymocytes has been alluded to. Direct evidence that both PGE$_1$ and adrenal corticosteroids effect the same population of thymocytes has recently been provided by competitive binding inhibition studies (77). Therefore, it is important to emphasize that PGE_1 and cortisone act independently, perhaps functioning in intrathymic and extrathymic feedback loops, respectively.

d. Effects of immunopotentiators and adjuvants. In addition to increasing the percentage of TdT^+ cells in mouse bone marrow (see Section 1c), the immunopotentiator, thiabendazole (TBZ) causes a marked increase in the percentage of lymphoblasts in the thymus cortex (36,37). The lymphoblasts are arranged as a dense band of large, pyroniophilic, TdT^+ cells in the subcapsular region; and are admixed with roughly equal proportions of small thymocytes in the deep cortex. The majority of these TdT^+ lymphoblasts incorporate ^{3}H-TdR after a 30 min pulse.

In contrast to the situation in bone marrow, stimulation of TdT^+ cortical thymocytes by TBZ is not dependent upon the presence of antigen. Indeed, the effect is prevented by the administration of DNFB, which when given alone causes marked atrophy of cortical thymocytes. It is not yet known if the stimulatory effect of TBZ on TdT^+ cortical thymocytes can be mediated by a humoral factor, such as that which stimulates the generation of TdT^+ bone marrow cells. Neither is it clear whether the accumulation of immature TdT^+ thymocytes results from direct stimulation of cortical thymocytes or from increased immigration of TdT^+ bone marrow cells nor whether the resulting cortical hyperplasia is associated with any change in the rate of production or release of functional T cells.

Of the commonly used adjuvants, BCG and Pseudomonas adjuvant cause a marked accumulation of large, pyroninophilic cortical thymocytes similar to that seen with TBZ (78). Presumably many of these cells are TdT^+. Pertussis vaccine, LPS, and Freunds adjuvant do not cause hyperplasia of cortical thymocytes.

A histological pattern indistinguishable from that seen in TBZ-treated normal mice occurs spontaneously in the thymus cortex of NZB/W F_1 strain mice (see Section 2c, above). Increased percentages of TdT^+ lymphoblasts are also seen in the thymus cortex of normal neonatal mice, adrenalectomized mice, and irradiated and bone marrow reconstituted mice. However, in these cases the maturational sequence of large (subcapsular) to small (deep) cortical thymocytes appears to be orderly.

3. Postthymic Cells

With the exception of the transient population of TdT^+ cells in the blood of infant rats and mice (79), and the possible release of TdT^+ cells from the thymus of adult mice bearing a syngeneic fibro-

sarcoma (80), TdT has not been found in postthymic T cells. Attempts
to induce TdT with thymosin and thymopoietin or by stimulation with
mitogens, antigens and lymphokines have been unsuccessful. Very
recently, Scheid, Rubenfeld and Silverstone are reported to have
induced TdT in murine spleen cell suspensions enriched for Thy 1^+
cells and cultured on feeder layers of human epidermal cells (unpublished observations cited in (56)). They have postulated that cultured epidermal cells may release a thymopoietin-like factor. Without additional information it is not possible to judge whether these
Thy 1^+ cells are postthymic cells (81) or prethymic cells (82).

CONCLUDING REMARKS

     The evidence that has been presented in this and a companion
paper (1) clearly shows that TdT^+ cells are represented in each of
the major compartments of T cell development: prethymic, thymic and
postthymic. Yet the exact role of TdT^+ cells in the development of
immunologically competent T cells is not known. Three major options
exist: i) TdT^+ cells generate a discrete lineage of T cells; ii)
TdT^+ cells generate all T cell subsets; or iii) TdT^+ cells comprise
a terminal, functionally sterile, branch of T cell development. The
distinction is of fundamental importance not only to an understanding
of normal thymocytopoiesis, but to an understanding of the possible
role of terminal transferase in T cell development and differentiation.

     The evidence that TdT^+ cells may constitute a specialized
lineage of T cells in rats and mice has been partly discussed elsewhere (1,7,53,83) and may be summarized as follows:

   i) TdT^+ thymocytes are restricted to the thymus cortex, where
       they constitute the major cell population.

  ii) TdT^+ cortical thymocytes and TdT^- medullary thymocytes have
       independent generative kinetics.

 iii) TdT^+ cortical thymocytes and TdT^- medullary thymocytes have
       markedly different phenotypes.

  iv) Cortical and medullary thymocytes are released independently
       to the peripheral lymphoid tissues.

   v) Two major subsets of prethymic cells exist in hemopoietic
       tissues; TdT^+ (or pre-TdT^+) cells and TdT^- (or pre-TdT^-)
       cells.

  vi) The counterparts of these prethymocyte subsets have been
       identified in the generative cell pools of thymus cortex and
       medulla, respectively.

vii) Selective ablation of TdT$^+$ (and possibly pre-TdT$^+$) bone marrow cells by low dose irradiation is followed by the selective regeneration of presumptive medullary thymocytes (TdT$^-$,
20-α-steroid dehydrogenase-positive).

viii) The kinetics of appearance of functional subsets of thymocytes during ontogeny suggest that they do not arise from a
common thymocyte precursor.

ix) TdT$^+$ postthymic cells in blood of neonatal rats and mice
have the antigenic (and possibly the functional) phenotype
of suppressor-cytotoxic T cells.

x) Some suppressor T cells in adult animals may also be derived
from TdT$^+$ cortical thymocytes.

The above evidence points to a duality of the T cell system at
all stages of development, one putative limb consisting of TdT$^+$ cells
and their functional progeny (suppressor T cells?), the other of
TdT$^-$ cells and their functional progeny (helper T cells) (see refs.
7,83,84). The ultimate proof of this hypothesis will require the
demonstration of selective generation of TdT$^+$ and TdT$^-$ T cell
lineages from purified subsets of thymocyte progenitors. Pending the
results of such studies it must be cautioned that there is no definitive evidence against the generally accepted single progenitor hypothesis of thymocytopoiesis, which has the rule of parsimony to commend
it.

For the purpose of discussion a tentative scheme of thymocytopoiesis based on the dual progenitor hypothesis is presented in
Figure 1. In this scheme, TdT$^+$ prethymic cells are generated from
TdT$^-$ precursors (pre-TdT$^+$ cells), which can be induced in vitro to
express both TdT and T cell surface antigens. The remaining prethymic cells can be induced to express T cell markers but not TdT.
Evidence that this dichotomy is not an artifact of the induction
assay has been presented elsewhere (7). Suffice it to say here that
at a minimum, the observation that a much higher proportion of prethymic cells can be induced to express TdT in normal than in
congenitally athymic mice, and that the specific activity of TdT in
bone marrow progressively decreases after adult thymectomy (and can
be restored by injection of thymosin), indicates that a feedback
loop exists by which the thymus regulates the commitment of early
prothymocytes to the TdT pathway. Other examples of perturbations
of this regulatory loop are found in adrenalectomized animals, in
the rebound from irradiation and cortisone treatment, in NZB/W mice,
and after the administration of thiabendazole. The nature of this
feedback loop is unknown, but among numerous possibilities it could
involve regulatory T cells, humoral thymus factors and other response
modifiers such as lymphokines, corticosteroids and PGE$_1$. A possibly
analogous feedback loop has been described by Cohen and Fairchild

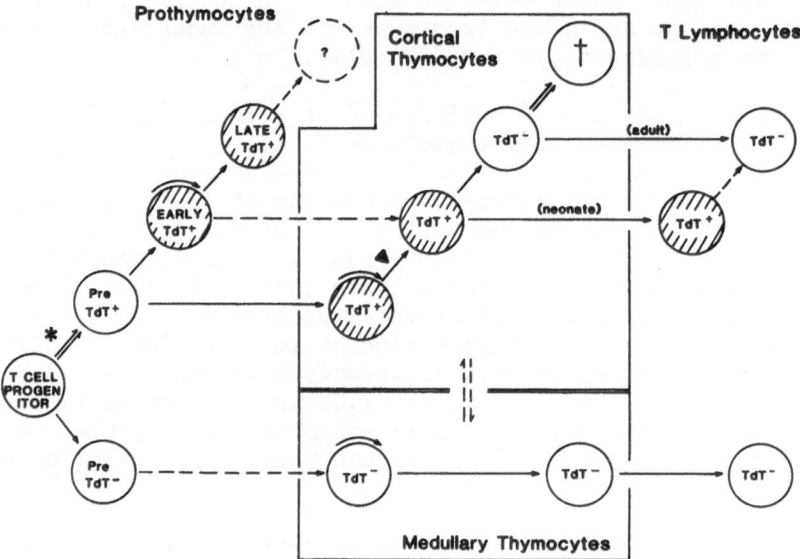

Fig. 1.   Scheme of thymocytopoiesis based on dual progenitor hypo-
          thesis (see text for details). Solid arrow, developmental
          pathway for which evidence exists; broken arrow, hypotheti-
          cal pathway; double arrow, major pathway; curved arrow,
          cell proliferation where known. *, extrathymic feedback
          loop; ∇, intrathymic feedback loop; †, intrathymic cell
          death.

(85).

An intrathymic feedback loop has been detected in the explanted fetal thymus. This loop appears to regulate the onset of proliferation of TdT$^+$ thymocytes in the subcapsular region of the thymus cortex. It can only be speculated that the conversion of proliferating TdT$^+$ thymocytes to non-proliferating TdT$^-$ thymocytes and the massive death of the latter in some way provides a signal for another wave of proliferation. It would be of considerable interest to know whether the intrathymic and extrathymic feedback loops are integrated in vivo and, if so, whether they are responsible for the gated entrance of prothymocytes into the thymus (86,87).

The possible significance of the massive intrathymic death of cortical thymocytes has intrigued many authors. Favored hypotheses include the elimination of self-reactive ("forbidden") clones and the generation and selection of non-self-reactive mutant clones of thymocytes (reviewed in (88)). The new observation provided by our study is that intrathymic death of cortical thymocytes appears to occur in a tightly regulated or programmed fashion in which the rapidly dividing TdT$^+$ cells enter G_0 and stop synthesizing TdT. Thus if TdT functions as a somatic mutagen (89,90), there would be ample opportunity for expression of and selection for (or against) the induced mutation. Alternatively it is possible that TdT might predispose "forbidden" clones of thymocytes to metabolic death by affecting the purine salvage pathway (91,92).

Although most TdT$^+$ bone marrow cells can be induced to express T cell differentiation antigens, our preliminary findings show that many TdT$^+$ cells do not function as thymocyte progenitors after being transferred into irradiated recipients (Greiner et al., unpublished). This apparent discrepancy may reflect the bias of the assay system itself, which preferentially detects those cells that can rapidly migrate to and undergo extensive proliferation in the thymus. Prethymic cells which lacked these properties at the time of injection might be missed. Thus, we have observed that TdT$^+$ bone marrow cells can be generated in vitro (42) by both TdT$^+$ and pre-TdT$^+$ precursors. This leaves open the possibility that, in vivo, pre-TdT$^+$ cells may preferentially function as prothymocytes under circumstances in which the demand for thymocytes is great, such as in the developing fetal thymus and in the regenerating adult thymus (1,15,52). However, under steady state conditions, TdT$^+$ bone marrow cells might function more efficiently than pre-TdT$^+$ cells as prothymocytes.

Alternatively, it is possible that some TdT$^+$ bone marrow cells may have functions aside from (or in addition to) prothymocyte activity. As an example, it has been suggested that some prethymic cells may act as natural killer (NK) cells (93), and about 50% of Gross virus-induced TdT$^+$ leukemias in the rat (94) have the antigenic phenotype (W3/25$^-$, OX8$^+$) of suppressor-cytotoxic T cells (R. Barton,

unpublished). We have recently found that the percentage of TdT^+
cells which bear the W3/13 antigen (46) is significantly greater in
the blood and spleen than in the bone marrow (65% versus 27%) of both
normal and congenitally athymic (nu/nu) rats. Inasmuch as $W3/13^+$ cells
are developmentally more mature than $W3/13^-$ cells, these results raise
the intriguing possibility that some TdT^+ bone marrow cells may normally bypass the thymus and migrate directly to spleen. There is
precedent for such an extrathymic pathway of T cell differentiation
in mice (82) and indirect evidence to suggest that some of the T cells
that are generated in this fashion may function as precursors of
cytotoxic lymphocytes (95,59).

Whatever their fate, the evidence strongly indicates that prethymic TdT^+ cells are heterogeneous with respect to antigenic phenotype
and responsiveness to thymic factors. Not only is there a difference
between the activity of thymosin and thymopoietin in this regard, but
between the several thymosin peptides as well (6,8). The latter observation is particularly intriguing because it has been suggested that
discrete thymosin peptides may act selectively to induce the differentiation of functionally predetermined subsets of thymocyte precursor
cells (2). Although it is difficult to draw in vivo correlates from
primarily in vitro data, the differential responsiveness of subsets of
prethymic cells to various thymic factors raises the possibility that
genotypic commitment to functional differentiation occurs at the prethymic, rather than the thymic, level of T cell development. This
notion is consistent with recent reports concerning the sequence of
appearance and representation of thymocyte subsets in the mouse (96,
97). Given the ability to isolate viable TdT^+ bone marrow cells on
the FACS (19) and to selectively generate prethymic TdT^+ cells in
vitro (42), it may be possible to directly test this hypothesis by
differential treatment with appropriate inducing agents.

In conclusion, it is evident that considerable progress has been
made in tracing the ontogeny of TdT^+ cells and in determining the
effects of biological response modifiers on their growth and differentiation. Yet fundamental questions remain about the developmental
potential of TdT^+ cells and about the role of terminal transferase
in this developmental process. Recent technological advances in the
induction, identification, purification and cloning of T and B cell
subsets should permit more definitive enquiries into the contribution of TdT^+ cells to the functional organization of the lymphoid
system.

REFERENCES

1. Goldschneider, I. Ontogeny of TdT-positive cells in
      rats and mice. (in press, these Proceedings)
2. White A. Chemistry and biological actions of products
      with thymic hormone-like activity. Biochem. Actions of
      Hormones. 7: 1 (1980).

3. Komuro, K., Goldstein, G., and Boyse, E.A. Thymus-
    repopulating capacity of cells that can be induced to differen-
    tiate to T cells in vitro. J. Immunol. 115: 195 (1975).
4. Silverstone, A., Cantor, H., Goldstein, G., and Baltimore, D.
    Terminal deoxynucleotidyl transferase is found in prothy-
    mocytes. J. Exp. Med. 144: 543 (1976).
5. Silverstone, A., Rosenberg, N., Baltimore, D., Sato, V.L.,
    Scheid, M.P., and Boyse, E.A. Correlating terminal
    deoxynucleotidyl transferase and cell-surface markers in the
    pathway of lymphocyte ontogeny. In Differentiation of Normal
    and Neoplastic Hematopoietic Cells (Clarkson, B., Marks, P.A.
    and Till, J.E., eds.), Cold Spring Harbor Laboratory. P. 433
    (1978).
6. Pazmino, N.H., Ihle, J.N., McEwan, R.N., and Goldstein, A.L.
    Control of dfferentiation of thymocyte precursors in the
    bone marrow by thymic hormones. Cancer Treat. Rep. 62: 1749
    (1978).
7. Goldschneider, I., Ahmed, A., Bollum, F.J. and Goldstein, A.L.
    Induction of terminal deoxynucleotidyl transferase and Lyt
    antigens with thymosin: Identification of multiple subsets
    of prothymocytes in mouse bone marrow and spleen. Proc. Nat.
    Acad. Sci. USA 78: 2469 (1981).
8. Pazmino, N.H., Ihle, J.N., and Goldstein, A.L. Induction in
    vivo and in vitro of terminal deoxynucleotidyl transferase by
    thymosin in bone marrow cells from athymic mice. J. Exp.
    Med. 147: 708 (1978).
9. Goldschneider, I., Gregoire, K.E., Barton, R.W., and Bollum, F.J.
    Demonstration of terminal deoxynucleotidyl transferase in
    thymocytes by Immunofluorescence. Proc. Nat. Acad. Sci. USA
    74: 734 (1977).
10. Rubenfeld, M., Silverstone, A.E., Knowles, D., de Sostoa, A.,
    Halper, J., and Edelson, R. Induction of T-cell differ-
    entiation by human epidermal cells. J. Invest. Dermatol. 74:
    252 (1980).
11. Rubenfeld, M., Knowles, D., Halper, J., Edelson, R., Silverstone,
    A. and de Sosta, A. The skin and T-cell differentia-
    tion. N. Eng. J. Med. 303: 1304 (1980).
12. Jager, G. and Lau, B. Expression of common acute lymphoblastic
    leukemia antigen and terminal deoxynucleotidyl transferase
    in normal mononuclear blood cells during diffusion chamber cul-
    ture (submitted for publication).
13. Janossy, G., Bollum, F.J., Bradstock, K.F., McMichael, A., Rapson,
    N. and Greaves, M.F. Terminal transferase-positive human bone
    marrow cells exhibit the antigenic phenotype of common acute
    lymphoblastic leukemia. J. Immunol. 123: 1525 (1979).
14. Bollum, F.J. Terminal transferase: Biological studies. in
    Advances in Enzymology (A. Meister, ed.), John Wiley, New York,
    Vol. 47. P. 347 (1978).

15. Gregoire, K.E., Goldschneider, I., Barton, R.W., and Bollum, F.J. Ontogeny of terminal deoxynucleotidyl transferase positive cells in lymphohemopoietic tissues of rat and mouse. J. Immunol 123: 1347 (1979).

16. Vines, R.L., Coleman, M.S., and Hutton, J.J. Reappearance of terminal deoxynucleotidyl transferase containing cells in rat bone marrow following corticosteroid administration. Blood 56: 501 (1980).

17. Hu, S.-K., Low, T.L.K., and Goldstein, A.L. In vivo induction of terminal deoxynucleotidyl transferase by thymosin in hydrocortisone acetate treated mice. Fed. Proc. 38: 107 (1979).

18. Goldschneider, I., Metcalf, D., Battye, F., and Mandel, T. Analysis of rat hemopoietic cells on the fluorescence-activated cell sorter. I. Isolation of pluripotent hemopoietic stem cells and granulocyte-macrophage progenitor cells. J. Exp. Med. 152: 419 (1980).

19. Goldschneider, I., Metcalf, D., Mandel, T., and Bollum, F.J. Analysis of rat hemopoietic cells on the fluorescence-activated cell sorter. II. Isolation of terminal deoxynucleotidyl transferase-positive cells. J. Exp. Med. 152: 438 (1980).

20. Whittum, J.A., Goldschneider, I., and Zurier, R.B. Effects of prostaglandin E_1 on terminal deoxynucleotidyl transferase-positive T cell progenitors in NZB/W F_1 and BALB/C mice. Fed. Proc. 39: 1130 (1980).

21. Basch, R. and Kadish, J.L. Hematopoietic thymocyte precursors. J. Exp. Med. 145: 405 (1977).

22. Greiner, D.L., Goldschneider, I., Barton, R.W. and Lubaroff, D.M. A quantitative assay system for thymocyte regeneration in the rat. Transplant. Proc. 13: 1457 (1981).

23. Goodwin, J.S. and Webb, D.R. Regulation of the immune response by prostaglandins. Clin. Immunol. Immunopath. 15: 106 (1980).

24. Strom, T.B., Lundin, A.P., and Carpenter, C.B. Role of cyclic nucleotides in lymphocyte activation and function. Prog. Clin. Immunol. 3: 115 (1977).

25. Parker, (1979). The role of intracellular mediators in the immune response. In Biology of the Lymphokines (S. Cohen, E.G. Pick, and J.J. Oppenheim, eds.). Academic Press, New York. P. 541 (1979).

26. Kuehl, F.A., Jr., Cirillo, V.J., and Oien, H.G. Prostaglandin-cyclic nucleotide interactions in mammalian tissues. In Prostaglandins: Chemical and Biological Aspects (S. Karim, ed.). University Park Press, Baltimore. P. 191 (1976).

27. Kurland, J. and Moore, M.A.S. Modulation of hematopoiesis by prostaglandins. Exp. Hemat. 5: 357 (1977).

28. Scheid, M.P., Goldstein, G., and Boyse, E.A. The generation and regulation of lymphocyte populations. Evidence from differentiative induction systems in vitro. J. Exp. Med. 147: 1727 (1978).

29. Bach, M.A. and Bach, J.F. Studies on thymus products. VI. The effects of cyclic nucleotides and prostaglandins on rosette-

forming cells. Interactions with thymic factor. Eur. J. Immunol. 3: 778 (1973).

30. Webb, D.R., Nowowiejski, I., Dauphinee, M., and Talal, N. Antigen-induced alterations in splenic prostaglandin and cyclic nucleotide levels in NZB mice. J. Immunol 118: 446 (1977).

31. Calvano, S.E., Mark, D.A., Good, R.A., and Fernandesz, G. Age-related changes in lymphoid tissue content of prostaglandins in (NZB x NZW)F$_1$ mice. Fed. Proc. 40: 974 (1981).

32. Fitzpatrick, F.A. and Stringfellow, D.A. Virus and interferon effects on cellular prostaglandin biosynthesis. J. Immunol 125: 431 (1980).

33. Tocco, D.J., Rosenblum, C., Martin, C.M. Absorption, metabolism and excretion of thiabendazole in man and laboratory animals. Toxicol. Appl.Pharmacol. 9: 31 (1966).

34. Lundy, J. and Lovett, E.J. Immunomodulation with thiabendazole: a review of immunological properties and efficacy in combined modality cancer therapy. Cancer Treat. Rep. 62: 1955 (1978).

35. Lovett, E.J. and Lundy, J. The effects of thiabendazole in a mixed leukocyte culture. Transplantation 24: 93 (1977).

36. Donskoy, E., Lovett, E.J., Conran, P.B., and Lundy, J.L. Morphologic studies with thiabendazole, an immunomodulator. Fed. Proc. 39: 1148 (1980).

37. Lundy, J., Donskoy, E., Elgebaly, S., and Goldschneider, I. Thiabendazole and lymphohematopoietic maturation. Cancer Res (in press).

38. Sharkis, S.J., Ahmed, A., Sensenbrenner, L.L., Jedrzjczak, W.W., and Sell, K.W. Thymic regulation of hematopoiesis. In Hematopoietic Cell Differentiation (D. Golde, M.J. Cline, D. Metcalf, and C.F. Fox, eds.). Academic Press, New York. P. 491 (1978).

39. Staber, F.G. and Metcalf, D. Humoral regulation of splenic hemopoiesis in mice. Exp. Hematol. (in press).

40. Dexter, T.M., Allen, T.D., and Lajtha, L.G. Conditions controlling the proliferation of haemopoietic stem cells in vitro. J. Cell. Physiol. 91: 335 (1977).

41. Schrader, J., Goldschneider, I., Bollum, F.J., and Schrader, S. In vitro studies of lymphocyte differentiation. II. Generation of terminal deoxynucleotidyl transferase positive cells in long term culture of mouse bone marrow. J. Immunol. 122: 2337 (1979).

42. Hayashi, J., Goldschneider, I., and Bollum, F.J. In vitro culture of terminal deoxynucleotidyl transferase-positive rat bone marrow cells. Fed. Proc. 39: 1133 (1980)

43. Sato, G.H. (1975). The role of serum in cell culture. In Biochemical Actions of Hormones (G. Litwack, ed.). Academic Press, New York p. 391 (1975).

44. Gospodarowicz, D. Purification of a fibroblast growth factor from bovine pituitary. J. Biol. Chem. 250: 2525 (1975).

45. Watson, J., Gillis, S., Manbrook, J., Mochizuki, D., and Smith, K.A. Biochemical and biological characterization of lympho-

cyte regulatory molecules. I. Purification of a class of
murine lymphokines. J. Exp. Med. 150: 849 (1979).

46. Mason, D.W., Brideau, R.J., McMaster, W.R., Webb, M., White,
R.A., and Williams, A.F. Monoclonal antibodies that define
T-lymphocyte subsets in the rat. In Monoclonal Antibodies
(Kennett, R.H., McKearn, T.J. and Bechtol, K.B., Eds.), Plenum
Press, New York. P. 251 (1980).

47. Mandel, T. Differentiation of epithelial mouse thymus.
Z. Zellforsch. 106: 498 (1970).

48. Ball, W.D. A quantitative assessment of mouse thymus differentiation. Exp. Cell Res. 31: 82 (1963).

49. Schlesinger, M. Antigens of the thymus. Prog. Allergy 16 : 214
(1972).

50. Owen, J.J.T. and Raff, (1970). Studies on the differentiation
of thymus-derived lymphocytes. J. Exp. Med. 132: 1216 (1970).

51. Borum, K. $_3$ Cell kinetics in mouse thymus studied by simultaneous
use of ^{3}H-thymidine and colchicine. Cell Tissue Kinet. 6: 545
(1973).

52. Goldschneider, I., Mandel, T., and Bollum, F.J. Cyclical
and reciprocal generation of terminal deoxynucleotidyl transferase positive and negative cells in explanted fetal mouse
thymuses. (In preparation).

53. Goldschneider, I., Shortman, K., McPhee, D., Linthicum, S.,
Mitchell, G., Battye, F., and Bollum, F.J. Identification
of subsets of proliferating low Thy 1 cells in thymus cortex
and medulla (submitted for publication).

54. Shortman, K. and Jackson, H. The differentiation of T lymphocytes.
I. Proliferation kinetics and interrelationships of subpopulations of mouse thymus cells. Cell. Immunol. 12: 230 (1974).

55. Fathman, C.G., Small, M., Herzenberg, L.A., and Weissman, I.L.
Thymus cell maturation. II. Differentiation of three "mature"
subclasses in vivo. Cell. Immunol. 15: 109 (1975).

56. Cayre, Y., De Sostoa, A., and Silverstone, A.E. Isolation of a
subset of thymocytes inducible for terminal transferase biosynthesis. J. Immunol. 126: 553 (1981).

57. Astaldi, G.C.B., Astaldi, A., Wijermans, Groenewovd, M., van
Bemmel, T., Schellekens, P.T.A., and Eijsvoogel, V.P. A thymus-
dependent serum factor active on precursors of mature T cells.
In Cell Biology and Immunology of Leukocyte Function (Quastel,
M.R., ed.), Academic Press, New York. P. 221 (1979).

58. Hu, S.-K., Low, T.L.K., and Goldstein, A.L. Modulation of terminal deoxynucleotidyl transferase (TdT) in vitro by thymosin
fraction 5 and purified thymosin α_1 in normal thymocytes. Fed.
Proc. 39: 1131 (1980).

59. Draber, P. and Kisielow, P. Identification and characterization
of immature thymocytes responsive to T cell growth factor. Eur.
J. Immunol. 11: 1 (1981).

60. Irle, C., Piguet, P.F., and Vassalli, P. In vitro maturation of
immature thymocytes into immunocompetent T cells in the absence
of direct thymic influence. J. Exp. Med. 148: 32 (1978).

61. Nagasawa, K. and Mak, T.W. Phorbol esters induce differentiation in human malignant T lymphoblasts. Proc. Nat. Acad. Sci. USA 77: 2964 (1980).

62. Dougherty, T.F. Effects of hormones on lymphatic tissue. Physiol. Rev. 32: 339 (1952).

63. Ishidate, M. and Metcalf, D. The pattern of lymphopoiesis in the mouse thymus after cortisone administration or adrenalectomy. Aust. J. Exp. Biol. 41: 637 (1963).

64. Blomgren, H. and Andersson, B. Characteristics of the immunocompetent cells in the mouse thymus: Cell population changes during cortisone-induced atrophy and subsequent regeneration. Cell. Immunol. 1: 545 (1971).

65. Jaffe, H.L. The influence of the suprarenal gland on the thymus. II. Direct evidence of regeneration of the involuted thymus following double supraadrenalectomy. J. Exp. Med. 40: 619 (1924).

66. Claesson, M.H. Diurnal variations in thymic lymphoid cell decay. Studies of intact, adrenalectomized, and adrenaline-treated mice. Acta Endocrinologica 70: 247 (1972).

67. Kung, P.C., Silverstone, A.E., McCaffrey, R.P., and Baltimore, D. Murine terminal deoxynucleotidyl transferase: Cellular distribution and response to cortisone. J. Exp. Med. 141: 855 (1975).

68. Ernstrom, V. Hormonal influences on thymic release of lymphocytes into the blood. in Hormones and the Immune Response, Ciba Foundation Study Group No. 36 (G.E.W. Wolstenholme and J. Knight, eds.), Churchill, London. p. 53 (1970).

69. Ritter, M.A. Embryonic mouse thymocyte development. Enhancing effect of corticosterone at physiological levels. Immunology 33: 241 (1977).

70. Horrobin, D.F., Manku, M.S., Oka, M., et al. Nutritional regulation of T lymphocyte function. Med. Hypoth. 5: 969 (1979).

71. Winkelstein, A. and Kelley, V.E. The pharmacological effects of PGE$_1$ on murine lymphocytes. Blood 55: 437 (1980).

72. Holmes, M. and Burnet, F.J. Thymic changes in NZB mice and hybrids. In The Thymus (G.E.W. Wolstenholm, and R. Porter, eds.), Little, Brown, Boston. P. 381 (1965).

73. De Vries, M.J. and Hijmans, W. Pathological changes of the thymic epithelial cells and autoimmune disease in NZB, NZW and B/W mice. Immunology 12: 179 (1967).

74. Dauphinee, M., Palmer, D.W., Talal, N. Evidence for an abnormal microenvironment in the thymus of New Zealand Black mice. J. Immunol. 115: 1054 (1975).

75. Talal, N. Autoimmunity and lymphoid malignancy in New Zealand Black mice. Prog. Clin. Immunol. 2: 101 (1975).

76. Zurier, R.B., Sayadoff, D.M., Torrey, S.B., and Rothfield, N.F. Prostaglandin E$_1$ treatment of NZB/NZW mice. I. Prolonged survival of female mice. Arth. Rheum. 20: 723 (1977).

77. Homo, R., Duval, D., Thierry, C. and Serrow, B. J. Steroid Biochem. (in press) (1980).

78. Khalil, A., Rappaport, H., Florentin, I., Bennoun, M., Davigny,
     M. and Mathe, G. Comparative study of the histologic reactions
     to intravenous injections of heat-killed Pseudomonas aeruginosa
     and of BCG. Biomedicine 30: 200 (1979).

79. Sasaki, R., Bollum, F.J., and Goldschneider, I. Transient
     populations of terminal transferase-positive (TdT$^+$) cells in
     juvenile rats and mice. J. Immunol.125: 2501 (1980).

80. Small, M., Lasser-Weiss, M., and Daniel, V. Release of immature
     cells from the thymus during solid tumor growth: Identifica-
     tion by assay of TdT activity. J. Immunol. 123: 259 (1979).

81. Stutman, O. Intrathymic and extrathymic T cell maturation.
     Immunol. Rev. 42: 138 (1978).

82. Roelants, G.E., London, J., Mayor Withey, K.S., and Serrano, B.
     Characterization of the Thy-1, Tla and Ig phenotype of peanut
     agglutinin-positive cells in adult, embryonic and nude mice
     using double immunofluorescence. Eur. J. Immunol. 9: 139 (1979).

83. Goldschneider, I. Early stages of lymphocyte development. Cur.
     Top. Develop. Biol. 14: 33 (1980).

84. Goldschneider, I. and Barton, R.W. Development and differentia-
     tion of lymphocytes. In The Cell Surface in Animal Embryogen-
     esis and Development (Poste, G. and Nicolson, G.L., eds.):
     Elsevier/North-Holland, Amsterdam. P. 599 (1976).

85. Cohen, J.J. and Fairchild, S.S. Thymic control of prolifera-
     tion of T cell precursors in bone marrow. Proc. Nat. Acad.
     Sci. USA 76: 6587 (1979).

86. Moore, M.A.S. and Owen, J.J.T. Experimental studies on the devel-
     opment of the thymus. J. Exp. Med. 126: 715 (1967).

87. LeDourin, N.M. and Jotereau, F.V. Tracing of cells of the avian
     thymus through embryonic life in interspecific chimeras. J.
     Exp. Med. 142: 17 (1975).

88. Shortman, K. The pathway of T-cell development within the thymus.
     Prog. Imunol. 3: 197 (1977).

89. Baltimore, D. Is terminal deoxynucleotidyl transferase a somatic
     mutagen in lymphocytes. Nature 248: 409 (1974).

90. Bollum, F.J. Terminal deoxynucleotidyl transferase: source of
     immunological diversity. Karl August Forster Lectures, Akad.
     Wiss. U. Lit. (Mainz), Steiner Verlog, P.1 (1975).

91. Bollum, F.J. and Goldschneider, I. Terminal deoxynucleotidyl
     transferase and lymphocyte differentiation. In Membranes,
     Receptors and the Immune Response (Kohler, H. and Cohen, E.D.,
     eds.): Alan R. Liss, New York. P. 189 (1980).

92. Hoffbrand, A.V. Function of terminal deoxynucleotidyl trans-
     ferase. (in press, these Proceedings).

93. Herberman, R.B., Timonen, T., Reynolds, C., and Ortaldo, J.R.
     Characteristics of NK cells. In Natural Cell-Mediated Immunity
     Against Tumors (R.B. Herberman, Ed.): Academic Press, New
     York. P. 89 (1980).

94. Barton, R.W., Tausche, F., and Goldschneider, I. Evidence for
     the cellular origin of Gross virus-induced leukemia in the rat:
     Description of a unique LDH isozyme sub-band in leukemic

lymphoid cells and lymphohemopoietic precursor cells. J. Immunol. 125: 2299 (1980).

95. Gillis, S., Union, N.A., Baker, P.E., and Smith, K.A. The in vitro generation and sustained culture of nude mouse cytolytic T-lymphocytes. J. Exp. Med. 149: 1460 (1979).

96. Scollay, R. and Weissman, I.L. T cell maturation: Thymocyte and thymus migrant subpopulations defined with monoclonal antibodies to the antigens Lyt-1, Lyt-2, and ThB. J. Immunol. 124: 2841 (1980).

97. Mathieson, B.J., Sharrow, S.O., Rosenberg, Y. and Hammerling, V. Lyt 1+23-cells appear in the thymus before Lyt 123+ cells. Nature 289: 179 (1981).

MURINE THYMOCYTE SUBPOPULATIONS: TWO AND THREE PARAMETER

ANALYSIS WITH VARIOUS MARKERS

Roland Scollay

The Walter & Eliza Hall Institute Med. Research
Post Office Royal Melbourne Hospital
Victoria, Australia

INTRODUCTION

The thymus has two main histologically defined subdivisions, the cortex and the medulla. Many markers have been used which apparently distinguish two major lymphocyte subpopulations (in suspension) and these appear to correlate with the histological subdivision; i.e. medullary cells have one set of markers, cortical cells another set. In our initial studies of thymocyte subpopulations with some of the more commonly used markers (1), it became clear that not all reagents distinguished the same subpopulations. As a result we began an extensive review of thymocyte subpopulations using simultaneous analysis of 3 parameters (on the fluorescence activated cell sorter) in order to assess the amount of agreement between markers. We have used various combinations of monoclonal antibodies to Thy 1, Lyt 1, Lyt 2, TL3 and H-2K and fluorescent peanut agglutinin, paying particular attention to dividing cells, since the most strongly TdT positive cells are among this category. The data are too extensive to show here, but a brief summary and a specific example may be useful.

SUMMARY

When non dividing cells are considered, correlation between markers is good and indeed the two populations defined in suspensions correlate well with the histological subdivisions. TL3 is an exception, correlating only very loosely. Terminal transferase as detected by immunofluorescence, correlates reasonably well.

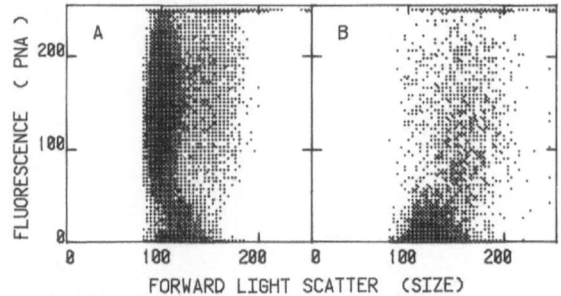

<u>Figure 1</u>.   3 dimensional fluorescence activated cell sorter
analysis. Normal thymocytes were stained with fluorescent PNA and
monoclonal anti H-2K (with a rhodaminated 2nd stage).   Analysis
was made of PNA and H-2K levels and cell size.   Box A shows a
contour plot of PNA levels against size for all thymocytes (i.e.
independent of H-2K levels). Box B shows the PNA against size
contour of cells selected for high H-2K levels.   Size increases
to the right, PNA increases upwards.   50,000 thymocytes were
analysed, 20% of these are included in Box B.   Of these, about
70% are PNA⁻, medium sized, 30% are PNA⁺, large size.

        When dividing cells are considered, the rules are different,
and none of the markers can, at this stage, be used to indicate
the source (cortex or medulla) or relationships of the cells.

Example

        H-2K is generally believed to be a good subpopulation marker
since it is low or negative on most cortical cells, but high on
medullary or cortisone resistant thymocytes (e.g. ref. 2).   But
if we select cells from total thymus solely on the basis of their
high H-2K levels, what other markers do the cells express?
Figure 1 shows the PNA and size characteristics of all thymocytes
(A), or of the cells in the thymus which have high H-2K levels
(i.e. levels above most cortical cells and equivalent to cortisone
resistant cells) (B). This selection includes 20% of the total
thymus.   Clearly the division contains 2 subdivisions;   medium
sized PNA negative cells (about 14% of total thymus and equivalent
to medullary cells) and large (i.e. dividing) PNA positive cells

(6% of total thymus and therefore mainly of cortical origin, since medullary dividing cells constitute less than 1% of total thymocytes (3)).

Amongst the total dividing population of the thymus, about 50% are high H-2, 25% low H-2 and 25% do not express detectable levels.

CONCLUSION

H-2K cannot be used to define the relationships or locality of dividing thymocytes. Most other markers show similar anomalies when large cells are considered. Particular care must be taken in correlating markers such as TdT with other subpopulation markers, since the cells of most interest are dividing cells.

REFERENCES

1. R. Scollay, S. Jacobs, L. Jerabek, E. Butcher and I. Weissman (1980) "T Cell Maturation: Thymocyte and Thymus migrant subpopulations defined with monoclonal antibodies to MHC region antigens" J. Immunol. 124, 2845.
2. J.T. Cerrotini and K.T. Brunner (1967) "Localisation of mouse isoantigens on the cell surface as revealed by immuno-fluorescence" Immunology 13, 395.
3. B.J. Bryant (1972) "Renewal and Fate in the Mammalian Thymus. Mechanisms and inferences of thymocytokinetics" Eur. J. Immunol. 2, 38.

TERMINAL TRANSFERASE POSITIVE CELLS IN THE HUMAN BONE

MARROW AND THYMUS

G.Janossy*, N.Tidman*, K.F.Bradstock*, A.V.Hoffbrand*
and F.J.Bollum[§]

*Departments of Immunology and Haematology, Royal Free
Hospital School of Medicine, Hampstead, London NW3
England; [§]Uniformed Services University of the Health
Sciences, Bethesda, Maryland 20014

INTRODUCTION

The original impetus for this study has been the urge for a
better understanding of human leukaemias, and the need for developing sensitive single cell assays to recognize individual leukaemic cells.   In order to achieve this aim, more had to be
learned about the normal human cell types (so-called 'normal
equivalent'cells) from which the leukaemias originate (1).   Terminal deoxynucleotidyl transferase (TdT) is one of the most useful
markers in this respect for the following reasons.   First, TdT
is present in the vast majority of acute lymphoblastic leukaemias
(ALL, ref.2,3); thus anti-TdT antibody, developed by Bollum (4),
can be used as a "pointer" to specifically recognize leukaemic
cells (e.g. residual malignant cells in treated patients) and
their normal counterparts. Second, being a *nuclear* enzyme it is
convenient to use anti-TdT antibody in combinations with various
antisera reacting with *membrane* antigens. With the help of these
combined assays a surprisingly extensive phenotypic profiles can be
assembled about the TdT positive cells in the bone marrow (BM)and
thymus. These studies represent the overture for studies which will
isolate these human cells and further analyse their functional
characteristics and development potential *in vitro*.

The establishment of these detailed phenotypic profiles has
been facilitated by the availability of monoclonal antibodies
(McAb). It will be demonstrated below that the combination
staining with anti-TdT and any given McAb is one of the most
important single tests which help to decide the clinical uses of
these McAb-s. This chapter represents a brief summary of the last
2-3 years work in the authors' laboratories and also includes some
observations from other collaborating laboratories as indicated in
the reference list.

METHODS

Leucocytes were separated and first stained with heterologous antisera or mouse McAb-s in suspension for membrane antigens. Most frequently indirect immunofluorescence with second layers coupled to tetraethylrhodamine-isothiocyanate (TRITC;red) was used. Stained cells were washed and cytocentrifuge spreads were made, fixed in cold methanol, washed in cold phosphate buffered saline and stained for nuclear TdT. A purified rabbit anti-calf TdT antibody (0.1 µg/µl) and, after washing, purified goat anti-rabbit Ig antibody coupled for fluorescein-isothiocyanate (FITC;green) was used. Other combinations, sometimes using three reagents in convenient arrangements (see ref.5-7), have been described elsewhere.

RESULTS

TdT$^+$ cells in human bone marrow (BM)

Fifteen BM samples from non-leukaemic neonates, children and young adults contained 0.3-15% TdT$^+$ cells (mean: 3.0%). Most of these cells expressed the cALL antigen, also expressed by leukaemic blasts of common ALL type, as well as Ia-like (HLA-DR) antigens detected by heterologous antisera (5,7). These cells were surface

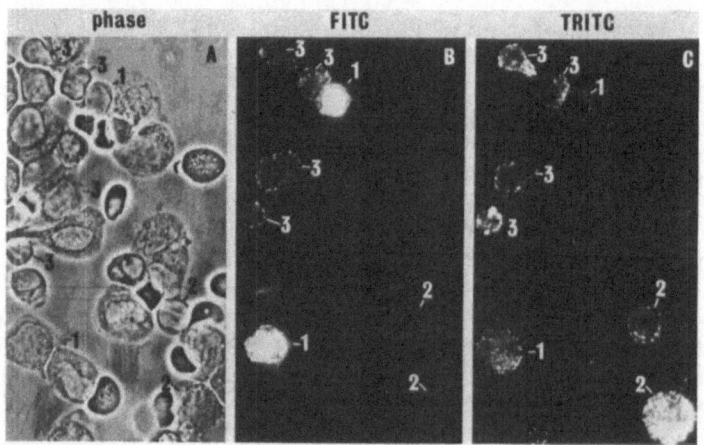

Fig.1. In normal and regenerating bone marrow, three different types of Ia$^+$ cells can be distinguished: (1) TdT$^+$,Ia$^+$,SmIg$^-$ non-T,non-B cells; (2) TdT$^-$,Ia$^+$,SmIg$^-$ cells that include 'lymphoblasts' and myeloblasts (in the corner) and (3) TdT$^-$,Ia$^+$,SmIg$^+$ B lymphocytes. Cells were first stained in suspension with chicken anti-HLA-DR serum plus sheep anti-chicken Ig-TRITC (for HLA-DR) and with goat anti-Hu-Ig-FITC (for SmIg). Smeared cells were restained with rabbit anti-TdT plus goat anti-rabbit-Ig-FITC (for nuclear TdT). The same area was photographed with phase-contrast and selective filters for FITC (nuclear TdT and SmIg) and TRITC (HLA-DR or Ia-like antigen). From ref.7 with permission.

membrane immunoglobulin negative (SmIg$^-$; Fig.1) and also negative
for human thymocyte/T lymphoid antigen (HuTLA$^-$; ref.5,7).   These
small cells were frequently small lymphocytes of non-T,non-B type
expressing the same phenotype as cALL blasts.   These studies have
clearly classified the cells of 'lymphoid' appearance into four
categories: 1) TdT$^+$ cells of non-T,non-B type (see above and cell
type 1 in Fig.1); 2) TdT$^-$ cells expressing HLA-DR antigens but no
lymphoid differentiation markers.   These cells have included at
least a proportion of identifiable myeloid precursors forming
granulocytic-monocytic colonies *in vitro* (type 2 in Fig.1);
3) B cells (TdT$^-$, but expressing HLA-DR and SmIg$^+$; type 3 in Fig.1);
and, finally, 4) T lymphocytes (TdT$^-$,HLA-DR$^-$,SmIg$^-$ expressing
HuTLA).   Other minor subpopulations of BM might also exist but are
difficult to observe in the absence of positive identification
markers.

     Recent observations with monoclonal antibodies (Mc Ab-s) have
confirmed and extended these findings (Fig.2; ref.10-15).   Two
Mc Ab-s react with the cALL antigen (MW 100.000) and also with
cALL$^+$ cells: J-5 and VIL-A1 (12,13).   An additional Mc Ab BA-2
also react with another antigen (MW 24.000) on an overlapping cALL$^+$
population (14).   A range of reagents detect HLA-DR (Ia-like)
antigens on BM cells: OKIa-1, DA-2, etc. (8).   Other reagents
(e.g. RFB-1 and OKT-10) have interesting reactivity pattern shown
in Fig.2.   Antigens are also detected by Mc Ab-s (e.g. PI 153/3;
ref.11) which are shared by cALL blasts and neuroblastoma cells.

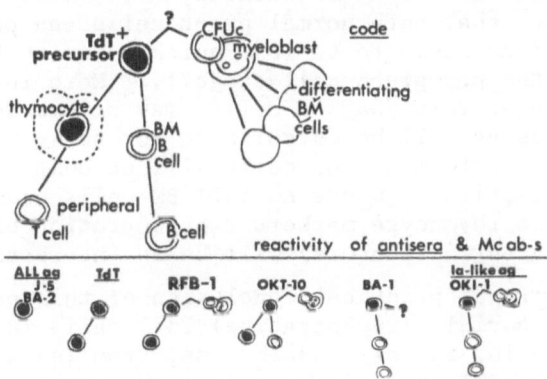

Fig.2.   Immunological characterization of human BM precursor
cells.   The reagents include (with the exception of anti-TdT) monoclonal antibodies.   These are used for membrane staining (with
TRITC, red) in combination with anti-TdT (on BM precursors and cortical thymocytes, depicted with full nucleus), with human T lymphoid antigen (HuTLA; on thymocytes and T cells) and with SmIg (on
B cells).   Myeloblasts can be recognized by staining for HLA-DR(Ia)
staining and morphology.   Myeloid colony forming cells can be
studied in cultures (CFUc) after labelling BM cells and separating
the antibody reactive population on cell sorter (16).   From ref.15
with permission.

In contrast, TdT$^+$ cells are negative with the wide range of anti-thymocyte/T cell Mc Ab-s analysed (OKT1,3,4,6,8,11,11A; see below) and react only with OKT10 (which is also reactive with myeloid precursors; Fig.2). TdT$^+$ cells are also negative with Mc Ab-s to myeloid differentiation antigens (OKM-1).

A further important conclusion of these studies is that the TdT$^+$ cells in normal BM and in the corresponding leukaemia, non-T, non-B ALL, have a very similar, if not identical, phenotype: cALL$^+$ (J-5$^+$,VIL-A1$^+$), HLA-DR$^+$ (DA-2$^+$,OKIa-1$^+$), BA-2$^+$,RFB-1$^+$,OKT10$^+$, PI153/3$^+$ but negative with OKT1,3,4,6, 8, 11 and 11A as well as with SmIg.

Minority populations amongst the BM TdT$^+$ cells

These analyses revealed considerable heterogeneity, and only 50-85% of TdT$^+$ cells expressed all the membrane antigens "necessary" to denote them as typical forms (cALL$^+$,Ia$^+$,RFB-1$^+$,OKT10$^+$, etc.). One particular heterogeneity has been emphasized before (7): a minute population of TdT$^+$ BM cells had cytoplasmic Ig (but no SmIg). These represented only 1.2% of the TdT$^+$ cells present in normal or regenerating marrow. These could be very early pre-pre-B cells (7,17),while the vast majority of pre-B cells (cyIg$^+$,SmIg$^-$ with blast-like morphology) are TdT$^-$. The reason for emphasizing the rarity of these cells is to contrast this observation with the facts that 20-30% of the cALL cases concomittantly contain both cytoplasmic IgM and TdT (pre-B leukaemias; 7,17-19). It was also interesting to note that both normal pre-B cells and pre-B ALL are RFB-1$^-$; this antigen seems to be lost quickly during B cell differentiation (at the pre-pre-B cell stage?). More interesting heterogeneity amongst this "bag of cells" may become detectable when functional assays will be established for early human lymphoid precursors and the performance of cells will be compared with their membrane characteristics. Since no TdT$^+$ BM cells have been observed to express thymocyte markers the generation of these in tests *in vitro* (by thymic factors) will be an interesting aim.

The most important practical conclusion of this part of the study is that the normal (regenerating) TdT$^+$ cells *in the marrow* are indistinguishable, in cell suspensions, from leukaemic cells, although >2% cyIg$^+$,TdT$^+$ cells in a patient who previously had pre-B ALL may be regarded as suspicious. Monitoring of cALL in the BM is not a practical possibility (20,21). (See also below).

TdT$^+$ cells in the human thymus

Analysis of labelling for TdT in sections of normal infant thymus confirmed the results seen in animals (22): TdT$^+$ cells represented ∿60-80% of cortical thymocytes and ∿3-6% of medullary thymocytes (22). Thus, the approximations that TdT$^+$ cells are cortical and TdT$^-$ lymphocytes represent medullary cells are close to the observed facts.

When anti-TdT is used in various combinations with conventional (rabbit and chicken) antisera to human thymocyte/T lymphocyte
antigens (HuTLA) and HLA-DR (Ia-like) antigens the immature (cortical) thymocytes and mature (medullary) thymocytes are clearly
defined as TdT$^+$,HuTLA$^+$,Ia$^-$ and TdT$^-$,HuTLA$^+$,Ia$^-$ respectively (6,7).
This is important because we have already pointed out above that
TdT$^+$,HuTLA$^+$,Ia$^-$ cells (of cortical thymocyte phenotype; Fig.4) are
absent in the normal human BM and, *vice versa*, cells of TdT$^+$,Ia$^+$,
HuTLA$^-$ (of BM types; Fig.1) are absent in the thymus. Consequently,

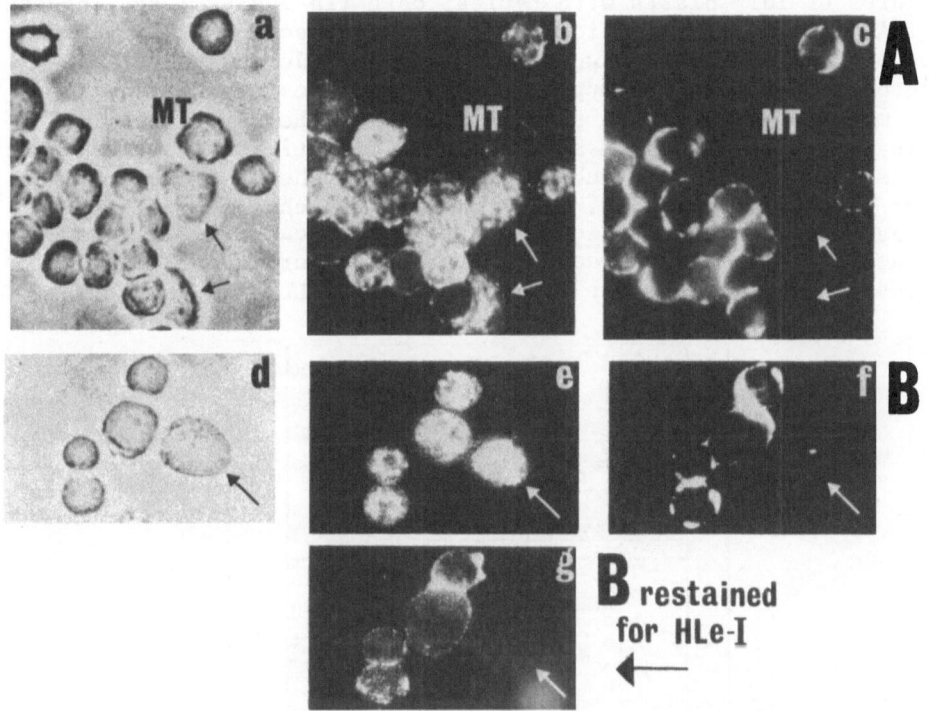

Fig.4.   Heterogeneity of human thymocytes demonstrated in triple-
labelling studies.   In *part A* the cells' membrane was stained for
HuTLA (rabbit anti-HuTLA) and the nucleus was stained for TdT
(rabbit anti-TdT plus goat anti-rabbit-FITC; green staining in b).
At the same time the cells were also stained with a monoclonal antibody (called NA1/34) detecting cortical thymocyte antigen HTA-1
(mouse reagent plus goat anti-mouse-TRITC, red staining in c).
Typical cortical thymocytes are TdT$^+$,HuTLA$^+$,HTA-1$^+$; medullary
thymocytes (MT) are TdT$^-$,HuTLA$^+$,HTA-1$^-$ (MT), and large oval blasts
are also seen which are TdT strongly positive, HuTLA$^+$,HTA-1$^-$
(arrows).   As shown in *part B*, these blasts are also negative
with another Mc Ab, HLe-I, which nevertheless strongly reacts with
cortical thymocytes (g) as well as with T cells and B cells (not
shown).   From ref.6 with permission.

leukaemic cells which show the TdT$^+$,HuTLA$^+$,Ia$^-$ phenotype can be
classified as *thymic* acute lymphoblastic leukaemia (Thy-ALL), as
previously suggested by other groups (23,24).   Admittedly, the
heterologous antisera used in these studies require meticulous
absorptions (reviewed in ref.25), and the leukaemias studied within
these individual groups show further considerable heterogeneity
(see below).   Nevertheless, it is striking that the common form
of ALL (which derives, in all probability, from a BM cell) and the
less frequent Thy-ALL (which derives from the thymus) are separate
disease entities.   Suspensions of leukaemic cells which contain
mixtures of TdT$^+$ blasts with typical cALL (Ia$^+$,HuTLA$^-$) and Thy-ALL
(HuTLA$^+$,Ia$^-$) characteristics are hardly ever seen (8,25) and one
type of these leukaemias does not fully change during the disease
progress to express the typical characteristics of the other.
This might be regarded as surprising, if one believes that TdT$^+$
cells in the childhood BM are very close relatives of prothymocytes,
because precedents for such "mixtures" and "phenotypic shifts" in
other leukaemias do exist.   For example, in chronic myelocytic
leukaemia acute exacerbations (blast crisis) can develop that
frequently contain *mixtures* of cells: a proportion of blast cells
express the cALL phenotype (lymphoid: TdT$^+$,cALL$^+$,Ia$^+$, sometimes

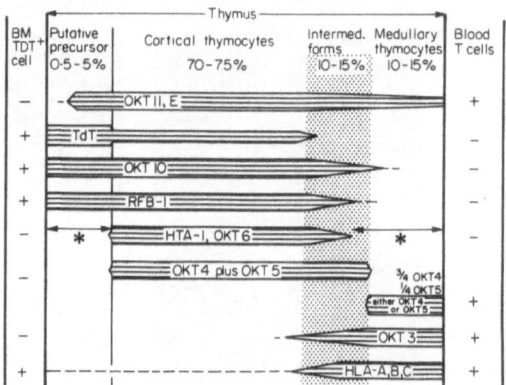

Fig.5.   Scheme of human thymocyte differentiation based on reactivity with monoclonal antibodies to   membrane antigens and with
nuclear terminal deoxynucleotidyl transferase.   Positivity of cell
populations is shown by horizontal bars;   dotted lines indicate
barely detectable or very weak positivity on a few cells.   These
cells simultaneously express both OKT4 and OKT5.   HTA-1$^-$ cells
(*asterisks*;   separated on immunoabsorbent columns) are strikingly
heterogeneous: the larger TdT$^+$ blasts in many respects resemble
bone marrow TdT$^+$ cells (which are also OKT10$^+$,RFB-1$^+$).   There are
a number of identifiable intermediary forms (see stippled area)
between cortical and medullary thymocytes.   Most cells in the medullary thymocyte population appear to have already been segregated
into inducer (OKT4$^+$,OKT5$^-$;   majority) and suppressor-cytotoxic cell
types (OKT5$^+$,OKT4$^-$;   minority).   Taken from ref.28 with permission.

even with cytoplasmic Ig) and other admixed blasts show myeloblast
features (TdT⁻,cALL⁻,Ia⁺).   Also, in this disease shifts are
seen; e.g. lymphoid blast crisis may progress into a myeloid blast
crisis (see revealing reports of both Philadelphia chromosome positive and negative cases in ref.26,27).   Thus, these indirect
clinical considerations suggest that TdT⁺ cells in the BM and thymus may not be very close relatives after all, or that the malignant TdT⁺,cALL⁺,Ia⁺ cells may not be able to enter the thymus and
change into Thy-ALL as readily as their *normal* counterparts might do.

A new range of monoclonal antibodies (Mc Ab-s) has further
helped to dissect the heterogeneity of normal thymocyte populations
(Fig.4 and 5).   A population of strongly TdT⁺ large thymic blasts
were observed which frequently had an oval nucleus containing 1-2
obvious nucleoli.   These large cells, 0.5-5% of thymocyte populations, do not express a number of thymocyte markers but apparently show a transitional phenotype (HuTLA⁺,Ia⁻,TdT⁺⁺,OKT11⁺, OKT10⁺,
RFB-1⁺,HTA-1⁻,OKT4⁻,OKT5⁻,OKT3⁻,HLA-A,B,C⁻) between TdT⁺ cells in
the BM (Fig.2) and the typical cortical thymocytes.   It is therefore possible that these blasts are precursors of cortical thymocytes
(6,28) but further studies are required to prove this point.

The obvious clinical relevance of these observations is that
the majority of Thy-ALL cases express the phenotype of these large
blasts and not that of the typical cortical thymocytes (6-8,11,29).
In other words, cases of Thy-ALL do not react particularly well with
a wide range of T-lineage specific Mc Ab-s such as NA1/34, OKT6,
OKT4, OKT5, OKT8 and OKT3.   Only OKT11 and OKT11A are, so far,
reliable markers of Thy-ALL blasts.   These detect sheep erythrocyte receptors and therefore give the same observations as E-rosetting (30).   Even this marker fails in a small proportion of E-
rosette negative Thy-ALL cases (31,32), indicating that further
search for additional anti-Thy-ALL reagents (or a return to rabbit
anti-HuTLA serum) is necessary.   Incidentally, it is interesting to
note that the large thymic blasts (putative prothymocytes; Fig.5)
isolated from normal thymus are OKT9⁻, while the leukaemic Thy-ALL
populations show a variable but frequently high number of OKT9⁺
blasts carrying transferrin receptor (29,33).   This is an interesting difference between leukaemic blasts and their "normal equivalent"
cells.

The features of typical small TdT⁺ cortical thymocytes are easy
to analyse because they represent the majority of thymocytes. These
cells carry an (almost) specific cortical thymocyte antigen HTA-1
(which also appears on the Langerhans cells of the skin (34)perhaps
in order to annoy the more pedantic scientists; 34). HTA-1 is detected
by two Mc Ab-s: NA1/34 (Fig.4) and OKT6 (Fig.5), which react with
two different epitopes on the same HTA-1 antigen.   These cells also
express both OKT4 and OKT5 (plus the equivalent OKT8), a finding
which is reminiscent of the observations originally made in mice

that cortical thymocytes simultaneously carry markers associated
with peripheral T lymphocytes of inducer type (OKT4) as well as
those of suppressor/cytotoxic type (OKT5, OKT8; ref.28,29).

TdT⁻ cells in the human thymus

When the observations shown in Fig.5 are put into a coherent
scheme, the inescapable suggestion, although not proof, is that
a) there is a continuous transition between cortical and medullary
thymocytes, and that b) during this process cortical cells lose
TdT, OKT6 and the double expression of OKT4 + OKT5,8 in this particular order.   Triple marker experiments show that before losing
TdT, some cells, perhaps already destined to become peripheral T
cells, acquire OKT3 and a more abundant expression of HLA-A,B,C
antigens (28).

The demonstration of double-labelled TdT$^+$,HTA-1$^+$ cells in the
medulla (22) is a further support for the migration of at least
some cortical cells into this region.   All these do not exclude
the possibility that a substantial proportion of medullary thymocytes may directly derive from TdT⁻ precursors of the BM (35)
and that their generation and proliferation is independent of
cortical thymocytes.

Monitoring of leukaemic cells using anti-TdT antibody

It was pointed out above that in treated patients with very low
numbers of leukaemic blast cells present the discrimination between
normal TdT$^+$ cells and leukaemic cells of common ALL type *within the
BM* is not feasible.   In contrast, the 'normal equivalent' cells
in Thy-ALL are restricted to the thymus: a combination staining
with anti-HuTLA and anti-TdT for identifying HuTLA$^+$,TdT$^+$ cells is
therefore operationally a leukaemia specific marker in both the BM
and the blood.   It has recently been demonstrated that this method
is capable of identifying individual Thy-ALL cells, which after
therapy in some patients are masquerading as small lymphocytes (36).
This technique is at least 10-times more sensitive for the identification of residual leukaemia than the conventional haematological
methods.

It is relevant here that *outside* the BM, in extramedullary
sites such as cerebrospinal fluid (37) or the testis, staining for
TdT alone is sufficient to identify ALL blasts of both common and
thymic type (see Janossy et al. this volume).

TdT in myeloid cells

Large blast cells with myeloblastic morphology and Ia-like
(HLA-DR) expression (e.g. cell 3 in Fig.1) react with conventional
anti-myeloid antisera (anti-My) made against ML-1 cell line and

extensively absorbed (using red cells 3x, B cell leukaemia 2x,
T cell leukaemia 6x and KM-3, an ALL line, 3x).   This reagent also
labels the whole myeloid lineage in the BM and monocytes in the
blood (38).   Similarly, cases of acute myeloblastic leukaemia and
myeloid blast crisis are My⁺.   When cells of lymphoid morphology
were analysed in normal and regenerating BM, TdT⁺ cells as well as
B lymphocytes (SmIg⁺) were My-antigen negative.   This paralleled
the finding of My-antigen negativity in cALL and B type CLL.

In a recent study Bradstock et al. (39) documented that in 3
out of 64 cases of AML the myeloblasts (identified by morphology
and cytochemistry) expressed the TdT enzyme.   This finding was
further supported in double-marker experiments on two of these
cases: My⁺ blasts expressed nuclear TdT.   These rare cases, as
well as earlier reports (40), represent examples of aberrant gene
expression.   The normal gene products (such as TdT, My-antigen
and myeloid enzymes) are synthesized by the leukaemic cells in
a combination not normally found in normal BM cells.   It is
important to emphasize that this "slipping control" of gene
expression can be most conveniently established by single cell
analysis with appropriate marker combinations, and that this phenomenon should not be confused with the more frequent situation
where the leukaemic populations consist of mixtures of blasts showing different characteristics (e.g. TdT⁺, peroxidase and Sudan
Black negative lymphoid blasts admixed to TdT⁻, peroxidase and
Sudan Black positive myeloblasts), but each reflecting "faithfully"
the features of normal equivalent cells   (Fig.6).

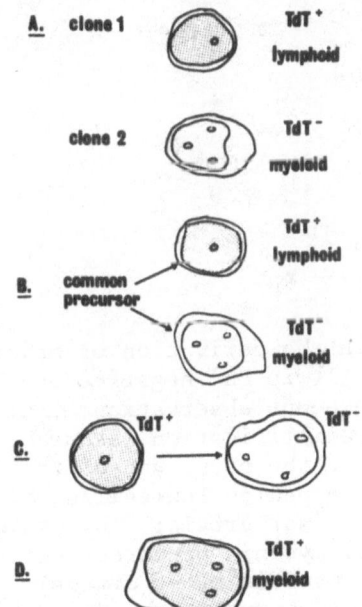

Fig.6.   Possible explanations
to account for the phenomenon of
TdT positive blasts observed in
cases of acute non-lymphocytic
leukaemia: (A) Co-existence of
two independent leukaemic clones
(rare).   (B) More than one pathway of differentiation open to
progeny of clonogenic leukaemic
cell (most frequent).   (C) Maturation of a TdT⁺ precursor cell
into a typical TdT⁻ myeloblast
(unlikely).   (D) Aberrant expression of TdT on myeloid cells
(rare).

DISCUSSION

In 1976 a differentiation sequence of human lympho-haemo-
poietic precursor cells has been suggested (Fig.7).   This was
regarded as a 'minimum-hypothesis' to render the then available
observations on human *leukaemias* (and their probable relationships
to the 'normal equivalent' cells) into a coherent scheme (41,42).
This hypothesis was based on the following tenets: a) most leukaemias form distinct subgroups which are identifiable by classical
haematological (cytochemical, ref.43) as well as immunological
criteria (Fig.7); b) these different subgroups probably reflect
the fact that correspondingly different 'normal equivalent' cells
are involved (1,44); c) in addition, there are peculiar disease
associations ("phenotypic shifts" and "mixed" cases) between different types of leukaemias, particularly between the myeloid and
cALL-type lymphoid blast crises in Philadelphia chromosome positive
chronic granulocytic leukaemia (26,44), and some of the TdT+ leukaemias show myeloid features (40,42).

As it has been shown above, this minimum hypothesis has been
useful to facilitate the investigation of the normal putative lymphoid precursor cells.   With the help of a highly specific anti-
TdT antiserum the normal equivalents of cALL blasts have been found

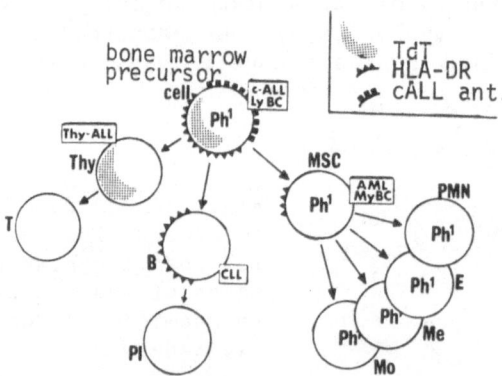

Fig.7.   Minimum hypothesis to explain the derivation of major
types of leukaemias from precursor cells in Ph' negative and Ph'
positive leukaemias (from ref.41).   Recent observations indicate
that this phenotypic chart is an oversimplification because TdT⁻
precursor cells of TdT+ forms exist(but the phenotype of this TdT⁻
pluripotent precursor, and their corresponding leukaemias, is still
unknown; see text).   Ly BC: lymphoid blast crisis; Thy: thymic;
CLL: chronic lymphoid leukaemia; MSC: myeloid precursor cells;
AML: acute myeloid leukaemia; My BC: myeloid blast crisis;
PMN: polymorphs; E: red cells; Me: megakaryocytes; Mo: monocytes.

in the BM and those of Thy-ALL were identified in the thymus.
Further Mc Ab-s have now contributed to the detailed description of
the phenotypes and heterogeneity of these cells (Fig.2 and 5).

On the other hand, this minimum hypothesis has been expected
to be an oversimplification - as has indeed turned out to be the
case in three important aspects. *First*, a subgroup of cALL with
pre-B cell features plus TdT expression has been identified (30%
of cALL cases) and a corresponding normal, but exceedingly rare,
$TdT^+,CyIg^+$ cell has been found (5,17-19). *Second*, animal studies
(45), and experiments *in vitro* (46) as well as observations on
patients following BM transplantation (47) strongly indicated that
the $TdT^+,Ia^+,cALL^+$ small cells may be lymphoid precursors but are
unlikely to include pluripotential stem cells of both the lymphoid
and myeloid pathways. TdT^+ cells seem to regenerate from TdT^-
precursors (45).   The *third* point is that since pluripotential
precursors are unlikely to be TdT^+ an explanation had to be found
for the phenomenon that TdT could be detected in some myeloid leukaemias.   This is provided by marker combination studies which show
that these leukaemias do not reflect *exactly* the characteristics of
normal cells: there is no identifiable normal equivalent for leukaemic TdT^+ cells of myeloid characteristics ("slipping" control of
normal genes in leukaemia).   Thus these rare cases of leukaemias
should not be used as an argument to support the possibility of TdT
enzyme in pluripotent precursor cells (Fig.6D).

The *main conclusions* from the studies summarized in this paper
are therefore as follows.   Anti-TdT antibody, a reliable strong
identification marker, has helped to focus attention on a class of
rare precursor cell.   This fact has helped to unravel some basic
problems about leukaemia derivation.   It has become feasible to
compare the malignant cells with their corresponding normal counter-
part.   Indeed, in some cases these comparative studies already
suggest abnormalities of the leukaemic cells.   As a result
("avalanche effect") more reagents can be found and standardized
which react with membrane antigens on TdT^+ cells.   These new antibodies can then be used for cell separation in order to compare, in
detail, the biochemical and genetic characteristics of isolated
normal and leukaemic cells.

One of the most interesting practical uses of anti-TdT antibody is that it provides a *simple assay* to detect newly produced
monoclonal antibodies (Mc Ab-s), which react with a class of precursor cell types.   With the help of three tests (combination
staining with TdT, analysis of morphologically identifiable
myeloblasts and functional assays for colony forming cells *in vitro*:
CFUc and BFUe) it has become possible to classify new Mc Ab-s into
two groups: 1. Anti-precursor cell Ab-s, and 2. antibodies against
differentiating cells (T cells, B cells, myeloid cells, etc.; see
Fig.2).   This classification of reagents is important because

reagents lacking reactivity towards precursor cells might be preferentially used in special therapeutic protocols. For example,
Mc Ab-s to T cells (such as OKT3) with no anti-precursor cell
activity may be selected for attempting the prevention of graft
versus host disease or to treat acute rejection crisis of trans-
planted organs (10,16).

Finally, these analytical studies may lead to the identification
of further Mc Ab-s with reactivity to the *pluripotential* (TdT⁻) cell
types. The importance of anti-TdT antibodies in human studies can
most convincingly be demonstrated by calling attention to the ign-
orance about the pluripotential precursor cells which so far *lack*
a good identity tag. It is hard to imagine that no leukaemic cells
express the characteristics of this pluripotential cell. It is
more likely that in the absence of a positive identification, leukaemic cells of pluripotential cell type (i.e. with an almost complete block of cellular differentiation) are masquerading in the
form of known (myeloid?) leukaemias. In fact, cases of acute
myeloid leukaemias with a few Sudan black positive cells (∿5%) as
well as another 5-10% TdT⁺ cells but with no sign of maturation in
the majority of blasts (85-90%) probably belong to the "pluripotential" leukaemia category. These patients appear to have a
particularly poor prognosis (39,48). When further reagents are
selected, with the help of anti-TdT and other antisera, a more
precise evaluation of these patients and further detailed studies
about haemopoietic stem cells (in isolation and tissue culture)
will become feasible.

ACKNOWLEDGEMENTS

This study was supported by Grants 186/A and 79/15 from the
Leukaemia Research Fund (to GJ, KFB and AVH) and Grant CA.23262
from the National Cancer Institute (to FJB). We are indebted to
Dr.G.Goldstein for the supply of OKT reagents, to Miss S.Granger
for preparation of fluorochromes conjugated with purified antibodies and Mrs.V.Lipton for secretarial help.

REFERENCES

1.   M.F. Greaves & G. Janossy: Patterns of gene expression and
cellular origins of human leukaemias, Biochem.Biophys.Acta 516:
193-230 (1978).
2.   R.P. McCaffrey, A. Harrison, B.S. Parkman & D. Baltimore:
Terminal deoxynucleotidyl transferase activity in human leukaemic
cells and normal thymocytes, N.Engl.J.Med. 292: 775-580 (1975).
3.   A.V. Hoffbrand. K. Ganeshaguru, G. Janossy, M.F. Greaves,
D. Catovsky & R.K. Woodruff: Terminal deoxynucleotidyl-transferase
levels and membrane phenotypes in diagnosis of acute leukaemia,
Lancet, ii, 520-523 (1977).

4.   F.J. Bollum: Antibody to terminal deoxynucleotidyl transferase Proc.Nat.Acad.Sci.U.S.A, 72: 4119-4112 (1975).

5.   G. Janossy, F.J. Bollum, K.F. Bradstock, A.McMichael, M. Rapson & M.F. Greaves: Terminal transferase positive human bone marrow cells exhibit the antigenic phenotype of non-T, non-B acute lymphoblastic leukaemia, J. Immunol, 123: 1525-1529. (1979).

6.   K.F. Bradstock, G. Janossy, G.Pizzolo, A.V. Hoffbrand, A. Michael, J.R. Pilch, C. Milstein, P. Beverley & F.J. Bollum: Subpopulations of normal and leukaemic human thymocytes: An analysis with the use of monoclonal antibodies, J.Nat.Cancer.Inst, 65: 33-42 (1980).

7.   G. Janossy, F.J. Bollum, K. Bradstock & J. Ashley: Cellular phenotypes of normal and leukaemic haemopoietic cells determined by selected antibody combinations, Blood, 56: 430-441. (1980).

8.   M.F. Greaves, D. Delia, R. Newman & L. Vodinelich: Analysis of leukaemic cells with monoclonal antibodies, In: Monoclonal Antibodies in Clinical Medicine (ed.A. McMichael and J. Fabre) Academic Press, London (1981).

9.   G. Janossy, G. Francis, D. Capellaro, A.H. Goldstone & M.F. Greaves: Cell sorter analysis of leukaemia-associated antigens on human myeloid precursors, Nature 276: 176-8 (1978).

10.   G. Janossy, N. Tidman, E.S. Papageorgiou, P.C. Kung & G. Goldstein: Distribution of T lymphocyte subsets in the human bone marrow and thymus - an analysis with monoclonal antibodies. J.Immunol. 126: 1608-13 (1981).

11.   M.F. Greaves: Monoclonal antibodies as probes for leukaemic heterogeneity and haemopoietic differentiation. In: Leukaemia Markers (ed. W . Knapp) Academic Press, New York. pp 19-32 (1981).

12.   J. Ritz, J.M. Pesando, Notis-McConarty, H. Lazarus & S.F Schlossman: A monoclonal antibody to human acute lymphoblastic leukaemia antigen, Nature 283: 583. (1980).

13.   K. Liszka, O.Majdic, P. Bettelheim & W. Knapp: A monoclonal antibody (VIL-Al) reactive with common acute lymphatic leukaemia cells. In: Leukaemia markers (ed. W. Knapp) Academic Press, London/new York, pp 61-64 (1981).

14.   T. Le Bien, R. McKenna, C. Abramson, K. Gajl-Peczalska, M. Nesbitt, P.Coccia, C. Bloomfield, R. Brunning & J. Kersey: The use of monoclonal antibodies, morphology and cytochemistry to probe the cellular heterogeneity of acute leukaemia and lymphoma.Cancer Research. In press.

15.   M.Bodger, G.E.Francis, D.Delia, J.A.Thomas, S.M.Granger & G.Janossy: A monoclonal antibody specific for immature human haemopoietic cells and T lineage cells. J.Immunol. in press (1981).

16.   D.H.Crawford, G.E.Francis, M.A.Wing, A.J.Edwards, G.Janossy, A.V.Hoffbrand, H.G.Prentice, D.Secher, I.McConnell, P.C.Kung & G.Goldstein: Reactivity of monoclonal antibodies with human myeloid precursor cells. Brit.J.Haematol., 49: in press (1981).

17.   L.B.Vogler, L.J.Preud'homme, M.Seligman, W.E.Gathings, W.M.Crist, M.D.Cooper & F.J.Bollum: Diversity of immunoglobulin expression in leukaemia cells resembling B lymphocyte precursors. Nature, 290: 339-341 (1981).

18.   J.Minowada, K.Koshiba, G.Janossy, M.F.Greaves & F.J.Bollum: A Ph' chromosome positive human leukaemia cell line (NALM-1) with pre-B cell characteristics. Leuk.Research, 3: 261-266 (1979).

19.   M.F.Greaves, W.Verbi, L.Vogler, M.Cooper, R.Ellis, K.Ganeshaguru, A.V.Hoffbrand, G.Janossy & F.J.Bollum: Antigenic and enzymatic phenotypes of the pre-B subclass of acute lymphoblastic leukaemia. Leuk.Research, 3: 353-362 (1979).

20.   M.F.Greaves, D.Delia, G.Janossy, C.Pain, S.Johnson, & T.A. Lister: Acute lymphoblastic leukaemia associated antigen. IV. Expression on non-leukaemia 'lymphoid' cells. Leuk.Research, 4: 15-32 (1980).

21.   K.F.Bradstock, G.Janossy, A.V.Hoffbrand, K.Ganeshaguru, P.Llewelin, H.G.Prentice & F.J.Bollum: Immunofluorescent and biochemical studies of terminal deoxynucleotidyl transferase in treated acute leukaemia. Br.J.Haematol., 47: 121-131 (1981).

22.   G.Janossy, J.A.Thomas, F.J.Bollum, S.Mattingly, G.Pizzolo, K.F.Bradstock, L.Wong, K.Ganeshaguru & A.V.Hoffbrand: The human thymic microenvironment: an immunohistological study. J.Immunol., 125: 202-212 (1980).

23.   B.Mills, L.Sen & L.Borella: Reactivity of antihuman thymocyte serum with acute leukemic blasts. J.Immunol., 115: 1038-1044 (1975).

24.   S.F.Schlossman, L.Chess, R.E.Humphreys & J.L.Strominger: Distribution of Ia-like molecules on the surface of normal and leukemic human cells. Proc.Natn.Acad.Sci.USA, 73: 1288-1299 (1976).

25.   G.Janossy: Membrane Markers. In: Methods in Haematology. The Leukaemic Cell. (Ed. D.Catovsky) Churchill Livingstone, Edinburgh/New York.

26.   E.N.Forman, T.Padre-Mendoza, P.S.Smith, B.E.Barker & P.Farnes: Ph'-positive childhood leukaemias. Spectrum of lymphoid-myeloid expressions. Blood, 49: 549-558 (1977)

27.   G.Janossy, R.K.Woodruff, A.Paxton, M.F.Greaves, D.Capellaro, B.Kirk, E.M.Innes, O.B.Eden, C.Lewis, D.Catovsky & A.V.Hoffbrand: Membrane marker and cell separation studies in Ph' positive leukemia. Blood, 51: 861-877 (1978)

28.   N.Tidman, G.Janossy, M.Bodger, S.Granger, P.C.Kung & G.Goldstein: Delineation of human thymocyte differentiation pathways utilizing double staining techniques with monoclonal antibodies. Clin.exp.Immunol., 45: 457-467 (1981).

29.   E.L.Reinherz, P.C.Kung, G.Goldstein, R.H.Levey & S.F.Schlossman: Discrete stages of human intrathymic differentiation: analysis of normal thymocytes and leukemic blasts of T lineage. Proc.Natn.Acad.Sci.USA, 77: 1588-1592 (1980)

30.   W.Verbi, M.F.Greaves, K.Koubek, G.Janossy, H.Stein, P.C.Kung & G.Goldstein: OKT11 and OKT11A: monoclonal antibodies with pan T reactivity which block sheep erythrocyte 'receptors' on T cells. Eur.J.Immunol. in press (1981).

31.   G.Janossy, A.V.Hoffbrand, M.F.Greaves, K.Ganeshaguru, C.Pain, K.F.Bradstock, H.G.Prentice & H.E.M.Kay: Terminal transferase enzyme assay and immunological membrane markers in the diagnosis of leukaemia - a multi-parameter analysis of 300 cases. Brit.J. Haematol., 44: 221-234 (1980).

32.   D.Catovsky, M.Cherchi, M.F.Greaves, G.Janossy, C.Pain & H.E.M.Kay: Acid-phosphatase reaction in acute lymphoblastic leukaemia. Lancet, i: 749-751 (1978).

33.   R.Sutherland, D.Delia, C.Schneider, R.Newman, J.Kemshead & M.F.Greaves: Ubiquitous cell surface glycoprotein on tumour cells is proliferation-associated receptor for transferrin. Proc.Natn. Acad.Sci.USA, in press (1981).

34.   E.Fithian, P.C.Kung, G.Goldstein, M.Rubenfeld, C.Fenoglio & R.Edelson: Reactivity of Langerhans cells with hybridoma antibody. Proc.Natn.Acad.Sci.USA, 78: 2541-2544 (1981).

35.   I.Goldschneider, A.Ahmed, F.J.Bollum & A.L.Goldstein: Induction of terminal transferase and Lyt antigens with thymosin: identification of multiple subsets of prothymocytes in mouse bone marrow and spleen. Proc.Natn.Acad.Sci.USA, 78: 2469-2475 (1981).

36.   K.F.Bradstock, G.Janossy, N.Tidman, E.S.Papageorgiou, H.G.Prentice, M.Willoughby & A.V.Hoffbrand: Immunological monitoring of residual disease in treated thymic acute lymphoblastic leukaemia. Leuk.Research, 5: 301-309 (1981).

37.   K.F.Bradstock, E.S.Papageorgiou & G.Janossy: Diagnosis of meningeal involvement in patients with acute lymphoblastic leukaemia using immunofluorescence for terminal transferase. Cancer, 47: 2478-2481 (1981).

38.   G.Janossy: Differentiation of human bone marrow cells and thymocytes. In: Leukaemia Markers (ed. W.Knapp) Academic Press, London/New York, pp 45-55 (1981).

39.   K.F.Bradstock, A.V.Hoffbrand, K.Ganeshaguru, P.Llewelin, K.Patterson, B.Wonke, G.Pizzolo, H.G.Prentice, M.Bennett & G.Janossy: Terminal deoxynucleotidyl transferase expression in acute non-lymphoid leukaemia - an analysis by immunofluorescence. Brit.J.Haematol., 47: 133-142 (1981).

40.   B.I.S.Srivastava, S.A.Khan & E.S.Henderson: High terminal deoxynucleotidyl transferase activity in acute myelogenous leukemia. Cancer Research, 36: 3847-3850 (1976).

41.   G.Janossy, M.F.Greaves, D.Capellaro, M.Roberts & A.H.Goldstone: Membrane marker analysis of 'lymphoid' and myeloid blast crisis in Ph' positive (chronic myeloid) leukaemia. In: Immunological Diagnosis of Leukemias and Lymphomas (Eds. S.Thierfelder, H.Rodt & E.Thiel) Springer-Verlag, Berlin/Heidelberg/ New York, pp 97-107 (1977).

42.   G.Janossy, M.F.Greaves, R.Sutherland, J.Durrant & C.Lewis: Comparative analysis of membrane phenotypes in acute lymphoid leukemia and in lymphoid blast crisis of chronic myeloid leukemia. Leuk.Research, 1: 289-300 (1977).

43.   G.Flandrin & M.T.Daniel: Cytochemistry in the classification
of leukemia.   In: The Leukaemic Cell (Ed. D.Catovsky) Churchill
Livingstone, Edinburgh/ New York, pp.24-48 (1981).
44.   G.Janossy, M.Roberts & M.F.Greaves: Target cell in chronic
myeloid leukaemia and its relationship to acute lymphoid leukaemia.
Lancet, ii: 1958-1969 (1976).
45.   R.L.Vines, M.S.Coleman & J.J.Hutton: Reappearance of terminal deoxynucleotidyl transferase containing cells in rat bone
marrow following corticosteroid administration.   Blood, 56:
501-509 (1980).
46.   H.A.S.Moore, H.E.Broxmeyer, A.P.C.Sheridan, P.A.Meyers,
N.Jacobsen & R.J.Winchester: Continuous human bone marrow culture:
Ia antigen characterisation of probable pluripotential stem cell.
Blood, 55: 682-690, (1980).
47.   B.Netzel, R.J.Haas, H.Rodt, H.J.Kolb, B.Belohradsky &
S.Thierfelder: Anti-leukaemic autologous bone marrow transplantation in childhood acute lymphoblastic leukaemia.   Transplant.
Proc., 13: 254-256 (1981).
48.   R.Mertelsmann: TdT in the clinical and pathophysiological
evaluation of human leukaemias.   This volume.

DISAPPEARANCE OF NUCLEAR TdT IN RPMI-8402 FOLLOWING TPA TREATMENT

N. Sacchi[1], U. Bertazzoni[2], D. Breviario[1], P. Plevani[3],
G. Badaracco[3] and E. Ginelli[1]

[1]Istituto di Biologia Generale, Facoltà di Medicina
Università di Milano, Italy
[2]Istituto di Genetica Biochimica ed Evoluzionistica
C.N.R., Pavia, Italy
[3]Istituto di Biologia Generale, E.U.L.O., Brescia,Italy

INTRODUCTION

Terminal deoxynucleotidyl transferase (TdT) is a useful marker
to study T-cell ontogeny and more generally lymphocyte differentiation.

In murine and rat systems the identification of T-cells by
means of developmental cell surface antigen markers (1,2) and TdT
(3) suggests the hypothesis of a double pathway of intrathymic T-
differentiation (see paper by Goldschneider in this book). According to this hypothesis TdT$^+$ and TdT$^-$ subsets of cells would be the
progenitors of two indipendent populations of T-functional
cells.

In man evidence of the segregation of two distinct thymocyte
populations from a common precursor cell comes from an extensive
study using monoclonal antibodies of the OK-T series (4). These subpopulations once exported in the peripheral compartment would represent the helper (OK-T4) and the cytotoxic/suppressor (OK-T8)
fractions.

In this kind of studies, based on the analysis of discrete subsets of cells, is never possible to follow the transition of one

thymocyte through the different stages. The possibility of _in vitro_ differentiation of human T-cell lines, that represent the leukemic transformation at a specific stage of intrathymic development may offer the opportunity to verify in a dynamic system the above mentioned hypothesis of T-cell ontogeny.

We report our studies of differentiation on the T-ALL line RPMI-8402 which for its phenotypic features can be related to an early thymocyte.

MATERIALS AND METHODS

Cell line

RPMI-8402 was originally derived from a 16-year-old female with T-ALL in relapse (we obtained the line from Dr. F.J. Bollum, USUHS, Bethesda, MD.) (5). Immunological and biochemical analysis showed that this line is positive for cALL, HuTLA, EAC rosettes, TdT, AP, and negative for Ia-like antigen, E rosettes, CyIg, SmIg. The cells were cultured in RPMI-1640 medium supplemented with 10% heat inactivated fetal calf serum, L-glutamine (2 mM) and penicillin-streptomycin (100 U/ml), at 37°C in 5% CO_2 atmosphere.

Inducers

Treatments of cells were performed for 2,4,6 days with not cytotoxic concentration of the following inducers: 1% dimethyl sulfoxide,   1% dimethyl formamide, 10^{-8}-10^{-5} M retinoic acid (Sigma), 0.5-1.5 mM butyric acid, 10^{-8}-10^{-7} M phorbol (P-L Biochemicals), phorbol 12,13 didecanoate (P-L Biochemicals), 10^{-8}-10^{-7} M 12-O-tetradecanoylphorbol 13-acetate (TPA) (P-L Biochemicals). All the phorbols were dissolved in 0.15 M NaCl.

TdT assay

Terminal transferase enzyme activity was measured according to the method described by Chang and Bollum (6). 1 U of enzyme activity corresponds to 1 nmole of dGTP polymerized in ·1 hr. TdT was tested also by an indirect immunofluorescence test (IF) (7). Cytospin preparations fixed in ice-cold methanol were coated with the rabbit anti-TdT antibodies, washed and labelled with FITC-coniugated goat anti-rabbit immunoglobulins as second layer. (The antibodies were kindly supplied by Dr. F.J. Bollum).

E-rosette assay

E-rosette forming cells were detected either by spontaneous
E-rosette formation with sheep red blood cells (SRBC) after 18 hrs
at 4°C or incubating cells with SRBC pretreated with neuraminidase
for 2 hrs at room temperature. Normal human lymphocyte isolated on
Ficoll/Hypaque served as control of SRBC samples.

DNA synthesis

DNA synthesis was evaluated by 60' pulses of ^{3}H thymidine at
various time intervals. The amount of incorporated radioactivity
was determined by trichloroacetic acid (TCA) precipitation on GF /C
filters.

RESULTS AND DISCUSSION

Treatments with dimethyl sulfoxide, dimethyl formamide, butyric
acid, retinoic acid, did not modify the phenotype of RPMI-8402 line.
Among the phorbol esters only the potent tumor promoter TPA showed
to be effective at the concentration of 10^{-8}-10^{-7} M.

The most striking change induced by TPA was the disappearance
of nuclear TdT in the treated cells. By means of the immunofluorescence test it was apparent that almost 90% of the cells grown in
the presence of TPA had a faint cytoplasmic fluorescence, while all
the control cells showed bright fluorescent nuclei (Fig. 1).

Likewise TdT activity decreased in the TPA-treated cell population (11 U/10^8 cells) in comparison to the control (115 U/10^8
cells). The residual activity would be due to the TdT$^+$ cells still
present after TPA treatment.

Concomitant to the loss of nuclear TdT we observed in the TPA
treated population a fraction of E-rosette forming cells (E$^+$) (no
rosettes were observed when sheep erythrocytes were pretreated with
neuraminidase) (Figs. 2 and 3). The occurrence of E$^+$ cells demonstrated that some TPA treated cells developed receptors for sheep
erythrocytes, peculiarity of mature thymocytes. On the contrary
cells with complement receptors (EAC rosettes) were not any more
detectable in the treated population.

DNA synthesis rate of treated cells was 50% of the control
after 24 hrs. This level of synthesis resulted to be constant at
the other time intervals (Fig. 4). The slow down of cell prolifera-

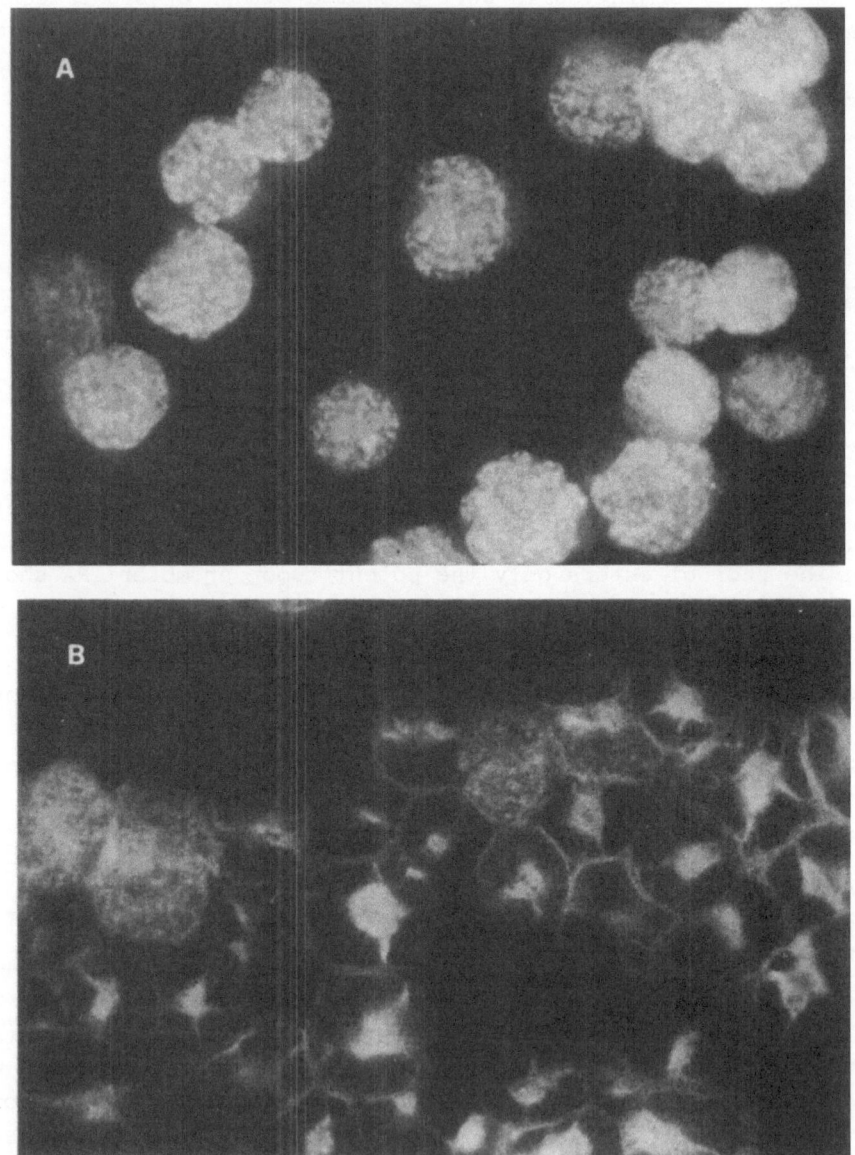

Fig. 1.   Immunofluorescence of 8402 cells grown in the absence (A)
          and in the presence (B) of 10^{-7} M TPA for 96 hrs.

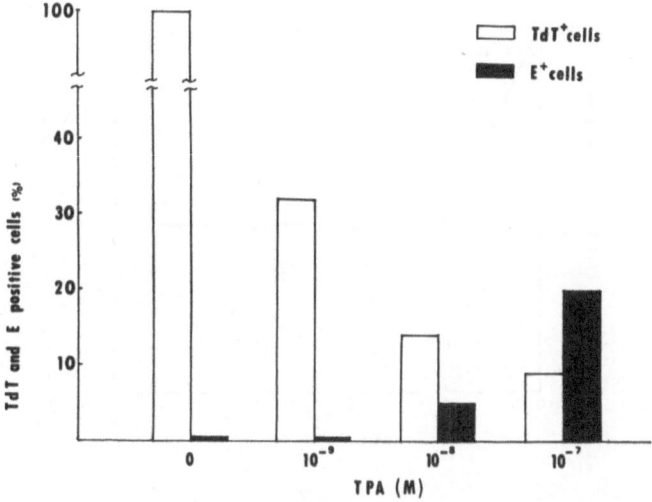

Fig. 2.   Induction of T markers in 8402 cell population after
          96 hrs of incubation with different concentrations
          of TPA.

tion was not accompanied by a decreased viability of the cells as
assessed by trypan blue dye exclusion. Therefore the drop in DNA
synthesis was not due to cytotoxicity of TPA, but rather to the
differentiation process.

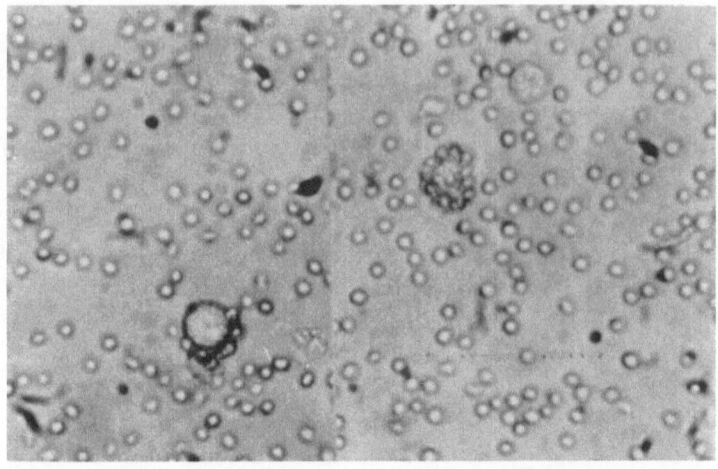

Fig. 3.   TPA treated 8402 cells with sheep erythro-
          cyte receptors (E^+ cells).

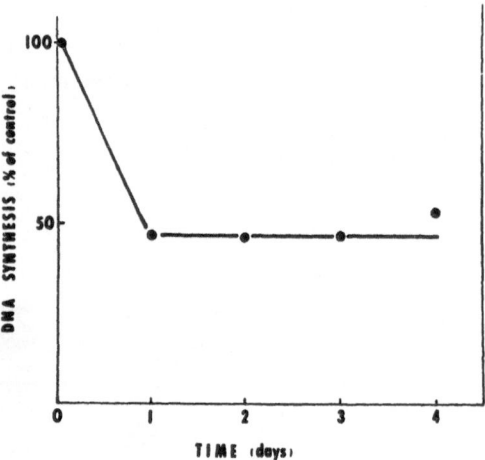

Fig.4. Effect of TPA 10^{-7}M on the DNA
synthesis rate of 8402 cells.

These findings taken together indicate that TPA can promote a
sequence of T- differentiation steps.The first forward step is represented by the loss of terminal transferase. Disappearance of nuclear TdT is generally considered a parameter of differentiation of
both T- and B- lineage prelymphocytes (8). A minor fraction of the
population undergoes a further step of maturation developing the
receptors for sheep erythrocytes.

The differential response within the 8402 cells to TPA induction may be due to the genetic heterogeneity of the cell population
as demenstrated by its chromosome distribution. Cytogenetic analysis showed that 8402 line was heteroploid with a chromosome distribution around a modal chromosome number of 79 (data not shown).

The extent of TPA- induced T maturation may also depend on the
stage of leukemic transformation of the different T lines.For in-
stance in the malignant T-cell line MOLT-3 (9),virtually all the
cells can be induced by TPA to become E$^+$. This line seems to be
more differentiated than the 8402 line since a percentage of E$^+$
cells is already present in untreated cell population.

TPA seems to be an efficient tool to switch on a maturation
process in malignant T lymphoblasts since it may somehow simulate

the physiological intrathymic stimuli promoting <u>in vitro</u> some
T differentiation steps.

    Presently differentiation systems as that one we have describ-
ed might be useful to study the still unsolved problem regarding
the possible processing of terminal transferase during differentiation suggested by the multiple forms found in human leukemic cells
(see papers by Coleman and Chang in this book).

ACKNOWLEDGEMENTS

    This work was supported by grant No. 80.01561.96 of the
Finalized Project "Control of Tumor Growth",C.N.R.,Italy.
This publication is contribution No.1810 of the Radiation Protec-
tion Programme (DG XII) of the European Community Commission.

REFERENCES

1. G. Nabel,M.Fresno,A.Chessman, and H.Cantor, Use of cloned popu-
    lations of mouse lymphocytes to analyze cellular differen-
    tiation, <u>Cell</u> 23:19 (1981)
2. Y.Caire,A.de Sostoa, A. E. Silverstone, Isolation of a subset
    of thymocytes inducible for terminal transferase biosynthesis,
    <u>J. Immunol.</u> 126:553 (1981).
3. K. Gregoire, I. Goldschneider, R. A. Barton and F.J. Bollum,
    Ontogeny of terminal deoxynucleotidyl transferase positive
    cells in lymphoemopoietic tissues in rat and mouse, <u>J. Immunol.</u>
    123:1347 (1979).
4. E. L. Reinherz , P. C. Kung, G. Goldstein, R.H. Levey, and S.F.
    Schlossman, Discrete stages of human intrathymic differentia-
    tion: analysis of normal thymocytes and leukemic lymphoblast
    of T-cell lineage, <u>Proc. Natl. Acad. Sci</u> 77:1588 (1980).
5. C. C. Huang, Y. Hou, L. K. Woods, G. E. Moore, and J. Minowada,
    Cytogenetic studies of human lymphoid T-cell lines derived
    from lymphocytic leukemia, <u>J. Natl. Cancer Inst.</u> 53: 655(1974).
6. L. M. S. Chang, and F. J. Bollum, Deoxynucleotide polymerizing
    enzymes of calf thymus gland, <u>J. Biol. Chem.</u> 246:909 (1971).
7. F. J. Bollum, Antibody to terminal deoxynucleotidyl transferase.
    <u>Proc. Natl. Acad. Sci.</u> 72:4119 (1975).
8. F. J. Bollum, Terminal Transferase: experienced biochemical rea-
    gent seeks biological assigments, <u>TIBS</u>, February issue: 41(1981)
9. K. Nagasawa, and T.W. Mack, Phorbol esters induce differentiation
    in human malignat T Lymphoblasts, <u>Proc. Natl. Acad. Sci.</u>
    77: 2964 (1980).

TERMINAL DEOXYNUCLEOTIDYL TRANSFERASE IN HUMAN FETAL TISSUES

Claudio Casoli**, Antonio Bonati* and Riccardo Starcich*

*Istituto di Patologia Speciale Medica
**Istituto di Microbiologia dell'Universita di Parma
  Parma, Italy

SUMMARY

The authors have investigated terminal deoxynucleotidil transferase (TdT) activity in the spleen, liver and thymus of a human foetus of 23 weeks. TdT significant levels have been found and two different molecular forms were defined in each tissue.

INTRODUCTION

Postnatal TdT levels have been studied by many authors both in normal (1-3) and neoplastic tissues (4-5). Some authors studied TdT in fetal tissues. Chang (6) found high TdT levels in cortical tymocytes of embryonic calf thymus gland.

Recently Sugimoto and Bollum (7) investigated TdT in the thymus and other organs of embryonic chickens with an immuno-fluorescent procedure. TdT positive cells in the thymus was found to attain a plateau about day 19, a few TdT positive cells were observed in the bursa of Fabricius at day 20, positive cells were extremely rare in both the spleen and bone marrow.

Gregoire et al. (8) examined TdT in haemopoietic tissues of rats and mice during fetal life and on the days after birth. TdT positive cells were not found in fetal liver, spleen or bone marrow, but appeared in bone marrow and spleen after birth. TdT positive cells in thymus and bone marrow decreased gradually, whereas TdT positive cells in spleen disappeared at 7 weeks of age. In rats the results of immunofluorescent valuation were confirmed by enzymatic activity.

The fact that the literature doesn't report studies on TdT levels in human fetal tissues prompted us to investigate TdT activity in the spleen, liver and thymus of a human fetus of 23 weeks.

METHODS

Human Fetal Cells

The thymus, liver and spleen of a human fetus of 23 weeks were studied. The foetus was obtained by spontaneous abortion. The gestational age was determined from crown to-heel length and expressed as weeks, calculated from the last menstrual period. The fetal thymus, spleen and liver, finely teased on stainless steel mesh, were passed through four layers of gauze to obtain a single-cell suspension of each organ. All manipulations were performed under aseptic conditions.

Cell suspensions, after morphologic and cytochemical analysis, were purified by Ficoll-Hypaque density-gradient centrifugation according to the method of Boyum (9). The mononuclear cells were washed once in phosphate buffered normal saline and counted.

Crude extracts from thymus, liver or spleen cells were obtained by detergent treatment of 3x freeze-thawed samples according to the procedure of McCaffrey et al. (10). Phenil methane sulfonylfluoride was included to reduce proteolysis.

Phosphocellulose chromatography

Crude extracts were centrifuged at 100,000 g for 1 h at 4°C and the supernatants applied at a flow rate of 10 ml/h to a 9 x 60 mm phosphocellulose columns equilibrated with 0,15 M KCl--TEGD. Columns were subsequently washed with 50 ml of this buffer. Enzyme activity was eluted with a 0,15-0,75 M KCl gradient in TEGD buffer. Fractions of approximately 2 ml were collected and 20 μl of each fraction were assayed in duplicate for TdT activity with and without 100 μM ATP, which has been shown to be a specific inhibitor for TdT without any inhibitory effect on other cellular DNA polymerase (11).

Assay System

Enzymes reactions were carried out in a final volume of 100 μl and contained the following components: 50 mM Tris-HCl ph 7.8, 1 mM DTT, 50 μg BSA, 0.5 mM MnCl$_2$, 0.5 μg oligo (dA)12-18, 20 μM (^{3}H)dGTP (120 cpm/ pmole). At the end of the incubation period (37°C, 1 h), sodium pyrophosphate and trichloroacetic acid were added for 10 min. The acid-insoluble fractions collected on GF/B filters were counted in a beta-counter.

An inhibition of at least 80% in the presence of 100 uM ATP was required for unequivocal demonstration of TdT activity.

One unit of TdT activity equalled 1nmole of (^3H)dGMP polymerized in 1 h. Specific activities were expressed as unit/10^8 cells. MOLT 4 cell line was valuated for standardizing our laboratory index: it is to 0.8 U/10^8 cells whereas the normal range for bone marrow is up to 0.01 U/10^8 cells.

RESULTS

Table 1 summarizes the levels of terminal transferase activity from thymus, liver and spleen cells. On the basis of our laboratory index these values seem to be significant. TdT levels in thymus cells are near to those found in spleen cells, the liver activity is 40% lower.

Table 1

TdT Levels in human fetal tissues

	thymus	0,75 units/10^8 cells
Foetus of	spleen	0,69 units/10^8 cells
23 weeks	liver	0,46 units/10^8 cells

Gradient elution of TdT from the phosphocellulose columns never yielded a unique activity peak (fig. 1). In our study two different forms of TdT can be defined in each tissue examined. These forms depend on the molarity of KCl at which they are eluted from the phosphocellulose: approximately in the thymus 0.20 and 0.35 M., in the spleen 0.25 and 0.35 M, in the liver 0.25 and 0.5 M.

DISCUSSION

We postulate two possible explanations for the presence of

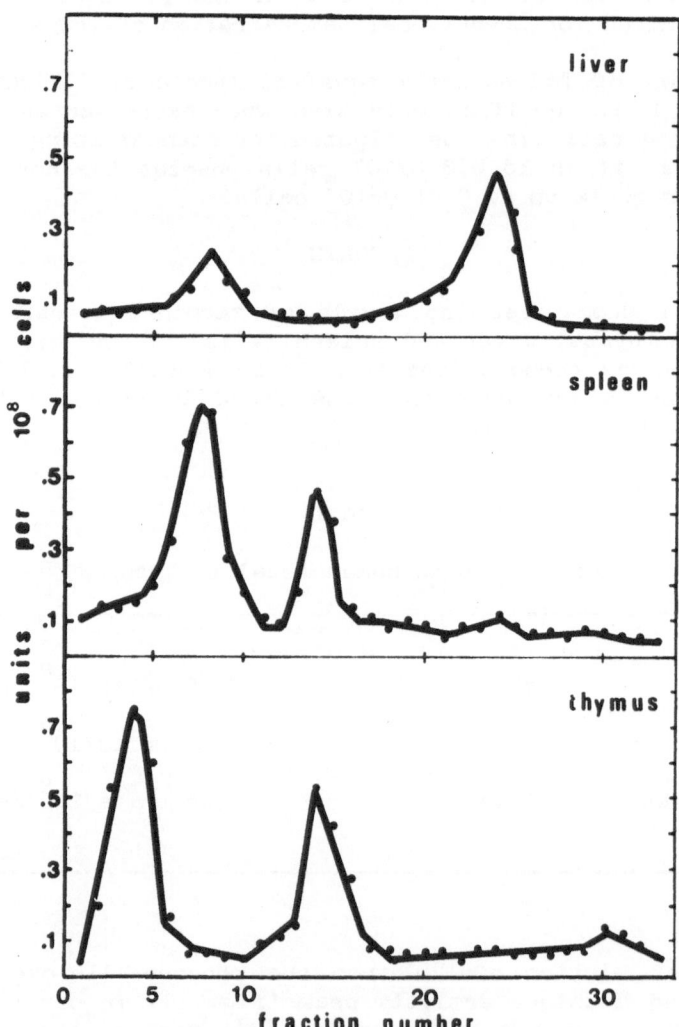

Figure 1.   Phosphocellulose chromatography of TdT. The gradient was from 0,15 to 0,75 M KCl. For technical details, see
methods.

TdT in fetal tissues.

The first hypothesis is that detectable TdT levels during fetal life in the thymus, liver and spleen could be expression of the activity of pre-thymic cells . Thus, Gregoire et al. (8), have found normal rate of TdT-positive cells in bone marrow and spleen from neonatally thymectomized rats and from congenitally athymic (nu/nu) NIH Swiss strain mice. The great majority then of TdT positive spleen cells from both nu/+ and nu/nu mice have been induced to express Lyt 1,2,3 alloantigens by a brief in vitro incubation with thymosin (12). Goldschneider et al. (13) too have induced terminal transferase and Lyt antigens in mouse bone marrow and spleen with thymosin. These studies suggest that the thymus is not necessary for the process of terminal transferase induction and define the nature of TdT positive cells as pre-thymic cells.

The second hypothesis considers TdT as an embryonic enzyme present in different tissues with a high proliferative ratio and perhaps active in maturative processes. Therefore, it is important to establish if there exists a relationship between TdT activity and CFU_s and CFU_c activity in haemopoietic fetal tissues. Because the literature shows that human fetal livers of 6-12 weeks gestational age contain CFU_c comparable to that found in adult marrow (14), we consider it would be very interesting to evaluate TdT levels at this gestational age.

Finally, the occurrence of several chromatographic forms of terminal deoxynucleotidil transferase has always been reported for mice, chickens and humans (15-17). Since no differences in catalytic properties were found between these forms, the interpretation of this observation is not clear. Experiments are in progress to study terminal transferase in human haemopoietic fetal tissues and the different physical-chemical characteristics of this enzyme.

ACKNOWLEDGEMENTS

The authors would like to thank Mrs. Romana Gardini for her expert technical assistance.

REFERENCES

1. GREENWOOD M.F., COLEMAN M.S., HUTTON J.J., LAMPKIN B., KRILL C., BOLLUM F.J. and HOLLAND P., Terminal deoxynucleotidyl transferase distribution in neoplastic and haemopoietic cells. J. Clin. Invest., 59 (1977) 889-899.

2. MEYSKENS F.L., KIETER C.A. and JONES J., Postnatal expres-
    sion of terminal deoxynucleotidyl transferase (TdT)
    in human thymus, Blood 52 suppl. 1 (1978) (Abstract)
    137.
3. YASMINEH W.G., SMITH B.M. and BLOOMFIELD C.D., DNA nucleo-
    tidylexotransferase of normal persons and leukemic
    patients, Clin. Chem. 26 (1980) 891-895.
4. HOFFBRAND A.V., GANESHAGURU K., JANOSSY G., GREAVES M.F.,
    CATOVSKY D. and WOODRUFF R.K., Terminal deoxynucleo-
    tidyl transferase levels and membrane phenotypes in
    diagnosis of acute leukaemia, Lancet 2 (1977) 520-523.
5. KUNG P.C., LONG J.C., McCAFFREY R.P., RATLIFF R.L., HARRI-
    SON T.A. and BALTIMORE B.S.D., Terminal deoxynucleo-
    tidyl transferase in the diagnosis of leukemia and
    malignant lymphoma, Am. J. Med. 64 (1978) 788-794.
6. CHANG L.M.S., Development of terminal deoxynucleotidyl tran-
    sferase activity in embryonic calf thymus gland, Bio-
    chem. Biophys. Res. Commun. 44 (1971) 124-131.
7. SUGIMOTO M. and BOLLUM F.J., Terminal deoxynucleotidyl
    transferase (TdT) in cick embryo lymphoid tissues, J.
    Immunol. 122 (1979) 392.
8. GREGOIRE K.E., GOLDSCHNEIDER I., BARTON R.W. and BOLLUM
    F.J., Ontogeny of terminal deoxynucleotidyl transfe-
    rase-positive cells in lymphohemopoietic tissues of
    rat and mouse, J. Immunol. 123 (1979) 1347.
9. BOYUM A., Separation of leucocytes from blood and bone
    marrow, Scand. J. Clin. Lab. Invest. 21 suppl. 97
    (1968) 1.
10. McCAFFREY F., SMOLER D.F. and BALTIMORE D., Terminal deo-
    xynucleotidyl transferase in a case of childhood acu-
    te lymphoblastic leukemia, Proc. Natl. Acad. Sci.
    U.S.A 70 (1973) 521.
11. BHALLA R.B., SCHWARTZ M.K. and MODAK M.J., Selective inhi-
    bition of terminal deoxynucleotidyl transferase (TdT)
    by adenosine ribonucleoside triphosfate (ATP) and its
    application in the detection of TdT in human leuke-
    mia, Biochem. Biophys. Res. Commun. 76 (1977) 1056.
12. HUTTON J.J., COLEMAN M.S., KENEKLIS T.P. and BOLLUM F.J.,
    Terminal deoxynucleotidyl transferase as a tumor cell
    marker in leukemia and lymphoma. Results from 1000
    patients. In Advances in Medical Oncology. Research
    and Education. Vol. 4: Biological Basis for Cancer
    Diagnosis. Edited by M. Fox. Pergamon Press, Oxford
    (1979) Pp. 165-175.
13. GOLDSCHNEIDER I., AHMED A.F., BOLLUM F.J. and GOLDSTEIN
    A.L., Induction of terminal deoxynucleotidyl transfe-
    rase and Lyt antigens in mouse bone marrow and spleen
    cells with thymosin: Identification of multiple sub-
    sets of prothymocytes, Manuscript submitted for pub-
    blication.

14. BARAK Y., KAROV Y., LEVIN S., SOROKER N., BARASH A., LANCET
    M. and NIR E., Granulocyte-macrophage colonies in
    cultures of human fetal liver cells: morphologic and
    ultrastructural analysis of proliferation and diffe-
    rentiation. Exp. Hematol. 8 (1980) 837.
15. McCAFFREY R., HARRISON T.A., PARKMAN R. and BALTIMORE D.,
    Terminal deoxynucleotidyl transferase activity in hu-
    man leukemic cells and in normal thymocytes. New
    England Journal of Medicine 292 (1975) 775-780.
16. KUNG P.C., SILVERSTONE A.E., McCAFFREY R.P. and BALTIMORE
    D., Murine terminal deoxynucleotidyl transferase: cel-
    lular distribution and response to cortisone. Journal
    of Experimental Medicine 141 (1975) 855-865.
17. PENIT C. and Chapeville F.: Developmental changes in termi-
    nal deoxynucleotidyl transferase of the chicken thy-
    mus. Biochem. Biophys. Res. Commun. 74 (1977) 1096.

EFFECT OF NUCLEOSIDES ON HUMAN T AND B LYMPHOBLASTOID CELL LINES

D. Breviario[1], P. Plevani[2], N. Sacchi[1], G. Badaracco[2],
U. Bertazzoni[3] and E. Ginelli[1]

[1]Istituto di Biologia Generale, Università di Milano
[2]Istituto di Biologia, EULO, Brescia
[3]Istituto di Genetica Biochimica ed Evoluzionistica
C.N.R., Pavia

INTRODUCTION

Human T and B lymphoblastoid cell lines have been used as model systems to study the biochemical basis of some immune disfunctions,
associated to specific purine metabolic enzyme deficiencies (1,2).

These studies showed that human T lymphoblasts have high level
of deoxyribonucleoside kinases and low levels of deoxyribonucleotide
dephosphorylating activities which both possibly contribute to their
enhanced ability to accumulate high concentrations of deoxynucleoside
triphosphates when exposed to exogenous nucleosides. As a consequence, T lymphoblasts are more sensitive to the inhibition of DNA
synthesis than B cells, and are selectively killed when treated
with high concentrations of nucleosides (3,4).

To explain the cytotoxicity of nucleosides in malignant T cells,
other biochemical differences between B and T cells have been considered. Recently, it has been reported that only lymphoblastoid cell
lines which contain terminal deoxynucleotidyl transferase (TdT) are
killed by elevated concentration of deoxyguanosine (5).

The physiological function of TdT is still unknown and its
possible in vivo product has not yet been identified. The finding
of TdT only in prelymphocyte cells (6,7) has suggested the possible

participation of the enzyme in creating DNA diversity by addition
of non-complementary nucleotides at 3'-OH termini (8).

In this regard the preferential nucleoside cytotoxicity of
TdT[+] cells could be explained by the presence and activity of the
enzyme during the processes of DNA replication or recombination
(5,9). This interpretation may be supported by other findings such
as the differential expression of d-Guanosine cytotoxicity in
different stages of T cell maturation (10) and the sensitivity of
a human pre-B TdT[+] lymphoblastoid cell line (NALM-1) to d-Guanosine
(5).

In a attempt to verify the involvement of TdT in this cytotoxicity phenomenon and to find variations in the DNA sequences
possibly linked to the enzyme activity in vivo, we have investigated
the effect of nucleosides on TdT[+] and TdT[-] cell lines derived from
the same patient (11).

RESULTS AND DISCUSSION

Figures 1 and 2 compare the cell growth and the extent of DNA
synthesis of human lymphoblastoid T (8402, TdT[+]) and B (8392, TdT[-])
cell lines (11) in response to different concentrations of d-Guanosine and Thymidine.

After a 24 hour treatment, the percentage of viable cells
was calculated as the ratio between the number of viable cells in
the presence of nucleosides and in control cultures. Concomitantly,
the extent of DNA synthesis was determined by incubating the cells
for 1 hr in the presence of ^{3}H-Thymidine or ^{3}H -d-Guanosine.

TdT[+] cells were much more sensitive than TdT[-] cells to DNA
synthesis inhibition by d-Guanosine and Thymidine. DNA synthesis
was almost completely inhibited in TdT[+] cells in the presence of
10 µM d-Guanosine whereas in TdT[-] cells the same effect was observed only by using one hundred higher fold nucleoside concentration (Figures 1 and 2).

In the presence of Thymidine, the maximum value of differential DNA synthesis inhibition between the two cell lines was
obtained at the concentration of 100 µM. Higher concentrations of
both nucleosides gave essentially the same effect (Figures 1 and 2).

Considering the cell viability, at high nucleoside concentrations we observed a 50% of viable TdT⁻ cells, that means a complete
arrest of cell proliferation. On the other hand the percentage of
viable TdT⁺ cells was less than 50%, indicating that a lethal effect
occurred (Figures 1 and 2).

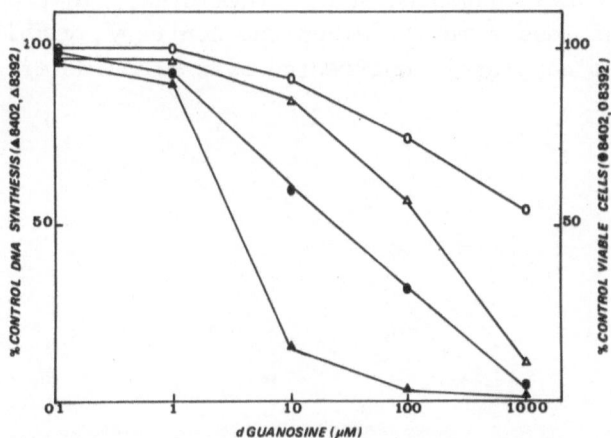

Fig.1. Effect of d-Guanosine on DNA synthesis and growth of
8402 (TdT⁺) and 8392 (TdT⁻) cell lines.

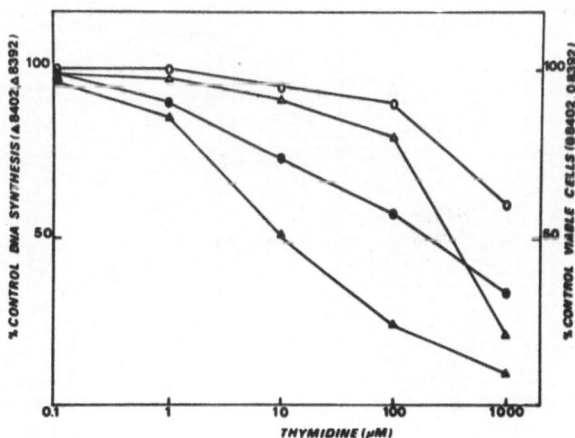

Fig.2. Effect of Thymidine on DNA synthesis and growth of
8402 (TdT⁺) and 8392 (TdT⁻) cell lines.

     T and B lymphoblast viability was also determined after
d-Guanosine and Thymidine removal. The cells, incubated for 24 hrs
with different concentrations of nucleosides, were washed extensively and inoculated at the same density after separation of viable
cells by centrifugation in 10% BSA. As shown in Figures 3 and 4,
the number of viable TdT[+] cells decreased by increasing the concentrations both of d-Guanosine and of Thymidine. In the latter case
the effect was less remarkable. On the contrary, viability of the
TdT[-] cells was completely unaffected by nucleoside treatment.

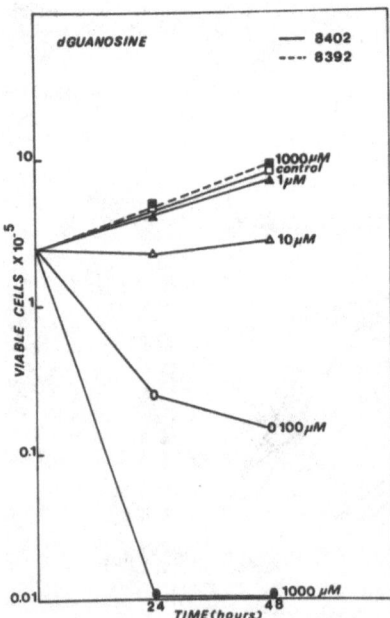

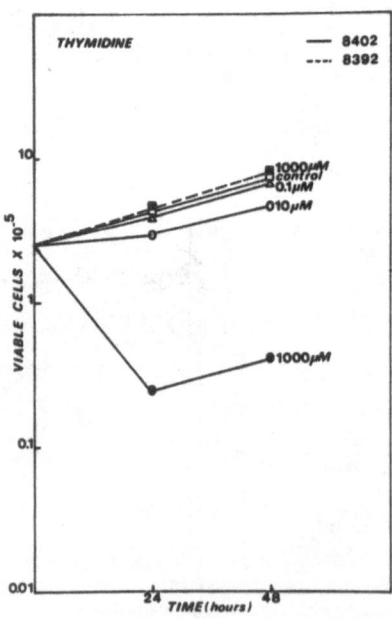

Fig.3. Growth curves of 8402
      (TdT[+]) and 8392 (TdT[-])
      cell lines after removal
      of d-Guanosine.

Fig.4. Growth curves of 8402
      (TdT[+]) and 8392 (TdT[-])
      cell lines after removal
      of Thymidine.

Comparable cytotoxic and cytostatic effects after treatment with d–Guanosine and Thymidine were obtained in a human lympho-blastoid pre–B cell line (NALM-1, TdT[+]) and in human promielocytic cell line (12) (HL-60, TdT[−])respectively (data not shown).

To investigate the mechanism of preferential sensitivity of TdT[+] cells to nucleosides we measured the size of deoxynucleosides triphosphate pool (13) in both cell lines after 24 hrs treatment with different concentrations of d–Guanosine and Thymidine (Table I). TdT[+] cells accumulated dGTP and dTTP faster than TdT[−] cells. At 100 μM nucleoside concentration, a 15–fold difference between the levels of dGTP and dTTP in TdT[+] and TdT[−] cells was observed.

These results suggest the existence of a positive correlation between elevated levels of triphosphates and selective inhibition of DNA synthesis in TdT[+] cells. This effect is probably due to the inhibition of the allosteric enzyme ribonucleotide reductase by dGTP and dTTP. The levels of the other triphosphates also showed positive or negative variations.

Table I. Accumulation of dNTPs in 8402 (TdT[+]) and 8392 (TdT[−]) cell lines treated with Deoxyguanosine and Thymidine

Nucleosides (μM)	Cell lines							
	8402				8392			
	dATP	dCTP	dGTP	dTTP	dATP	dCTP	dGTP	dTTP
	(pmoles/10^6 cells)				(pmoles/10^6 cells)			
Deoxyguanosine								
0	29	37	12	22	6	28	13	15
1	29	33	12	15	−	−	−	−
10	87	40	30	22	9	28	12	19
100	72	61	454	20	12	34	37	15
1000	−	−	−	−	10	39	150	16
Thymdine								
0	29	37	12	22	6	28	13	15
1	22	30	12	23	−	−	−	−
10	42	26	16	114	7	29	13	14
100	127	18	54	307	8	28	16	22
1000	−	−	−	−	12	27	23	198

Since the lethal effect of nucleosides on TdT$^+$ cells could be interpreted as the result of incorporation of non complementary nucleotides into DNA by TdT, we have tried to detect the presence of possible changes at DNA level in TdT$^+$ cells by analyzing the molecular organization of some regions of the human genome after nucleoside treatment.

In preliminary experiments, nuclear DNA was extracted from TdT$^+$ cells treated for 24 hrs with either 10 μM d-Guanosine or Thymidine, digested with Eco RI and fractionated in 1% agarose gel elecrophoresis. The DNA fragments were subsequently prepared for molecular hybridization either <u>in situ</u> or on nitrocellulose filter (14,15). Hybridation of filters and gels were done with ^{32}P-nick translated plasmids containing mouse rDNA, human β globin DNA and a human middle repetitive sequence (16-18). No difference was observed in sequence organization of DNA extracted from TdT$^+$ treated and untreated cells.

These data would suggest that cytotoxicity in TdT$^+$ cells is not a consequence of macroscopic alteration of DNA. It is possible, however, that TdT activity is restricted to very limited regions of DNA, so that it could not be detected in our experimental conditions.

AKNOWLEDGEMENTS

The work was supported in part by grant no. 80.01561.96 of the Finalized Project "Control of Tumor Growth" C.N.R., Italy. This publication is contribution no. 1810 of the Radioprotection Programme (DGXII) of the European Community Commission.

REFERENCES

1. E. R. Giblett, J. E. Anderson, F. Cohen, B. Pollara and H. F. Meuwissen, Adenosine deaminase deficiency in two patients with severely impaired cellular immunity, <u>Lancet</u> 2:1067 (1972)
2. E. R. Giblett, A. J. Amman, D. W. Wara and L. K. Diamond, Nucleoside phosphorylase deficiency in a child with severely impaired T-cell immunity and normal Bcell immunity, <u>Lancet</u> 1:1010 (1975).
3. D. A. Carson, J. Kaye, S. Matsumoto, J. E. Seegmiller and L. Thompson, Biochemical basis for the enhanced toxicity of deoxyribonucleosides toward malignant human T cell lines, <u>Proc.Natl.Acad.Sci.</u> 76: 2430 (1980).

4. R. L. Wortmann, B. S. Mitchell, N. L. Edwards and I. H. Fox,
    Biochemical basis for differential deoxyadenosine toxi-
    city to T and B lymphoblasts: Role for 5'-nucleotidase,
    Proc.Natl.Acad.Sci (USA) 76: 2434 (1979).

5. F. J. Bollum and I. Goldschneider, Membranes, Receptors and
    the immune response, in "Progress in Clinical and Biolo-
    gical Research" (Cohen, E. R. and Köhler H. eds.) Alan R.
    Liss Inc., New York (1980) vol. 42, pp. 189-202.

6. F. J. Bollum, Terminal deoxynucleotidyl transferase as a hema-
    topoietic cell marker, Blood 54:1203 (1979).

7. K. E. Gregoire, I. Goldschneider, R. W. Barton and F. J. Bollum,
    Ontogeny of terminal deoxynucleotidyl transferase positive
    cells in lymphopoietic tissues of rat and mouse,
    J. Immunol. 123:1347 (1979).

8. F. J. Bollum, Terminal deoxynucleotidyl transferase: Biologi-
    cal studies, in "Advance in Enzymology" (Meister A. ed.)
    Wiley Interscience, New York, pp. 347 (1978).

9. F. J. Bollum, Terminal transferase: experienced biochemical
    reagent seeks biological assignment. TIBS, February: 41
    (1981).

10. A. Cohen, J. W. W. Lee, H. M. Dosch and E. Gelfand, The expres-
    sion of deoxyguanosine toxicity in T lymphocytes at
    different stages of maturation. J. Immunol. 125: 1578
    (1980).

11. C. C. Huang, Y. Hou, L. K. Woods, G. E. Moore and T. Minowada,
    Cytogenetic study of human, lymphoid T cell lines
    derived from lymphocytic leukemia, J. Natn. Cancer Inst.
    53: 665 (1974).

12. S. J. Collins, R. C. Gallo and R. E. Gallagher, Continous
    growth and differentiation of human myeloid leukemic
    cells in suspension, Nature, 270: 347 (1977).

13. L. Kuehl, Isolation of skeletal muscle nuclei, Exp. Cell Res.
    91: 441 (1975).

14. T. M. Shinnick, E. Lund and O. Smithes, Hybridization of
    labelled RNA to DNA in agarose gels. Nucleic Acids Res.
    2: 1911 (1975).

15. E. M. Southern, Detection of specific sequences among DNA
    fragments separated by gel electrophoresis. J. Mol. Biol.
    98: 503 (1975).

16. S. Ottolenghi, B. Giglioni, P. Comi, A. M. Gianni, E. Polli,
    C. T. A. Acquaye, J. H. Oldham and G. Masera, Globin
    gene deletion in HPFH, $\delta^o{}'\beta^o$ thalassaemia and Hb lepore
    desease, Nature 278: 654 (1979).

17. D. C. Tiemeier, S. M. Tilghman and P. Leder, Purification and
    cloning of a mouse ribosomal gene fragment in coliphage
    lambda. Gene 2: 173 (1977).

18. E. Ullu and M. L. Melli, unpublished results.

20α-HYDROXYSTEROID DEHYDROGENASE (20αSDH) - A NEW ENZYMATIC

MARKER FOR PRE T AND T LYMPHOCYTES

Yacob Weinstein

Department of Hormone Research
The Weizmann Institute of Science
Rehovot 76100, Israel

ASSOCIATION OF 20αSDH WITH T LYMPHOCYTES AND MEDULLARY THYMOCYTES

The enzyme 20αSDH which reduces progesterone to 20α-hydroxy-progesterone is found in various lymphatic organs (1,2). The following evidence suggests that within the mouse lymphocyte population, 20αSDH is confined to T lymphocytes (1,2): 20αSDH activity is found in thymocytes, splenocytes and lymph node lymphocytes from all the normal mouse strains, but not in splenocytes from congenitally athymic or neonatally thymectomized mice; it is associated with spleen cells bearing the θ antigen; T-cell mitogens such as Con A and PHA, alloantigens (3), but not the B cell mitogen LPS, are able to induce <u>in vitro</u> an increase in 20αSDH activity (2). 20αSDH is found in thymocytes and T cells from all mouse strains. In the thymus, the 20αSDH activity is located within the hydrocortisone resistant, PNA negative, medullary cells (2,4,5).

20αSDH ACTIVITY IN BONE MARROW CELLS

The Effect of Age and Strain on 20αSDH Activity

Low levels of activity are found in 2-week-old mice, and increase between 2 weeks to 2 months of life (6,7). This increase in 20αSDH is strain specific: BALB/c, SJL, NZW and NZB mice have high enzymatic activity while C57BL/6, DBA/1 and DBA/2 mice show low activity. This property is not linked to the major histocompatibility complex (6).

Distribution of 20αSDH Activity among Subpopulations of Bone Marrow Cells

The pattern of the cell size distribution determined by light scattering with a Fluorescence Activator Cell Sorter (FACS), indicates that there are two distinct subpopulations of nucleated cells. The small cells are the size of a small lymphocyte (about 7 microns) and the large cells are the size of blast cells (10-14 microns). More than 75% of the 20αSDH activity was found in the large cell population (4,6). Almost no 20αSDH activity was found in newborn liver cells from BALB/c, SJL and in spleens of BALB/c nude mice ten days after lethal irradiation and bone marrow reconstitution (6). These organs contain stem cells, erythroid and myeloid cells at various stages of differentiation (8). About 1% of the liver cells are pre B-lymphocytes (9). These results suggest that the enzyme is not located in erythroid, myeloid or pre B cells. Macrophages are also devoid of 20αSDH activity (5).

POSSIBLE LOCALIZATION OF 20αSDH IN MARROW PRE T LYMPHOCYTES

Stimulation of bone marrow cells from congenitally athymic nude or normal mice with PHA in conditions which are known to stimulate pre T lymphocytes (10) caused stimulation of 20αSDH activity as well as an increase in ^{3}H thymidine incorporation after 4 days in culture (6,7). Moreover, at least part of the bone marrow cells which were stimulated by PHA to express θ-antigen also have 20αSDH activity (6). This was found by labeling the PHA stimulated marrow cells with fluorescent anti θ, isolation of the labeled cells on FACS and measurement of the 20αSDH activity in the labeled cells (6). These results indicate that some pre T cells have 20αSDH activity.

TESTOSTERONE EFFECT ON THE THYMUS AND MARROW 20αSDH ACTIVITY

Thymus

It has been shown that testosterone influences the size of the thymus (11). Orchidectomy causes hypertrophy of the thymus, which leads to the population of the thymus with additional young, hydrocortisone-sensitive, cortical thymocytes (11). Since cortical thymocytes lack 20αSDH activity (4) the total 20αSDH activity of such thymus cells is low. Testosterone implant (in Silastic capsule) causes atrophy of the thymus cortex in the orchidectomized mice within 10 days. The high levels of 20αSDH and the vigorous response of the residual thymocytes from testosterone treated mice to PHA and Con A, show that they belong to the mature hydrocortisone resistant medullary thymocytes. The effective testosterone concentration which affects lymphocyte proliferation in vitro is 3000 ng/ml. The concentration of androgens in the blood of orchidectomized mice, normal males and orchidectomized mice with testosterone implant are about 1, 3 and 15 ng/ml, respectively. It is unlikely that changes in testosterone concentrations which are 200-3000 fold below these concentrations are sufficient to explain thymic growth after orchidectomy or thymus atrophy in the presence of testosterone implant. From these we conclude that testosterone affects the thymic cell population indirectly probably via the bone marrow.

Bone Marrow

Orchidectomy causes a decrease in the relative number of the large marrow cells concomitantly with a decrease in the marrow 20αSDH activity (4). Testosterone replacement therapy reverses this effect and induces an increase in marrow 20αSDH activity and in the relative number of large marrow cells. Since most of the 20αSDH activity resides in the large cells, the increase in marrow 20αSDH activity caused by testosterone could be explained by the accumulation of these (large) cells in the presence of the androgen.

CONCLUSIONS

It was suggested that the cortical and medullary thymocytes belong to separate lineages of T cell differentiation (12), and that there are multiple subsets of prothymocytes (13). Our results support these hypotheses and indicate that in the bone marrow there are separate lineages of pre-medullary and precortical thymocytes.

Since the small decrease in blood testosterone levels after orchidectomy is enough to increase the small bone marrow population and the number of cortical thymocytes, it is conceivable that the precortical cells will be among the small marrow cell population. Moreover, increase in testosterone blood levels causes a decrease in the small marrow cells, and a decrease in the thymic cortical cells. In view of the likelihood that 20αSDH activity is a property of pre T lymphocytes (4,6), and the fact that most of the 20αSDH activity is located in the medullary thymocytes (4), the marrow cells that express 20αSDH activity are likely to be the premedullary thymocytes.

The enzyme terminal deoxynucleotidyl transferase (Tdt) is found in the marrow pre-T cells and in thymocytes. Tdt is located in the small marrow cells and in the thymic cortical PNA positive cells. Peripheral T cells have no Tdt activity (14). Neonatal thymectomy causes a marked decrease in marrow Tdt activity (15), while it increases the marrow 20αSDH activity (6). We suggest (Fig. 1) that the medullary and cortical thymocytes belong to two separate lineages of differentiation. The large marrow cells with 20αSDH activity are the precursors of the medullary thymocytes and the small marrow

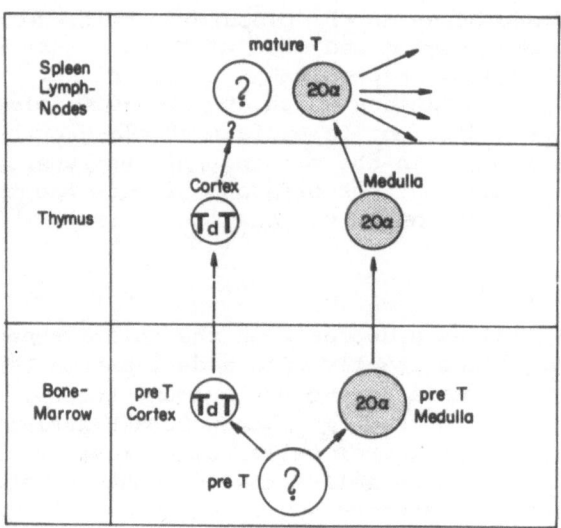

Fig. 1.   Proposal for the ontogeny of T lymphocytes: 20αSDH activity
          is located in marrow pre T lymphocytes, medullary thymo-
          cytes and mature T lymphocytes. Tdt activity is located in
          the small marrow cells and cortical thymocytes. The large
          marrow cells with 20αSDH activity are the precursors of the
          medullary thymocytes. The precortical thymocytes originate
          from the small marrow lymphocytes with Tdt activity.

cells with Tdt activity are the precursors of the cortical thymocytes.

REFERENCES

1. Y. Weinstein, H.R. Lindner, and B. Eckstein, Nature, 266:632
    (1977)
2. Y. Weinstein, J. Immunol., 119:1223 (1977)
3. Y. Weinstein, and E. Okon, J. Natl. Cancer Inst., 64:89 (1980)
4. Y. Weinstein, and Z. Bercovich, J. Immunol., 126:998 (1981)
5. L. Pepersack, J.C. Lee, R. McEwan, and J.N. Ihle, J. Immunol.,
    124:279 (1980).
6. Y. Weinstein, Thymus, in press (1981)
7. A.S. Fuks, and Y. Weinstein, J. Immunol., 23:1266 (1979)
8. D. Metcalf, and M.A.S. Moore, in "Haemopoietic Cells", North
    Holland Publishing Co., Amsterdam, London, 1971.
9. F. Melchers, Eur. J. Immunol., 7:482 (1977)
10. O.W. Press, C. Rosse, and J. Clagett, J. Exp. Med. 146 (1977)
11. E.J. Castro, Proc. R. Soc. Lond. E., 185:425 (1974)
12. K. Shortman, and E. Jackson, Cell Immunol., 12:230 (1974)
13. I. Goldschneider, A. Ahmed, F.J. Bollum, and A.L. Goldstein,
    Proc. Natl. Acad. Sci., in press (1981)
14. D. Baltimore, A.E. Silverstone, P.C. Kung, T.A. Harrison, and
    R.P. McCaffrey, Cold Spring Harbor Symp., 41:63 (1976)
15. N.H. Pazmino, J.N. Ihle, and A.L. Goldstein, J. Exp. Med. 147:
    708 (1978).

HYBRIDS BETWEEN MYELOMA CELLS AND THYMOCYTES - AN ATTEMPT TO

APPROACH THE BIOLOGICAL FUNCTION OF TERMINAL TRANSFERASE

Catherine Transy and Pierre Rouget*

Institut de Recherche en Biologie Moléculaire
C.N.R.S. et Université Paris 7
2, place Jussieu - 75251 PARIS CEDEX 05 - FRANCE

INTRODUCTION

   Terminal deoxynucleotidyl transferase (TdT ; nucleoside triphosphate : DNA deoxynucleotidyl-transferase, E C 2.7.7.31) is an
enzyme able to catalyze the polymerisation of deoxynucleotides onto
a primer without requiring a template (1). During the years following its discovery (2), many data on its enzymatic properties were
described, most of them being extensively reviewed by Bollum (3, 4).
The TdT cellular distribution among normal adult tissues appears to
be restricted to the thymus and bone marrow (5, 6). The presence of
TdT was also described in peripheral cells derived from acute lymphoblastic leukemias and from lymphomas ; in these cases, the cells
were characterized as precursors of T or B lymphocytes (7-11). More
recently, TdT was detected in transient populations of peripheric
cells in young rats and mice (12). The biological function of TdT
remains unknown although its biochemical properties and its cellular
distribution have suggested a possible role in the differentiation
of lymphoid cells and in the generation of the diversity of immunoglobulins (4, 13).

   In fact, a vast amount of experimental data are now available from many laboratories concerning both the amino acid sequences
of the immunoglobulin (Ig) and the DNA structure of the corresponding genes (14-19).

   At first, these data show that the genes which specify the
variable (V) and the constant (C) regions of immunoglobulins are

----

*To whom correspondence should be addressed.

distant in germ—line DNA and brought into a relative proximity during
the lymphoid system development (15). This modification occurs
through a junction between a (V)-DNA segment (coding for 97 to 97
amino acids) and one of the joining (J)-segments coding for the remaining amino acids of the variable Ig region. For the heavy chains, another DNA segment, called D for diversity (16, 18) is involved in a
V-D-J junction and seems to correspond to a part of the third hypervariable (HV-3) region. During these rearrangements, an intervening
sequence between the J and C-DNA segments remains and will be further
spliced from a pre-mRNA.

     In the process of DNA rearrangements two mechanisms can be in-
volved in the diversification of immunoglobulin genes : choice of the
(J) segment to be joined to the (V) region and some variations in the
exact sites of V-J or V-D-J reconbinations. These events at least can
affect the DNA segment coding for the third hypervariable region.

     Nevertheless, these observations did not rule out the possible
involvement of somatic mutations in the generation of diversity. In
fact, several authors (20, 21) have observed the occurence of somatic
mutations by comparing the sequences of germ-line V genes and of the
corresponding gene in myeloma or hybridoma lines. Most of them are
located in the hypervariable regions. Other authors have shown spontaneous mutations arising in cultured lymphoid cells (22, 23).

     Therefore the question of the possible role of TdT in the gene-
ration of the diversity of Ig or of T-cells receptors for antigens
remains of actual interest but also remains to be established. So far,
two important observations agree with this hypothesis : the observation of TdT in peripheric cells of young rats and mice (12) and the
report of a correlation between a virtual absence of TdT-positive
cells and a lower Ig-heterogeneity in mice affected by an autoimmune
disease (24). The observations of sequences homologies in the DNA
segments of the three HV regions (21) and of sequence homologies
upstream to the HV segments of λ and κ chains (25) can also be taken
in account with this hypothesis. Indeed, it seems very likely that if
TdT is involved in somatic mutations, it can only work after the
action of at least an endo-exonuclease system and very probably under
the form of a multienzymatic complex. The above repeated sequences
might be sites of recognition for this complex, allowing it to work
in a specific way.

     One of our attempts to examine TdT as a possible generator of
immunological diversity has been based on the following methodological
approach : a) construction of somatic hybrids between TdT-positive
cells and cells which secrete monoclonal antibodies such as myeloma
or hybridoma lines ; b) analysis of the stability or variations of the
antibodies secreted by these hybrids, tentatively related to the expression of TdT. For this study we have used so far thymocytes and

pre-T lines as parental TdT-positive cells. As parental B-cells, we have taken lines secreting IgA, κ which have the advantage of being well-defined regarding their antibody activity (anti-levan) and their allotypic and idiotypic determinants. Although our results do not yet allow any answer to the question of TdT function, the partial analysis of the Ig secreted by the hybrids would be presented. The question of the expression of T-specific markers and of TdT would be discussed. The further construction of other types of hybrids (pre-B-B, pre T-T, ...) would also be discussed in an attempt to analyse the regulation of the expression of T and B specific characters and to approach the TdT function.

MATERIAL AND METHODS

- Cell lines. - The B parental cells were derived from the ABPC-48 myeloma (26) and kindly provided by P. Legrain and G. Buttin (27). They were grown in Eagle medium supplemented with 10 % horse serum. The AS-3 line is HGPRT⁻ and the ASO 1.(12).7 cells are both HGPRT⁻ and ouabain-resistant, this last character being codominant. Both lines secrete the ABPC 48 protein which is an IgA, κ with an anti-levan activity. This immunoglobulin also can be characterized by its allotypic and idiotypic determinants using monoclonal antibodies obtained from G. Buttin laboratory.

        The TdT-positive parental cells were either thymocytes of several mice strains (Balb/c, B10-S, AKR) or continuous cloned lines (R 33.2) derived from a thy-1 positive lymphoma induced by M.MuLV in B10.14R mice.

        Other hybridoma lines obtained from P. Legrain and G. Buttin were used as a source of monoclonal antibodies against ABPC-48. The A11-2 line secretes an antibody directed against an allotypic determinant of the heavy chain of ABPC-48 and the IDA lines serie secrete antibodies against the idiotypic determinants of this Ig. The SP-2 line is a nonsecretory myeloma used as control in many experiments.

- Cell fusion, selection and cloning of the hybrids. - Fusions were carried out using 45 % PEG-1000 (Merck) in serum free medium according to a filter procedure described by Buttin et al. (28). The hybrids were selected in a medium containing 10^{-5} M azaserine, 5.10^{-5} hypoxanthine and, according to the experiments, 0.5×10^{-3} M ouabain. To be cloned, the selected hybrids were submitted to limit dilution on feeder layers of thymocytes in microtiters plates. Caryotypes were determined by counting the chromosomes from 20 different metaphase plates.

- Detection of surface immunoglobulines. - These assays were derived from the rosette method described by Juy et al. (29). For the detec-

tion of surface immunoglobulins with anti-levan activity, we followed
the formation of rosettes between the hybrids (or the control cells)
and levan coated sheep red blood cells (levan-SRBC).

- Rosette inhibition assay for the detection of secreted immunoglobulins. - The method described by Juy et al. (28) was used. The
SRBC were coated with ABPC-48 and allowed to form rosettes with A11-2
or IDAs cells. The culture supernatant of the hybrids was assayed for
the inhibition of the rosette formation in order to check the presence of the allotypic or idiotypic determinants.

- Enzyme linked immunosorbent assay (ELISA) for the detection of
secreted immunoglobulins. - The antibodies secreted by A11-2 cells
(anti-allotypic) or by IDAs cells (anti-idiotypic) were coated at
various dilutions on microtitration plates in 0.1 M sodium bicarbo-
nate ; the wells were washed twice with PBS containing 0.1 % Tween-20
and were saturated with the same buffer added with 0.1 % BSA. After
washing they were incubated with 10-20 ng β-galactosidase conjugated
ABPC-48, in the above buffer. These experiments were performed in the
absence or presence of various amounts of the culture supernatant
from the hybrids to be assayed. The wells were washed 4 times with
the above PBS-Tween solution. Then, the enzymatic reaction of bound
β - gal-ABPC-48 was allowed to take place for 15 h at 37° C in 0.1 M
phosphate (pH 7.0) containing 10^{-3} M $MgSO_4$, 2×10^{-4} M $MnSO_4$, 2×10^{-3} M
Mg-Titriplex, 0.1 M 2-mercaptoethanol and 2.5×10^{-3} M paranitrophenol
β-D galactopyranoside. Then the reaction was stopped by addition of
0.1 M sodium carbonate and the optical density was measured at 405 nm
in a Titertek Multiskan Photometer.

- Assays for H-2 and thy-1 alloantigens. - For the detection of these
alloantigens we used an indirect immunofluorescence method : 10^5 to
10^6 cells were centrifuged 10 min at 200 g and washed with ER medium
containing 10^{-2} M Hepes pH 7.4 and 5 % foetal calf serum. They were
incubated with the concerned antibodies for 30 mn at 4° C. After several washings with the above solution, cells were incubated with fluorescein conjugated goat IgG against mouse IgG or IgM. After washing,
the cells were mounted in 80 % (w:v) glycerol.

- Assays for Terminal deoxynucleotidyl transferase. - Extraction of
and quantitative determination of TdT were carried out under conditions similar to that described by McCaffrey et al. (8). In a total
volume of 100 μl, the incubation mixture contained 50 mM Tris-HCl pH
7.9, 2 mM DTT, 2.5×10^{-4} M $MnCl_2$, 5 μg/ml $d(pA)_{10}$ (collaborative research), 2×10^{-5} M dGTP and 2 μCi [^{3}H]-dGTP.

       For the immunofluorescence assays, the antibody against calf
TdT was prepared as described for chicken TdT (30) except that the
affinity chromatography was performed through a calf TdT-Sepharose
column. The cells were treated according to the procedure previously

described (30) with slight modifications. After cytocentrifugation and fixation with paraformaldehyde (4 % in 0.2 M sodium cacodylate, pH 7.4), the cells were treated with acetone for 5 min at 4° C. They were washed with PBS and incubated with rabbit purified antibody against calf TdT (0.02 mg/ml in PBS containing 2 mg/ml bovine serum albumine), for 30 min at room temperature. After extensive washing with PBS, the preparation were incubated for 1 h at room temperature with fluorescein conjugated sheep or goat IgG against rabbit IgG. Subsequently, they were washed with PBS and mounted in 80 % (w/v) glycerol.

RESULTS

1. Production of T-B hybrids. Hybridation frequency. - We fused AS-3 or ASO-1(12)7 myeloma cells with thymocytes of 4 weeks-old Balb/c, B10-S and AKR mice. The fusion procedure on filters allowed to obtain for each experiment several hybrid populations arising from independant fusion events. Table 1 shows that the hybridation frequency was about 10^{-5} either for syngeneic or allogenic fusions. Control experiments of fusion of myeloma cells and splenocytes showed a similar hybridation frequency.

2. Verification of the hybrid status. - The above selected lines were characterized as hybrids according to 3 criteria :

- These cells grew in an ERH medium containing 5.10^{-5} M hypoxanthine and 10^{-5} M azaserine whereas parental cells died. Furthermore the fused cells did not grow in an ERH medium containing 10^{-5} M azaserine alone. This observation shows that the $HGPRT^+$ phenotype of the thymocytes was expressed and that the obtained lines did not result from a selection of mere azaserine resistant variants.

- Each caryotype was examined on 20 different metaphasic plates and showed significant and regular differences with the parental cells.

- In the case of allogenic fusions, the hybrid character was also verified by the expression of the two parental H-2 haplotypes using indirect immunofluorescence assays.

The concordance between these 3 criteria shows that the selected cells were really hybrids.

3. Surface immunoglobulins on hybrids. - The parental B cells, i.e. AS-3 and ASO 1(12)7 synthetise the ABPC-48 Ig which is present on their surface and which also is secreted. We have determined whether this Ig was still conserved in the hybrids. The following experiments were carried out both on 30 uncloned different hybrid populations and on 18 cloned hybrids. The hybrids were allowed to give rise to rosettes with levan-coated SRBC. Results reported in Table 2 showed that most of the hybrids presented an active anti-levan antibody on their surface.

Table 1. Fusion between thymocytes and myeloma cells

Fusion	Origin of thymocytes	Cells per filter Thymocytes	B cells	Filter number	Viable cells after fusion before selective medium	Number of inoculated wells in selective medium	Number of positive wells in selective medium
A	Balb/c (H2-d,Thy1-2)	$4,5 \times 10^7$	4×10^6 (AS0-1(12)-7)	1	5×10^6	48	48
				2	1.5×10^6	36	36
				3	6.5×10^6	51	51
B	Balb/c (H2-d,Thy1-2)	1.8×10^7	4.2×10^6 (AS0-1(12)-7)	1	5×10^5	15	15
				2	5×10^5	15	15
				3	3.5×10^5	8	8
				4	4×10^5	8	8
				5	7.5×10^5	16	16
				6	4×10^5	8	8
				7	3×10^5	8	8
				8	6×10^5	15	15
C	B10-s (H2-s,Thy1-2)	2×10^7	3×10^6 (AS0-1(12)-7)	1	8.5×10^5	16	16
				2	5×10^5	19	19
				3	5.5×10^5	62	62
				4	5×10^5	18	18
D	AKR (H2-k, Thy1-1)	2.8×10^7	3×10^6 (AS0-1(12)-7)	1	1.0×10^6	19	19
				2	1.6×10^6	24	24
				3	1.5×10^6	22	22
				4	3.2×10^6	45	45
E	Balb/c (H2-d,Thy1-2)	10^7	3×10^6 (AS3)	1	2×10^6	24	11
				2	1×10^6	12	7
				3	4×10^6	48	21
				4	3×10^6	48	23
				5	4.5×10^6	48	26
				6	4.5×10^6	48	26

Table 2. Expression of Anti-levan Surface Immunoglobulins
and of Thy-1.2 in myeloma-thymocytes hybrids.

Fusion	Filter number	Hybrids	Anti-levar surface immunoglobulins	Thy 1-2
A	1	A1	+	−
		A2	+	−
		A3	+	−
		A4	+	−
	2	A5	+	−
		A6	+	−
		A7	+	−
		A8	+	−
	3	A9	+	−
		A10	+	−
		A11	+	−
		A12	+	−
B	1	B1	+	−
	2	B2	+	−
	3	B3	+	−
	4	B4	+	−
	5	B5	+	−
C	1	C1	+	−
		C2	+	−
	2	C3	+	−
		C4	+	−
	3	C5	+	−
		C6	+	−
	4	C7	+	−
		C8	+	−

4. Secretion of Ig with allotypic and idiotypic determinants of
ABPC-48. - The secretion of immunoglobulins by the above hybrids was
determined also both on uncloned and cloned hybrids populations. It

was first determined by assaying the culture supernatant for inhibition of rosette formation between ABPC-48 coated SRBC and A11-2 (anti-
allotype) or IDA-20 (anti-idiotype) cells. Results reported in Table 3
indicate that most of the hybrids inhibited these rosettes.

These observations were confirmed in a more quantitative way
through ELISA experiments. The wells of microtitration plates could
be coated with the Ig secreted by A11-2 (anti-allotype) and IDA-20
(anti-idiotype). After coating these proteins remained able to bind
β-galactosidase conjugated ABPC-48 Ig. The measurement of β-galactosidase activity allowed a quantitative determination of ABPC-48 bound
to A11-2 and IDA-20 proteins. When these experiments were performed
in the presence of supernatant of cells secreting Ig with the allotypic or idiotypic determinants of ABPC 48, the binding of β-gal-
ABPC-48 suffered a quantitative inhibition. The Fig. 1 shows examples
of results from these experiments and Table 3 indicates that most of
the hybrids secrete Ig with the allotypic and idiotypic determinants
recognized on ABPC-48 by A11-2 and IDA-20 antibodies.

Table 3. Analysis of the immunoglobulins synthetized
by the myeloma-thymocytes hybrids.

	Hybrids	Antilevan surface Ig	Secreted Ig with the allotypic determinant recognized by ALL-2	Secreted Ig with the idiotypic determinant recognized by IDA-20
Cloned hybrids	A44	+	+	+
	A4-26	–	–	–
	A4-27	–	–	–
	A4-29	+	+	+
Uncloned hybrids from independent fusion events	D1	+	+	+
	D2	+	+	+
	D3	+	+	+
	D4	+	+	+
	D5	+	+	+
	E14	+	+	+
	E30	+	+	+
	E71	+	+	+
	E85	+	+	+

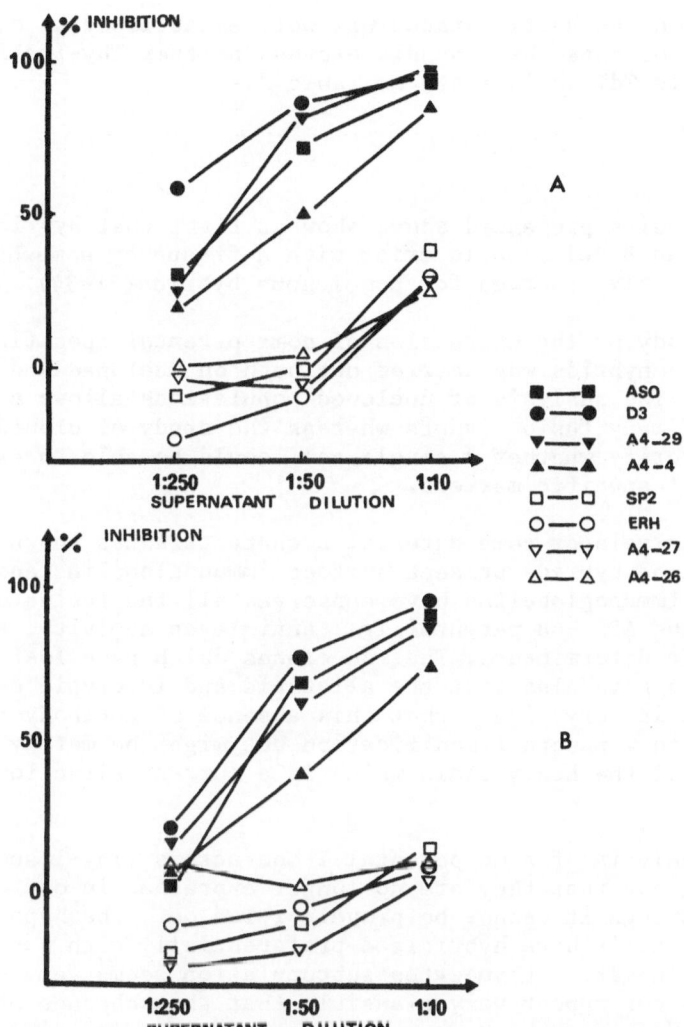

Fig. 1.   Inhibition of the binding of ABPC-48 immunoglobulin to the
          All-2 anti-allotype (A) or to the IDA-20 antiidiotype (B)
          by the immunoglobulins secreted by some hybrids. ASO
          corresponded to the positive control and SP2 to the nega-
          tive controls.

5. <u>Apparent extinction of the T-specific parental character</u>. - Most of the parental thymocytes expressed the Thy-1 (Thy 1.1 or Thy 1.2 according to the mice strains) and more than 70 % were TdT-positive.

Although the hybrid status was well established as described in § 1, none of these hybrids did express neither Thy-1 alloantigens nor detectable TdT as indicate in Table 2.

DISCUSSION

The results presented above show at first that hybrids between thymocytes and B cells could arise with a frequency somewhat similar to that currently observed for homologous hybrids(31-34).

The study of the expression of some parental specific charac- ters by these hybrids was carried out both on uncloned and cloned populations. The analysis of uncloned populations allows a rapid screening of many fusion events whereas the study of cloned hybrids permits to verify whether a single cell could be able to express both parental specific markers.

The analysis of some parental B characters has shown that most of the obtained hybrids present surface immunoglobulins and secrete them. These immunoglobulins have conserved all the tested characte- ristics of the ABPC-48 parental Ig : anti-levan activity, allotypic and idiotypic determinants. The two clones which have lost the anti- body activity have also lost the allotypic and idiotypic determinants. Therefore it is very likely that this absence of antibody activity is not related to a punctual modification but might be merely related to the loss of the heavy chain which is a current situation in hybri- domas (31).

The analysis of some parental T characters (Thy-1 and TdT) strongly suggest that they are no longer expressed in our obtained hybrids. Although it cannot be provely ruled out, the hypothesis that myeloma cells have hybridized preferentially with a small (Thy-1, TdT)-negative thymocytes subpopulation seems very unlikely. It also does not appear very plausible that this absence of Thy-1 and TdT was due to a chromosome segregation since many independant fusion events have been assayed. In fact, the T-B fusions examined by other authors so far have shown some discrepancy. In some cases the Ig expression was conserved and Thy-1 extincted (32) whereas other experiments lead to an inverse situation with even the re- expression of the Thy-1 allele of the parental B-cell(33, 34).

Although we have undertaken the establishment of other hybrids between TdT-positive T cell lines and myeloma cells, the co-expres- sion of TdT and immunoglobulins in T-B hybrids does not seem very easy to obtain. However, with this hope, we are carrying out experiments in two ways : submitting these hybrids to thymic hormones with the aim to induce TdT expression and modifying the gene ratios by rehybriding the hybrids with TdT-positive cells in order to induce a different modulation of genes expression.

Another, maybe easier, tentative approach could be the construction of hybrids between TdT-positive pre-B cells and myeloma lines and, in a somehow parallel way, the establishment of hybrids between TdT-positive pre-T lines and T-cells bearing antigen specific receptors.

ACKNOWLEDGEMENTS

We are grateful to Drs. G. Buttin, F. Chapeville, P. Legrain and L. Phalente for very helpful discussions and gift of cell lines. Part of this work was supported by the ATP "Biologie Moléculaire du Gène", from C.N.R.S.

REFERENCES

1. F.J. Bollum, Oligodeoxynucleotide-primed reactions catalysed by calf thymus polymerase, J. Biol. Chem. 237:1945 (1962).
2. F.J. Bollum, Claf thymus polymerase, J. Biol. Chem. 235:2399 (1960).
3. F.J. Bollum, Terminal deoxynucleodityl transferase, in : "The Enzymes", 3rd ed., P.D. Boyer, Ed., Academic Press, New-York, vol. 10:145 (1974).
4. F.J. Bollum, Terminal deoxynucleotidyl transferase, Biological studies, in : "Advances in Enzymology", A. Meister, Ed., Wiley-Interscience, New-York, vol. 47:347 (1978).
5. L.M.S. Chang, Development of Terminal deoxynucleotidyl transferase activity in embryonic calf thymus gland, Biochem. Biophys. Res. Commun., 44: 124 (1971).
6. K.E. Gregoire, I. Goldschneider, R.W. Barton, and F.J. Bollum, Proc. Natl. Acad. Sci. USA, 74:3993 (1977).
7. M.S. Coleman, J.J. Hutton, P. De Simone and F.J. Bollum, Terminal deoxynucleotidyl transferase in human leukemia, Proc. Natl. Acad. Sci. USA, 71:4404 (1974).
8. R.P. McCaffrey, T.A. Harrison, R. Parkman and D. Baltimore, Terminal deoxynucleotidyl transferase activity in human leukemia cells and in normal human thymocytes, New-England J. Med., 292: 775 (1975).
9. A.E. Silverstone, H. Cantor, G. Goldstein and D. Baltimore, Terminal deoxynucleotidyl-Transferase is found in prothymocytes, J. Exp. Med., 144:543 (1976).
10. J.C. Brouet, J.L. Preud'Homme, C. Pénit, F. Valensi, P. Rouget and M. Seligmann, Acute lymphoblastic leukemia with Pre-B-cell

characteristics, <u>Blood</u>, 54:269 (1979).

11. L. Boumsell, A. Bernard, H. Coppin, Y. Richard, C. Pénit, P.
    Rouget, J. Lemerle and J. Dausset, Human T-cell differentiation
    antigens and correlation of their expression with various markers
    of T-cell maturation, <u>J. Immunol</u>. 123:2063 (1979).

12. R. Sasaki, F.J. Bollum and I. Goldschneider, Transient popula-
    tions of Terminal transferase positive (TdT[+]) cells in juvenile
    rats and mices, <u>J. Immunol</u>. 125:2501 (1980).

13. D. Baltimore, Is terminal deoxynucleotidyl transferase a somatic
    mutagen in lymphocytes ? Nature 248:409 (1974).

14. E.A. Kabat, The structural basis of antibody complementarity,
    <u>in</u> : Adv. Prot. Chem., C.B. Anfinsen, J.T. Edsall, F.M. Richards,
    Eds., Academic Press, New-York, vol. 32:1 (1978).

15. O. Bernard, N. Hozumi, and S. Tonegawa, Sequences of Mouse immuno-
    globulin light chain genes before and after somatic changes,
    <u>Cell</u>, 15:1133 (1978).

16. H. Sakano, R. Maki, Y. Kurosawa, W. Roeder and S. Tonegawa, Two
    types of somatic recombination are necessary for the generation
    of complete immunoglobulin heavy-chain genes, Nature, 286:676
    (1980).

17. V.G. Seidman, E.E. Max and P. Leder, A K-immunoglobulin gene
    is formed by site-specific recombination without further somatic
    mutation, <u>Nature</u>, 280:370 (1979).

18. P. Early, H. Huang, M. Davis, K. Calame and L. Hood, An immuno-
    globulin heavy chain variable region gene is generated from
    three segments of DNA : V_H,D and J_H, <u>Cell</u>, 19:981 (1980).

19. R.P. Perry, D.E. Kelley, C. Coleclough and J.F. Kearney, Organi-
    zation and expression of immunoglobulins genes in fetal liver
    hybridomas, <u>Proc. Natl. Acad. Sci</u>. USA, 78:247 (1981).

20. N.M. Gough and O. Bernard, Sequences of the joining region genes
    for immunoglobulin heavy chain and their role in generation of
    antibody diversity, <u>Proc. Natl. Acad. Sci. USA</u>, 78:509 (1981).

21. S. Tonegawa, A.M. Maxam, R. Tizard, O. Bernard and W. Gilbert,
    Sequence of a mouse germ-line gene for a variable region of
    an immunoglobulin light chain, <u>Proc. Natl. Acad. Sci. USA,</u> 75:
    1485 (1978).

22. D.S. Secher, C. Milstein and K. Adetugbo, Somatic mutants and
    antibody diversity, Immunological Rev. 36:51 (1977).

23. C. Milstein, K. Adetugbo, N.J. Cowan, G. Köhler, D.S. Secher
    and C.D. Wilde, Somatic cell genetics of antibody-secreting
    cells : studies of clonal diversification and analysis by cell
    fusion, <u>Cold Spring Harbor Symp. Quant. Biol.</u>, 16:793 (1976).

24. K.S. Landreth, K. McCoy, J. Clagett, F.J. Bollum and C. Rosse,
    Deficiency in cells expressing terminal transferase in auto-
    immune (motheaten) mice. <u>Nature</u>, 290:409 (1981).

25. T.T. Wu, E.A. Kabat and H. Bilofsky, Some sequence similarities
    among cloned mouse DNA segments that code for λ and κ kight
    chains of immunoglobulins, Proc. Natl. Acad. Sci. USA, 76:
    4617 (1979).

26. M. Potter, Antigen-binding myeloma proteins of mice, Adv. in Immunol., 25:141 (1977).
27. B. Goud, P. Legrain, J.C. Antoine, S. Avrameas and G. Buttin, Cross-linking of surface receptors and endocytosis of antigens are not sufficient to suppress antibody production of two hybridoma cell lines, J. Receptor Research (in press).
28. G. Buttin, C. Le Guern, L. Phalente, E.C.C. Lin, L. Medrano and P.A. Cazenave, Production of hybrid lines secreting monoclonal anti-idiotypic antibodies by cell fusion on membrane filters, Curr. Topics in Microbiol. and Immunol., 81:27 (1978).
29. D. Juy, P. Legrain, P.A. Cazenave and G. Buttin, A new rapid rosette-forming cell micromethod for the detection of antibody synthesizing hybridomas, J. Immunol. Methods, 30:269 (1979).
30. P. Rouget and C. Penit, Terminal deoxynucleotidyl transferase during the development of chicken thymus, cell differentiation, 9:329 (1980).
31. G. Köhler, Immunoglobulin chain loss in hybridoma lines, Proc. Natl. Acad. Sci. USA, 77: 2197 (1980).
32. G. Köhler, T. Pearson and C. Milstein, Fusion of T and B cells, Somatic Cell Genetics, 3:303 (1977)
33. G.M. Iverson, R.A. Goldsby and L.A. Herzenberg, Expression of Thy 1.2 antigen on hybrids of B. Cells and a T lymphoma, Curr. Topics in Microbiol. and Immunol. 81:192 (1978).
34. M.J. Taussig, A. Holliman and L.J. Wright, Hybridization between T and B lymphoma cell lines, Immunology, 39:57 (1980).

THE BOLLUM ENZYME IN LEUKEMIA AND LYMPHOMA CELLS:

THE FIRST DECADE

Ronald McCaffrey
Richard Bell
Anne Lillquist
George Wright[+]
Earl Baril[++]

Section of Medical Oncology of
the Evans Memorial Department of Clinical Research
Boston University Medical Center
the Department of Medicine
Boston University School of Medicine
and the Hubert H. Humphrey Cancer Research Center
Boston University, Boston, MA.

[+]Department of Pharmacology
University of Massachusetts Medical School
Worcester, MA.

[++]Worcester Foundation for Experimental Biology
Shrewsbury, MA.

INTRODUCTION

It is now almost a decade since the original report of the presence of the Bollum enzyme (terminal deoxynucleotidyl transferase, TdT) in leukemia cells[1]. That discovery was a fortuitous event which occurred in the course of experiments characterizing what at first appeared to be tumor virus reverse transcriptase in the blast cells of a child with acute lymphoblastic leukemia[1,2]. The restricted expression of terminal transferase in normal cells had been reported one year earlier by Lucy Chang, working in Fred Bollum's laboratory[3]. We interpreted our observation in the light of this restricted expression, postulating that TdT-positive leukemic cells were clonal expansions of TdT-positive normal cell compartments[1,4].

In this review we will focus on the significance, both conceptually and in terms of therapeutic strategy, of TdT expression in malignant hematopoietic cells. Some preliminary data on the potential utility of 6-anilinouracils as specific cytocidal agents for TdT-positive leukemias will also be included.

ORIGINAL DOSCOVERY OF TdT IN LEUKEMIA CELLS

Shortly after the discovery by Baltimore[5] and Temin and Mizutani[6] of the retrovirus DNA polymerase, reverse transcriptase, we began a study which sought to determine if human leukemia cells contained this unique viral "footprint". Using embryonic mouse bone marrow fibroblasts as a model tissue, we showed that cell lines infected with Moloney leukemia virus contained a distinct DNA polymerase which functioned with the template·primer combination poly(C)·oligo(dG), whereas uninfected companion lines had no such polymerase. Solubilized reverse transcriptase from several RNA tumor viruses also had this poly(C)·olido(dG)-accepting activity. Constitutive cell DNA polymerases (pol α, pol β, and pol γ) and bacterial DNA polymerases all failed to function with poly(C)-·oligo(dG)[7]. Having established this viral specificity for this template·primer combination we embarked upon a survey of human leukemia cells for the presence of a DNA polymerase which would synthesize poly(dG) on a poly(C) template.

Our first sample for study came from a 5 1/2 year old girl with acute lymphoblastic leukemia. Blast cells from the peripheral blood of this child were processed identically to the embryonic mouse bone marrow fibroblasts. Like the virally infected model cells, her cells contained a distinct polymerase which functioned with poly-(C)·oligo(dG). A preliminary report was written, declaring the discovery of reverse transcriptase in these cells[2]. Further characterization of this poly(C)·oligo(dG) activity showed that interpretation to be incorrect: a series of "leaving out" and "unrelated substrate" experiments showed this to be a template-independent, end addition enzyme[1]. A similar enzyme was present in normal human thymus, but not in other cells or tissues assayed[1]. We discussed the biologic significance of our observation in terms of the Chang survey[3] for TdT in normal tissues; we hypothesized that TdT-positive leukemic cells might be normal TdT-positive cells "frozen" at a defined point in their ontogeny, and that perhaps TdT would be a useful marker for studying cells "at risk" for leukemogenic transformation[1,4,8,9,].

A search for an authentic, reverse transcriptase-like, poly-(C)·oligo(dG) utilizing polymerase in human leukemia cells was continued for 15 additional cases; none contained such an enzyme[10]. We therefore shifted the focus of our study from reverse transcrip-

tase to an exploration of the biologic significance of TdT expression in leukemic cells.

TdT EXPRESSION IN MALIGNANT HEMATOPOIETIC CELLS

Survey studies of the expression of TdT in malignant cells from
a wide spectrum of hematopoietic neoplastic processes have been
reported from several centers[11-15]. A summary of the data on the
first several hundred patients surveyed by us is given in Table 1.

Table 1. Occurrence of Terminal Transferase in Hematologic
Malignancy

Clinical Diagnosis	No. of Cases	No. TdT(+)
Acute lymphoblastic leukemia	300	290
Acute myeloblastic leukemia	120	10
Acute undifferentiated leukemia	30	16
Blastic chronic myelogenous leukemia	100	38
Post-polycythemia vera leukemia	15	3
Post-myeloid metaplasia leukemia	16	10
Post-chemo/radiotherapy leukemia	9	2
Stable phase chronic myelogenous leukemia	30	0
B-cell chronic lymphocytic leukemia	15	0
T-cell chronic lymphocytic leukemia	3	0
Sézary syndrome	6	0
Hairy cell leukemia	9	0
Multiple myeloma	7	0
Hodgkin's disease	7	0
Lymphoblastic lymphoma	15	15
Nodular lymphoma	6	0
Diffuse, poorly differentiated lymphocytic lymphoma	9	0
Diffuse histiocytic lymphoma	6	0

Specimens of blood, bone marrow, or neoplastic nodes were tested by
either the fluorescent antibody technique[37] or biochemical assay[1],
or both. In the FA assay, positive samples showed more than 20%
of cells staining; negative results indicate less than 0.5% of cells
were stained. In the biochemical assay, positive samples contained
from 0.3 to 10 units of TdT per 10^8; negative samples contained
less than 0.003 TdT units per 10^8 cells.

This distribution of enzyme-positive and -negative cases within the various diagnostic categories in this series is similar to that which has been reported in surveys performed by other groups. Thus, even acknowledging the minor exceptions of TdT-negative lymphoblastic disease and TdT-positive myeloblastic disease, it is clear that TdT activity reliably distinguishes between lymphoblastic and myeloblastic leukemia with about 95% confidence. However, among the less common variants of acute leukemia there is an apparent randomness to blast cell TdT expression. As shown in Table 1, TdT can be present in blast cells from a spectrum of apparently diverse, non-lymphoid neoplastic processes - blastic chronic myelogenous leukemia, undifferentiated acute leukemia, and leukemia following a variety of dyspoietic states. At least two possible interpretations of this "randomness" are possible. One is to agree that randomness, or chance, accounts for TdT expression in leukemic cells; that TdT expression in the leukemias is haphazard and a reflection of a chaotic metabolic cellular state. A second interpretation is to consider the expression of TdT in leukemic blast cells to be an indication of a 'lymphoid' origin of these cell populations. We have favored this second interpretation. Data from a variety of sources have emerged, which increasingly support such an interpretation, as discussed below.

CLINICAL UTILITY OF BLAST CELL TdT ASSAYS

The utility of TdT assays in the formulation of clinical decisions derives from the hypothesis that TdT-positive blast cells are lymphoid in nature, irrespective of their morphology. (The use of the term 'lymphoid' in this non-morphologic manner is obviously confusing. We have used 'lymphoid' in this manner, not to fracture the English language, but to convey a meaning for which there is no suitable alternative term). We tested this "TdT-lymphoblast" hypothesis directly in a therapeutic trial using the agents vincristine and prednisone in blastic chronic myelogenous leukemia (CML)[16]. This variant of acute leukemia was selected for study because it had been traditionally classified as a variant of acute myeloblastic leukemia; yet fully one third of cases have TdT-positive blast cells. Vincristine and prednisone were selected as therapeutic agents because of the high rate (80%) of responsiveness to these agents in authentic lymphoblastic leukemia and the low rate (15%) in authentic myeloblastic leukemia. We argued that a high rate of responsiveness to this drug combination among the TdT-positive patients would partially validate the hypothesis that such cases were "lymphoblastic" in nature. An added impetus to this study was the similarity between the incidence of TdT positivity and the proportion of cases reported by Canellos in 1971 to be responsive to vincristine and prednisone therapy[17]. Our study sought to determine whether TdT-positivity and vincristine-prednisone responsiveness would coincide.

A total of 30 patients with Philadelphia-chromosome positive blastic CML was studied in a multi-institutional cooperatve clinical trial (Table 2). Blast crisis was defined by the presence of at least 30% blast cells in the bone marrow or peripheral blood. Three patients had de novo blast crisis, the remaining 27 had a well-defined preceeding chronic phase, ranging in duration from 8 months to 10 years. Patients' ages ranged from 3 to 78 years; 13 were below the age of 50 years. By Wright-Giemsa morphologic criteria, 13 of the 30 patients had lymphoblastic disease; the remaining 17 were myeloblastic. Sixteen were TdT-positive; 14 were TdT-negative. (This series was biased toward TdT-positive cases: a random series of 30 blastic CML patients would contain approximately 10 TdT-positive cases).

Therapy was limited to vincristine sulfate, 1.5 mg/m^2 (with a maximum dose of 2 mg) given intravenously each week, and prednisone, 60 mg/m^2 given by mouth each day. In all cases, at least two doses of vincristine and 14 days of prednisone therapy were administered before a patient was considered to be nonresponsive. The results of the TdT assay were not known to the physicians caring for the patients until the chemotherapy had been completed. We defined complete response as the total elimination of blast cells from the peripheral blood, return of peripheral blood values to normal, and return of normal marrow cellularity with less than 5% blast cells.

Table 2. Blastic CML Study - Patient Summary

Age	3-78 years
TdT-positive	16
TdT-negative	14
De Novo Ph^1+ acute leukemia	3
Preceeding stable phase CML, Ph^1+	27
Lymphoblastic	13
Myeloblastic	17

Responsiveness to vincristine-prednisone therapy is summarized in Table 3. Only one of the 14 TdT-negative cases responded, whereas 11 of the 16 TdT-positive cases achieved remission (p< 0.008). The association of several patient characteristics with responsiveness was analyzed. TdT-positivity alone predicted a 67% response rate (11 of 16 TdT positive patient responded). Only 1 of 14 (7%) TdT-negative patients responded. When age and TdT status were considered together, 78% of TdT-positive patients under age 50 years responded, whereas only 11% of TdT-negative patients under age 50 years responded. Thus, age and TdT status considered together were extremely significant. Morphology alone was not significantly predictive for either responsiveness or TdT status, a finding which requires emphasis (Table 4).

Table 3. Efficacy of Vincristine-Prednisone Therapy in Blastic CML

Terminal Transferase Negative Cases	
Number of patients	14
Responses	1
Failures	13

Terminal Transferase Positive Cases	
Number of patients	16
Responses	11
Failures	5

The co-presence with TdT of independent, lymphoid related surface markers in cells from blastic CML patients also supports the "TdT-lymphoblast" hypothesis. We first showed this association in 1973 in a small group of blast crisis patients studied simultaneously in London and Boston for TdT expression and common ALL antigen (CALLA) status[18]. Additional reports[19,20] on larger numbers of patients have substantiated this relationship.

Table 4. Blastic CML Study - Characteristics of Responders and Non-Responders to Vincristine-Prednisone Therapy

	TdT(+) (16 cases)		TdT(-) (14 cases)	
	Responders (11 cases)	Failures (5 cases)	Responders (1 case)	Failures (13 cases)
Morphology				
Lymphoblastic	6	2	1	4
Myeloblastic	5	3	-	9
Age (Years)				
Range	3 - 78	23 - 75	47	5 - 62
Median	27	51	47	52

A practical consequence of these observations is that TdT status (and/or other lymphoid-related markers) can be used a priori to recognize blast crisis patients who are likely to respond to vincristine-prednisone therapy. It is also probable that other forms of TdT-positive leukemia (Table 1), which are not usually considered to be lymphoblastic in nature, would be similarly vincristine-prednisone responsive.

Among the lymphomas, TdT expression is confined to those with lymphoblastic morphology. (A minor subset - probably 5% - of patients with 'histiocytic' lymphoma also have TdT-positive malignant cells). Our data and those of others[21-23] on TdT expression in the lymphomas suggest that the finding of TdT-positive cells in lymphomatous tissue is highly suggestive of lymphoblastic lymphoma. Since the unequivocal establishment of the diagnosis of lymphoblastic lymphoma is not a trivial matter for the pathologist[24], the use of TdT as a reliable marker in this setting represents an important diagnostic advance.

Likewise, the recognition of TdT-positive cells in extramedullary sites has facilitated the recognition of leukemic cells in cerebrospinal fluid and testicular biopsies[25,26]. The immunofluorescent assays for TdT have been particularly useful in this regard.

Several attempts have been made to monitor remission status of TdT-positive leukemia patients with serial marrow TdT assays.

Despite the development of elegant methodology, there has been a notable lack of success in adapting the assay to this use[27,28]. Bone marrow from patients with leukemia can have two types of TdT-positive cells: residual TdT-positive leukemia cells and TdT-positive normal marrow cells. The TdT-positive normal marrow cell population can fluctuate dramatically in response to a variety of stimuli, including fever, chemotherapy, and certain viral ill-nesses[29]. Thus, the relative contribution of either class of TdT-positive cell, normal or malignant, to the activity observed in remission marrow cannot be assessed with current assay systems. Only when marrow is largely replaced by leukemic cells, or when leukemic cells are obtained from peripheral blood can the observed TdT activity be termed leukemia-associated. Thus, our present inability to distinguish normal marrow cell TdT from TdT which is leukemia cell-associated, limits the use of TdT assays of bone marrow to monitor remission status in TdT-positive leukemia.

MOLECULAR FORM OF TdT

Several chromatographic species of TdT can be identified using phosphocellulose columns to analyze either crude or post-DEAE cellulose purified cell extracts[30,31]. Homogenates of acute lym-phoblastic leukemia cells, harvested from either bone marrow or peripheral blood, contain TdT which resolves into two forms on phosphocellulose: Peak I elutes at 0.3M salt, and Peak II at 0.4M salt (Figure 1A). In a few instances virtually all the activity is confined to the Peak II area. No examples of TdT activity eluting exclusively in the Peak I area have been encountered in over 500 chromatograms of acute lymphoblastic leukemia samples. This double-peak pattern does not appear to be an artefact of initial chromato-graphy: pre-cycling on DEAE cellulose, or re-chromatography on phosphocellulose of the isolated peaks, does not alter their chromatographic behavior (Figures 1B and 1C). Human, calf, mouse, and chicken thymus show a similar two-peak phosphocellulose TdT profile.

The determinants of this two-peak pattern require definition. Biochemically, as assessed by substrate-initiator preferences, Peak I and Peak II are indistinguishable[30]. By glycerol gradient isokinetic sedimentation, each has an apparent molecular weight of 52,000. In preliminary experiments we have determined that phos-phorylation-dephosphorylation may be the major factor defining the phosphocellulose profile. Dephosphorylation of Peak I (using alkaline phosphatase treatment) causes it to elute at a lower salt concentration from the phosphocellulose, whereas phosphorylation (using brief exposure to protein kinase) causes Peak I to shift to the Peak II area. Similarly, Peak II, after alkaline phosphatase exposure, shifts to the Peak I area and after protein kinase treatment elutes, on rechromatography, at 0.5M salt.

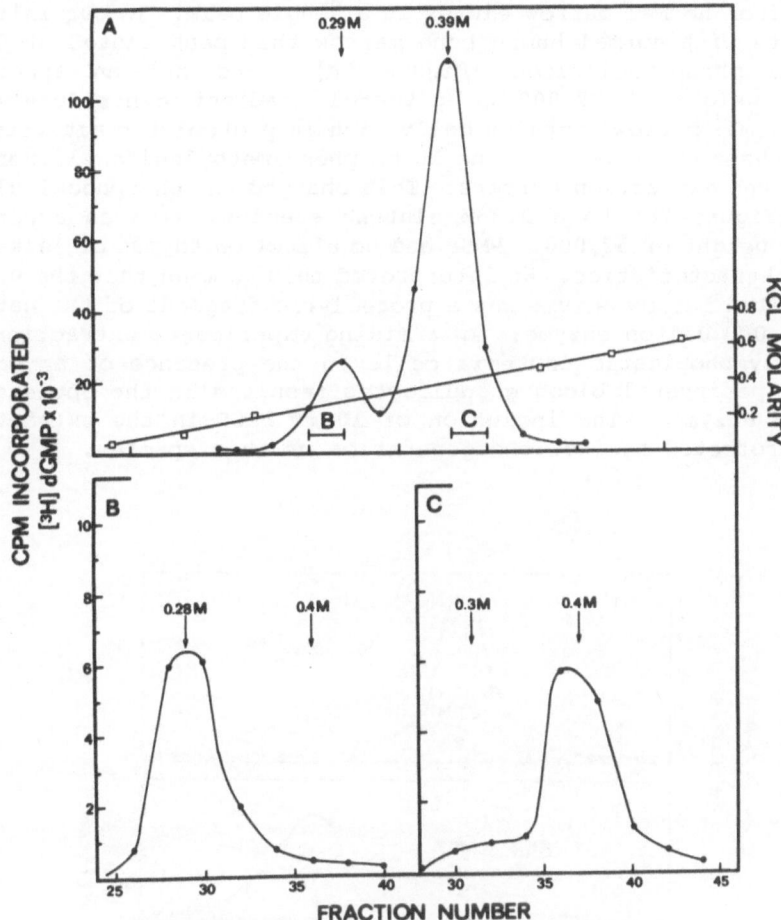

Figure 1. TdT activity from blast
cells from a patient with acute
lymphoblastic leukemia. Panel A
shows activity recovered on initial
phosphocellulose chromatography, and
Panels B and C activities recovered
after rechromatography on phosphocellulose of pooled fractions from
initial column (indicated on Panel A
as "B" and "C"), after dialysis and
passage through a DEAE cellulose
column.

TdT from normal marrow elutes as a single peak. In our initial experiments with normal human bone marrow this peak eluted at 0.2M salt from phosphocellulose (Figure 2A) and had an apparent molecular weight of 32,000 on glycerol gradient centrifugation. Because normal marrow contains cells rich in proteolytic activity we repeated these experiments using 10 mM phenylmethylsulfonylfluoride (PMSF) in the extraction buffers. This changed the phosphocellulose profile (Figure 2B) to a 0.33M eluting species, with an apparent molecular weight of 52,000. PMSF had no effect on thymus or leukemic cell TdT characteristics. We interpreted this to mean that the 0.2M, 32,000 dalton marrow enzyme was a proteolytic fragment of the native 0.33M, 52,000 dalton enzyme. In a mixing experiment, extraction of TdT from lymphoblastic leukemia cells in the presence of an equal number of peripheral blood granulocytes resulted in the appearance of a 0.2M enzyme. The inclusion of 10 mM PMSF in the extraction buffers protected against the generation of this species.

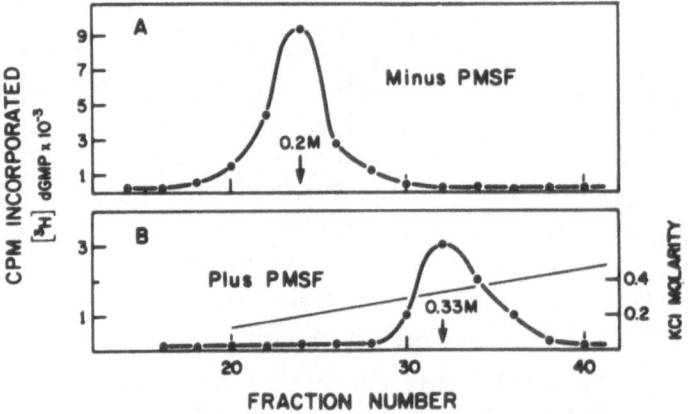

Figure 2. TdT activity from normal human bone marrow. Panel A shows the chromatographic profile obtained in the absence of PMSF in the extraction buffers. Panel B shows the profile obtained in the presence of PMSF. Phosphocellulose chromatography performed as described in reference 9.

Since TdT-positive marrow cells are the presumed precursors of TdT-positive thymocytes, the single-peak 0.33M TdT enzyme may represent a "less differentiated" species than the two-peak enzyme. This speculation is supported by the phosphocellulose pattern seen in murine thymus following recovery from cortisone-induced thymic involution (Figure 3). Here, the ratio of Peak I to Peak II changes dramatically over the week following involution. In the first few days of thymus repopulation, presumably with primitive, actively proliferating cells, Peak I predominates. The pre-cortisone ratios of Peak I to Peak II are gradually re-established around day 6. This fluctuation is reminiscent of the variant patterns seen in acute, fulminant relapse[9]. The explanation for this chromatographic heterogeneity may be that during "explosive" events, such as thymic repopulation or acute leukemic relapse, the regulation of phosphorylation-dephosphorylation of TdT is altered.

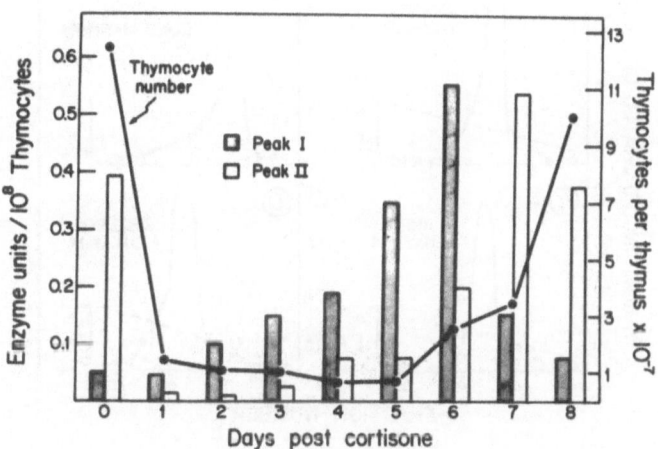

Figure 3. Effect of cortisone treatment on the number of cells and TdT activity in murine thymus. A single 150 mg/kg dose of cortisone acetate was administered to C57Bl/6J mice intraperitoneally on day zero. The number of cells per thymus and TdT activity per 10^8 thymocytes were determined sequentially thereafter[31].

Data on TdT expression in the murine leukemias also support
this interpretation.   Normal murine thymocyte TdT in the phosphocellulose chromatography system elutes, as mentioned, as two peaks
similar to human thymocytes (Figure 4A).   Normal mouse bone marrow
TdT is, however, predominantly Peak I, as is normal human bone marrow
(Figure 4B).   The viral lymphoblastic lymphoma of AKR mice, which
begins anatomically in the thymus[32], is characterized by a two-peak
TdT pattern (Figure 4C).   However, the thymic-sparing leukemia,
which develops in mouse bone marrow in response to Abelson leukemia
virus infection[33], also has, when well established, the "thymocyte"
TdT phosphocellulose pattern (Figure 4D).   This duplicates the
observations we have made in human lymphoblastic leukemia, where the
phosphocellulose TdT profile is almost always "thymocyte", whereas
only a minority (about 20%) of patients have disease with "thymocyte" (T-cell) characteristics assessed by other means.

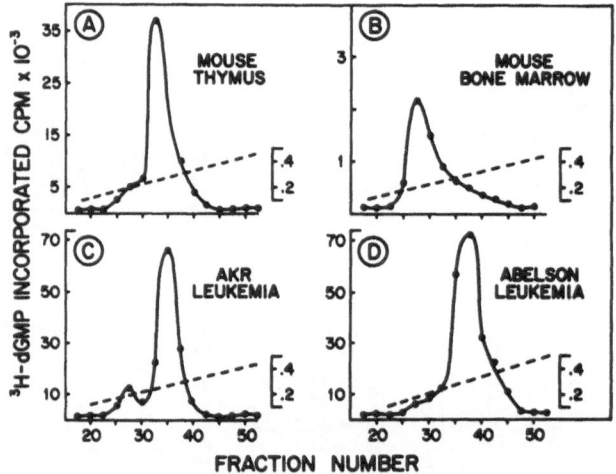

Figure 4.   TdT in murine tissues.
Panel A: phosphocellulose chromato-
gram of normal mouse thymus.   Panel B:
normal mouse bone marrow.   Panel C:
AKR leukemic blast cells.   Panel D:
Abelson-virus-induced leukemic blast
cells.

TdT heterogeneity at the molecular level thus occurs both as a proteolytic artefact during purification and as a result of metabolic modification (phosphorylation-dephosphorylation). Cell line-specific forms of TdT[34], or anatomic compartment-specific forms[35], should be analyzed in the light of these considerations.

CONTROL OF TdT EXPRESSION

Data from several sources are now consistent with a model which states that TdT-positive marrow cells may be initially capable of either B- or T-cell differentiation[29,36]. Surface Ig-positive cells are TdT-negative; and we were not able to demonstrate TdT in bursa of chickens[37]. TdT-positive marrow T-cell precursors exit and migrate to the cortical thymus. Such migration has been reliably observed only during the neonatal period in rats[29]. Presumably, traffic of TdT-positive cells between marrow and thymus after the neonatal period occurs with a frequency too low to be detected. No cells have been identified exiting from the thymus while still expressing TdT[38]. The occurrence of TdT positivity in leukemia cells with myeloid characteristics[39] suggests that TdT may be present in an early stem cell compartment.

The factors regulating the expression of TdT in the bone marrow and thymus are largely unknown. It is likely, by analogy with control mechanisms operative in other arms of the cellular hematopoietic system that humoral factors are involved. Stimulated by the observations of Pike et al.[40] on the induction of E-rosetting cells in bone marrow from a child with severe combined immunodeficiency disease by short-term incubation of bone marrow cells on normal thymic epithelial monolayers, we attempted a similar experiment with TdT-positive lymphoblastic leukemia cells. Our reasoning was that these TdT-positive cells (which were simultaneously E-rosette positive) might represent a clonal expansion of thymocytes "frozen" at the TdT-positive stage, yet capable of "thawing" into a more differentiated state (i.e. capable of becoming TdT-negative), if provided with appropriate thymic "signals". We, therefore, in association with Dr. Erwin Gelfand and colleagues, exposed 4×10^7 TdT-positive, E-rosette positive leukemia cells to a normal thymic epithelial monolayer for 6 hours, harvested the cells, and assayed for TdT. An identical aliquot of cells exposed to a skin fibroblast monolayer, initiated from the donor of the thymic epithelial monolayer, served as control. As shown in Figures 5A and 5B, exposure to the thymic monolayer resulted in a dramatic reduction in TdT activity. E-rosetting remained positive, and the cells had maintained their viability. Other parameters of lymphoid maturation[41] were not measured in these experiments. Further experiments of this type are now in progress in our laboratory.

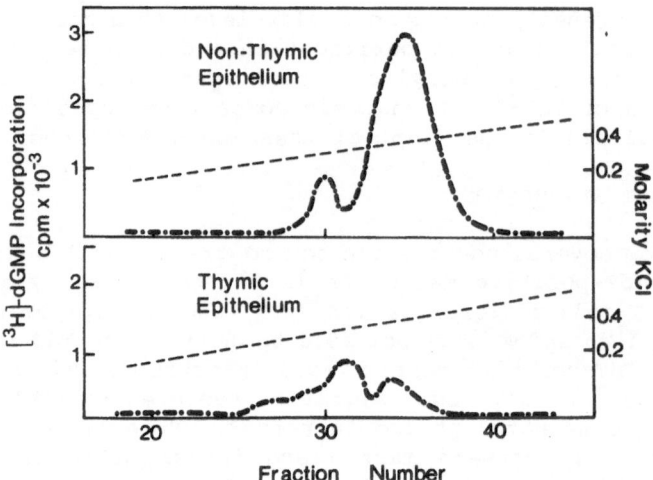

Figure 5. TdT expression in T cell acute lymphoblastic leukemia cells: Effect of exposure to thymic
epithelial monolayers

TdT AS A THERAPEUTIC TARGET

     The physiologic function of TdT in the cells in which it is
expressed, either leukemic or normal, is presently unknown. Its
strict limitation in normal animals to lymphoid cells during the
early phases of their differentiation suggests that it may play a
critical role in this process. Presumably TdT subserves a similar
role in leukemia cells, although the process of differentiation is
itself obviously disturbed. Whatever its role might be, we have
asked whether inhibition of TdT in leukemic cells might constitute a
lethal event to such cells. We thus recently began a search for
specific TdT inhibitors to explore this question. Our strategy for
the identification and exploitation of such inhibitors is summarized
in Table 5.

     Stimulated by the work of Wright, Baril and Brown with 6-
anilinouracils as inhibitors of authentic DNA polymerases[42-44], we
systematically analyzed the effect of a large series of uracil
analogues on human leukemic and calf thymus TdT. Two compounds
showed significant and specific inhibition (Table 6 and 7).

Table 5. Strategy for Identification and Exploitation of Selective
          TdT Inhibitors

Screen candidate compounds using biochemical assays.

Test for specific growth inhibitory effects
on cell lines in culture.

Determine acute animal toxicity and pharmacokinetics.

Establish effects on normal and neoplastic
cell populations in vivo.

Table 6. Effect of Uracil Analogues on TdT Activity

Designator	Name	nmoles ^{3}H-dGMP inc.
Control		1.22
GW-9E	6-anilinouracil	1.21
GW-7B	6-(benzylamino)uracil	1.03
GW-7C	6-(phenetylamino)uracil	1.31
GW-11D	6-(p-butylanilino)uracil	1.12
GW-22E	6-(p-hydroxyanilino)uracil	1.21
GW-16C	6-(p-acetamidobenzylamino)uracil	1.43
GW-18B	6-(cyclohexylamino)uracil	1.31
GW-20B	6-(cyclohexylmethylamino)uracil	1.33
GW-18E	6-(n-pentylamino)uracil	1.20
GW-22A	6-(iso-pentylamino)uracil	1.21
GW-17B	6-(3',4'-trimethyleneanilino)uracil	1.23
GW-28A	6-(d-naphthylamino)uracil	1.24
GW-33E	5-(p-methoxybenzyl)-6-aminouracil	1.20
GW-17E	6-(p-methoxyanilino)uracil	0.51
GW-18C	6-(p-aminoanilino)uracil	0.69

TdT activity was assayed biochemically, as previously described[1],
using homogeneously purified leukemic terminal transferase. All
uracil analogues were used at 200 µmolar final concentration.

Table 7.   Specificity of Inhibition of TdT by GW-17E and GW-18C

	CPM ^{3}H-dNMP incorporated		
Enzyme	Control	+17E	+18C
TdT	12,130	4,900	3,100
pol α	15,040	15,217	15,417
pol β	5,221	5,492	5,218
pol γ	7,490	7,223	7,165

Effect of 400 μmolar inhibitor on activity of homogeneously purified human leukemic TdT and HeLa cell DNA polymerases.   TdT was
assayed as described in reference 1; pol α and pol β as described
in reference 44; and pol γ as described in reference 7.

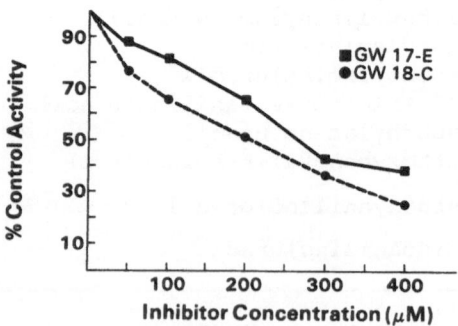

Figure 6.   Inhibition of TdT by 6-(p-methoxy-
anilino)uracil (GW-17E) and 6-(p-aminoanilino)
uracil (GW-18C).

As shown in Figure 6, a dose-dependent inhibition curve could be demonstrated for both compounds. Preliminary studies comparing growth of TdT-positive with TdT-negative cell lines in the presence of either compound suggests that indeed inhibition of TdT may be a lethal event (Table 8). Should these preliminary findings be confirmed with other TdT-positive and negative lines, we will extend these studies as outlined on Table 5. The concentrations of GW 18-C and GW 17-E required to inhibit TdT are relatively high. Modified compounds are now being synthesized to probe the influence of structure on inhibition and to design inhibitors effective at lower concentrations.

Table 8. Effect of GW-17E on Cell Proliferation

Cultured Lines	TdT Status	% of Control
HeLa	Negative	96
L1210	Negative	92
EL-4	Positive	26

Cells were grown in RPMI medium with 10% fetal calf serum, incubated in 37°C in moist air with 5% CO_2. The inhibitor was diluted in growth medium to achieve a variety of concentrations. The effect of the inhibitor on cell proliferation was calculated, at 72 hours, by comparing cell numbers at the end of the incubation period (72 hours) in treated and control cultures. All experiments were done in triplicate. Results shown are with inhibitor at 400 µmolar concentration.

CONCLUSION

In the eight years since the initial findings of TdT in acute leukemia, it has become established as a useful marker in the classification of leukemia and lymphoma. In addition, in certain circumstances, it is of use in the selection of therapy, its presence indicating a high likelihood of response to vincristine and glucocorticoids. Its use in the monitoring of remission status is severely compromised by the fluctuating normal TdT-positive subpopulation of the marrow. The analysis of molecular forms of TdT, in the hope of finding compartment cell line or leukemia-specific TdT types, has not yet yielded a means to discriminate between normal and malignant TdT bearing populations.

The regulation of TdT expression is of more than academic interest, as defects in such regulatory mechanism(s) may be responsible for the disordered proliferation state which results in leukemia or lymphoma. Our preliminary data implicate thymic factors in these regulatory mechanisms.

The function of TdT remains unknown two decades after its discovery. The development of potent and specific TdT inhibitors will provide a critical tool in the dissection of the biological role of TdT. Such compounds may be of therapeutic use in TdT-positive malignant states as definitive or adjunctive therapy, or possibly in the in vitro destruction of TdT-positive malignant cells in bone marrow prior to autologous grafting.

ACKNOWLEDGEMENT

We are indebted to our many colleagues who supplied clinical samples for analysis.

Supported by Grants CA18662, A-105877, FR-128, CAL-19514, CA15187 and GM21747 from the U.S. National Institutes of Health. Ronald McCaffrey was the recipient of Research Career Development Award CA00099 from the U.S. National Institute of Health.

REFERENCES

1. R. McCaffrey, D. F. Smoler, and D. Baltimore, Terminal deoxynucleotidyl transferase in a case of childhood acute lymphoblastic leukemia, Proc Natl Acad Sci USA 70:521 (1973).
2. R. P. McCaffrey, and D. Baltimore, Detection of reverse transcriptase in cell extracts, Blood 40:933 (1972).
3. L. M. S. Chang, Development of terminal deoxnucleotidyl transferase activity in embryonic calf thymus gland, Biochem Biophys Res Commun 44:124 (1971).
4. R. P. McCaffrey, D. F. Smoler, and D. Baltimore, DNA polymerases in lymphoid cells, in: "Modern Trends in Human Leukemia", R. Neth, R.C. Gallo, S. Spiegelman, and F. Stohlman, eds., J.F. Lehmans Verlag, Munich, Germany (1974).
5. D. Baltimore, RNA-dependent DNA polymerase in virions of RNA tumor viruses, Nature 226:1209 (1970).
6. H. M. Temin, and S. Mizutani, RNA-dependent DNA polymerase in the virions of Rous sarcoma virus, Nature 226:1211 (1970).
7. D. Baltimore, R. McCaffrey, and D. F. Smoler, Properties of reverse transcriptase, in: "Virus Research: Proceedings of the 1973 I.C.N.-U.C.L.A. Symposium on Molecular Biology", C. F. Fox, and W. S. Robinson, eds., Academic Press, New York (1973).

8. R. P. McCaffrey, D. F. Smoler, T. A. Harrison, and D. Baltimore,
    A thymus-specific enzyme in acute lymphoblastic leukemia
    cells, in: "Advances in the Biosciences", T. M. Fliedner,
    and S. Perry, eds., Pergammon Press, New York (1975).

9. R. P. McCaffrey, T. A. Harrison, P. C. Kung, R. Parkman,
    A. E. Silverstone, and D. Baltimore, Terminal deoxy-
    nucleotidyl transferase in normal and neoplastic hemato-
    poietic cells, in: "Modern Trends in Human Leukemia II",
    R. Neth, ed., J. F. Lehmans Verlag, Munich, Germany
    (1976).

10. R. P. McCaffrey, and D. Baltimore, unpublished observations.

11. P. C. Kung, J. C. Long, R. P. McCaffrey, R. L. Ratliff,
    T. A. Harrison, and D. Baltimore, Terminal deoxynucleotidyl
    transferase in the diagnosis of leukemia and malignant
    lymphoma, Am J Med 64:788 (1978).

12. M. F. Greenwood, M. S. Coleman, J. J. Hutton, B. Lampkin,
    C. Krill, F. J. Bollum, and P. Holland, Terminal deoxy-
    nucleotidyl transferase distribution in neoplastic and
    hematopoietic cells, J Clin Invest 59:889 (1977).

13. R. Mertelsmann, B. Koziner, D. Filippa, S. Gupta, B. D.
    Clarkson, R. A. Good, and F. P. Siegal, Characterization
    of malignant lymphoma in leukemic phase by multiple
    differentiation markers of mononuclear cells, Am J Med
    63:556 (1977).

14. M. F. Greaves, Cell surface characteristics of human leukaemic
    cells, in: "Essays in Biochemistry", P. N. Campbell, and
    R. D. Marshall, eds., Academic Press, New York (1979).

15. F. J. Bollum, Terminal deoxynucleotidyl transferase as a
    tumor cell marker in leukemia and lymphoma: Results
    from 1000 patients, in: "Tumor Markers", Mihich and Baserga,
    eds., Pergamon Press, Oxford (1979).

16. S. M. Marks, D. Baltimore, and R. P. McCaffrey, Terminal
    transferase as a predictor of initial responsiveness
    to vincristine and prednisone in blastic chronic myelogenous
    leukemia, N Engl J Med 298:812 (1978).

17. G. P. Canellos, V. T. DeVita, J. Whang-Peng, and P. P. Carbone,
    Hematologic and cytogenetic remission of blastic trans-
    formation in chronic granulocytic leukemia, Blood 38:671
    (1971).

18. R. McCaffrey, M. Greaves, T. A. Harrison, T. Revezs, M.
    Beard, and D. Baltimore, Biochemical and immunologial
    evidence for lymphoblastic conversion in chronic myelogenous
    leukemia, Blood 46:1044 (1975).

19. A. V. Hoffbrand, K. Ganeshaguru, G. Janossy, M. F. Greaves,
    R. K. Woodruff, and D. Catovsky, Terminal deoxynucleotidyl
    transferase levels and membrane phenotypes in diagnosis
    of acute leukemia, Lancet ii:520 (1977).

20. G. Janossy, F. J. Bollum, K. F. Bradstock, and J. Ashley,
     Cellular phenotypes of normal and leukemic hematopoietic
     cells determined by analysis with selected antibody
     combinations, Blood 56:430 (1980).

21. J. C. Long, R. P. McCaffrey, A. C. Aisenberg, S. M. Marks,
     and P. C. Kung, Terminal deoxynucleotidyl transferase
     positive lymphoblastic lymphoma, Cancer 44:2127 (1979).

22. J. A. Donlon, E. S. Jaffe, and R. C. Braylan, Terminal deoxy-
     nucleotidyl transferase activity in malignant lymphomas,
     N Engl J Med 297:461 (1977).

23. P. Vezzoni, F. Campagnari, G. DiFronzo, and L. Clerici,
     Terminal deoxynucleotidyl transferase in human lymphomas:
     Possible existence of forms with high and low molecular
     weights, Br J Cancer 43:312 (1981).

24. B. N. Nathwani, H. Kim, and H. Rappaport, Malignant lympho-
     blastic lymphoma, Cancer 38:964 (1976).

25. G. Janossy, Personal communication.

26. K. F. Bradstock, E. S. Papageorgiou, G. Janossy, A. V. Hoff-
     brand, M. L. Willoughby, P. D. Roberts, and F. J. Bollum,
     Detection of leukaemic lymphoblasts in CSF by immunofluo-
     rescence for terminal nucleotidyl transferase, Lancet
     i:1144 (1980).

27. K. F. Bradstock, G. Janossy, A. V. Hoffbrand, K. Ganeshaguru,
     P. Llemellin, H. G. Prentice, and F. J. Bollum, Immuno-
     fluorescent and biochemical studies of terminal deoxy-
     nucleotidyl transferase in treated acute leukaemia,
     Br J Haem 47:121 (1981).

28. M. Greaves, A. Paxton, G. Janossy, C. Pain, S. Johnson,
     and T. A. Lister, Acute lymphoblastic leukaemia associated
     antigen. III Alterations in expression during treatment
     and in relapse, Leukaemia Res 4:1 (1980).

29. F. J. Bollum, Terminal deoxnucleotidyl transferase as a
     hematopoietic cell marker, Blood 54:1203 (1979).

30. R. P. McCaffrey, T. A. Harrison, R. Parkman, and D. Baltimore,
     Terminal deoxynucleotidyl tansferase activity in human
     leukemia cells and in normal human erythrocytes, N Engl
     J Med 292:775 (1975).

31. D. Baltimore, A. E. Silverstone, P. C. Kung, T. A. Harrison,
     and R. P. McCaffrey, What cells contain terminal deoxy-
     nucleotidyl transferase? in: "The Generation of Antibody
     Diversity: A New Look", A. J. Cunningham, ed., Academic
     Press, New York (1975).

32. R. Siegler, and M. A. Rich, Influence of thymic mass on
     murine viral leukogenesis, Nature 209:313 (1966).

33. H. T. Abelson, and L. S. Rabstein, Virus-induced thymic
     independent disease in man, Cancer Res 30:2213 (1970).

34. M. R. Deibel Jr, and M. S. Coleman, Purification of a high
     molecular weight human terminal deoxynucleotidyl transferase,
     J Biol Chem 254:8634 (1979).

35. M. R. Deibel Jr, M. S. Coleman, K. Acree, and J. J. Hutton,
     Biochemical and immunological properties of human terminal
     deoxynucleotidyl transferase purified from blasts of
     acute lymphoblastic and chronic myelogenous leukemia,
     J Clin Invest 67:725 (1981).
36. G. Janossy, F. J. Bollum, K. F. Bradstock, A. McMichael,
     N. Rapson, and M. F. Greaves, Terminal transferase-positive
     human bone marrow cells exhibit the antigenic phenotype
     of common acute lymphoblastic leukemia, J Immunol 123:1525
     (1979).
37. D. Baltimore, A. E. Silverstone, P. C. Kung, T. A. Harrison,
     and R. P. McCaffrey, Specialized DNA polymerases in lymphoid
     cells, in: "Cold Spring Harbor Symposium on Quantitative
     Biology XLI: Origins of Lymphocyte Diversity", p.63,
     (1977).
39. J. C. Long, and R. P. McCaffrey, Mediastinal mass, N Engl
     J Med 299:296 (1978).
40. K. W. Pyke, H.-M. Dosch, M. M. Ipp, and E. W. Gelfand, De-
     monstration of an intrathymic defect in a case of severe
     combined immunodeficiency disease, N Engl J Med 293:424
     (1975).
41. E. L. Reinhertz, P. C. Kung, G. Goldstein, and S. F. Schlossman,
     Separation of functional subsets of human T cells by
     a monoclonal antibody, Proc Natl Acad Sci USA 76:4061
     (1979).
42. G. E. Wright, and N. C. Brown, Inhibitors of Bacillus subtilis
     polymerase III. Structure-activity relationships of
     6-(phenylhydrazino)uracils, J Medicinal Chem 20:1181
     (1977).
43. N. C. Brown, J. Gambino, and G. E. Wright, Inhibitors of
     Bacillus subtilis DNA polymerase III. 6-(arylalkylamino)-
     uracils and 6-anilinouracils, J Medicinal Chem 20:9 (1977).
44. G. E. Wright, E. F. Baril, and N. C. Brown, Butylanilinouracil:
     A selective inhibitor of HeLa cell DNA synthesis and
     HeLa cell DNA polymerase alpha, Nucleic Acid Res 1:99
     (1980).

PROGNOSTIC SIGNIFICANCE OF TERMINAL TRANSFERASE ACTIVITY AND

OTHER FACTORS IN CHILDHOOD ACUTE LYMPHOBLASTIC LEUKEMIA

John J. Hutton, Mary Sue Coleman, and Steven Moffitt

Department of Medicine, University of Texas, San Antonio
Department of Biochemistry, University of Kentucky,
Lexington, and Department of Biometry, Emory University
Atlanta, Georgia

Terminal deoxynucleotidyl transferase (TDT) is an intracellular biochemical marker of certain types of lymphoid precursors. Clinically, TDT serves as a useful marker of malignant lymphoblasts. TDT activity is elevated in 85-95% of cases of non-B, non-T and T-marked acute lymphoblastic leukemias, 5-10% of cases of acute myelogenous leukemia, and 30-40% of cases of chronic myelogenous in blastic phase (reviewed by Coleman and Hutton, 1981). The presence of TDT in leukemic cells is correlated with a favorable response to therapy with drugs cytocidal to lymphoblasts. There is marked variation in the activity of TDT in leukemic cells at diagnosis of acute lymphoblastic leukemia. In our experience values ranged from 0 to 694 units/10^8 nucleated cells in bone marrow from 118 children and from 0 to 1790 units/10^8 nucleated cells in peripheral blood from 51 children with acute lymphoblastic leukemia (Hutton et al, 1979). There are no studies testing the prognostic significance of these quantitative variations in level of TDT at diagnosis. Similarly, there are no studies testing whether quantitative measurements of TDT activity during remission of leukemia can predict relapse of disease.

In an attempt to answer these questions we measured TDT activity in cells from the bone marrow and peripheral blood of 206 children with acute lymphoblastic leukemia. Details of our methods of assay of TDT and the activities obtained have been published (Greenwood et al, 1977; Coleman and Hutton, 1981). The studies were done in collaboration with Dr. F.J. Bollum, Uniformed Services University of the Health Sciences; Drs. Martha Greenwood and Phillip Holland, University of Kentucky; Dr. Beatrice Lampkin, Childrens Hospital,

Cincinnati, Ohio; and Dr. Carl Krill, Childrens Hospital, Akron, Ohio. Markers of leukemic blasts and clinical characteristics of most patients at diagnosis have been published (Coleman et al, 1976; Greenwood et al, 1977). Cells from peripheral blood and bone marrow were obtained at diagnosis and repetitively during remission of disease. Other data recorded for subsequent analysis of prognostic factors included: institution where treatment was given, treatment group, age, peripheral leukocyte count at diagnosis, presence or absence of a mediastinal mass, presence or absence of surface immunoglobulin on blasts, whether blasts formed rosettes with sheep erythrocytes, whether remission was achieved, duration of complete remission, and survival. We have examined whether the quantitative activity of TDT at diagnosis predicts achievement of complete remission, duration of complete remission, or survival. We have also tested whether changes in TDT activity during remission predict relapse of disease.

Patients were entered on study from January 1973 through December 1978. Follow up data were obtained through September, 1979. There were 206 children ranging in age from less than one year to 16 years. The mean age was 77 months. Complete remission was obtained in 96% of these patients. 43% relapsed during the course of our studies. The median duration of complete remission was 1400 days. Children were generally treated on protocols of either the Pediatric Hematology/Oncology Group or the Childrens Cancer Study Group. Treatment protocols were complex and the details are not relevant. However, children were stratified by risk, generally received vincristine and prednisone if they were low risk, and vincristine, prednisone, cyclophosphamide followed by asparaginase, if they were high risk. There were several changes in the treatment protocols during the course of the study so treatment was not uniform throughout. Maintenance chemotherapy was generally continued for 3 years after complete remission was achieved or until relapse occurred.

Table 1. Terminal Transferase Activities at Diagnosis of Childhood Acute Lymphoblastic Leukemia

| Group | Bone Marrow | | Peripheral Blood | |
	Activity	Specimens	Activity	Specimens
All Patients	91 ± 143[a]	157	54 ± 159	163
Exclude TDT				
Values > 120	29 ± 29	120	20 ± 27	145
Normal Children	5.9 ± 19	198	0.5 ± 0.7	51
Normal Adults	2.7 ± 3.7	160	1.0 ± 3.1	16

[a]Units/10^8 nucleated cells. Mean $\pm$ SD.

Table 1 shows terminal transferase activities at the time of diagnosis of acute lymphoblastic leukemia. While all patients had terminal transferase measured in their leukemic cells, not all patients had measurements made on cells from both the bone marrow and peripheral blood. Approximately 25% of the patients had cells examined from only one of the two sources. For the 157 patients with terminal transferase measurements in bone marrow, the mean TDT activity was $91 + 143$ units/10^8 cells. For peripheral blood, this figure was $54 + 159$ units/10^8 cells in a set of 163 specimens. The large standard deviation reflects the wide range of activities seen in patients with acute lymphoblastic leukemia. A significant proportion of the variation is due to a relatively small number of patients with extremely high activities. For example, while the mean was 91 units the values ranged from 0 to approximately 1000 units, when all patients were considered. If we exclude the high outliers and set the upper limit of values at 120 units, then the mean activity in bone marrow is $29 + 29$ units for 120 specimens and $20 + 27$ units for the 145 specimens of peripheral blood (Table 1). Setting the upper range at 120 units/10^8 cells excludes 24% of values obtained for bone marrow and 11% of values obtained for peripheral blood. The bone marrow from 198 children without acute leukemia contained 5.9 units of transferase activity per 10^8 cells with a standard deviation of 19. We have pointed out in prior publications the wide range of normal in children (Greenwood et al, 1977). Particularly high values are seen in patients with fungal diseases and idiopathic thrombocytopenic purpura. The normal mean value in adults is about half the normal mean value in children. If we exclude values of terminal transferase greater than 120 units, then the average value in children with acute lymphoblastic leukemia is 5 times the normal value. The elevation of transferase in lymphoblastic leukemia is dramatic, but there is significant overlap with values obtained in children without a malignancy so that measurements of TDT cannot diagnose malignancy.

We plotted the terminal transferase activity in cells from bone marrow against the terminal transferase activity in cells from peripheral blood for a series of patients in whom both measurements were made (Figure 1). There is a weak, but statistically significant, correlation between the two measurements. In a set of 77 measurements, the correlation coefficient is 0.19 ($p < 0.04$).

As a first step in analyzing whether the quantitative value of terminal transferase obtained in either bone marrow or peripheral blood is a prognostic indicator, we plotted the distribution of values seen in patients who are alive vs. those who are dead (Figure 2). Values in peripheral blood were plotted separately from those in bone marrow. The small letter "A" indicates the patient is still alive and the small letter "D" indicates the patient is dead. There is no clustering of dead patients at any particular range of transferase activities. For clarity we have only shown the distribution

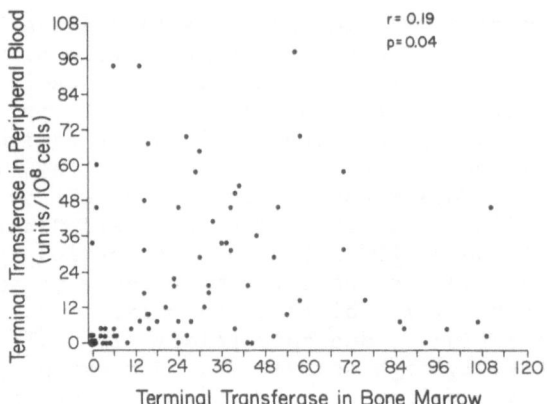

Fig. 1.   Activities of terminal transferase in nucleated
          cells from the bone marrow and peripheral blood
          of children with acute lymphoblastic leukemia.
          Each point represents a different patient.

from 0 to 120 units of TDT/10^8 cells, but when the plot is extended
to 1000 units/10^8 cells there is still no obvious difference between
the distribution of values in dead versus living patients.

A significant number of children had low levels of TDT activity
in their leukemic lymphoblasts (Figure 2). Because the activity of
TDT is generally high in blasts from children with ALL, we wondered
whether the low values were due to technical problems with the
collection of cells, difficulties with the enzymatic assay, or the
existence of a subtype of ALL with distinctive clinical characteristics. Table 2 shows data on 4 patients selected at random from
among those with TDT less than 1 unit/10^8 cells in bone marrow. The
number of cells assayed ranged from 0.3 X 10^7 to 12 X 10^7. We generally consider 1 to 2 X 10^7 cells as necessary for reliable assay of
transferase, if the specific activity of the ^{3}H-dGTP used in the
assay is in the vicinity of 100 cpm/pmol. Sensitivity of the assay
can be increased by increasing the specific activity of the isotope,
but at great monetary cost. In 2 of the 4 cases presented in Table
2, the number of cells extracted for the assay was lower than

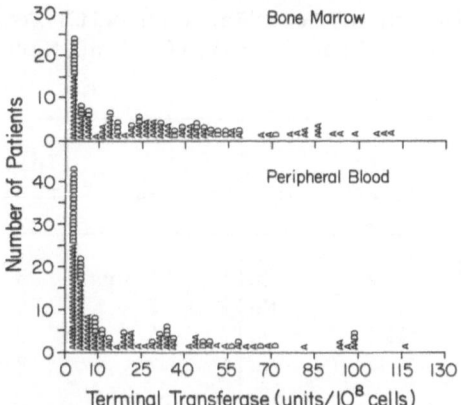

Fig. 2. Distribution of terminal transferase activities
         in patients at diagnosis. Each symbol represents
         a different patient. Patients who were still alive
         at the termination of the study are represented by
         "A", those who were dead "D".

optimal. Because cells were collected at 5 different institutions,
we did not have ideal control of the way they were handled prior to
assay in our laboratories. In our experience, terminal transferase
is stable at -30°C for several months and at -70° for several years.
When we looked carefully at data for patients with low values of
terminal transferase, we could not discern any particular pattern
in the number of days specimens were frozen. In 2 of the 4 cases
shown, samples were frozen for several months. In the majority of
cases included in this report, the cells were frozen less than 1
month so we do not believe that handling or freezing the specimens
is a major cause of low activities of transferase. Many of the
children with low TDT were younger than average. In all cases, the
blast cells were examined for their ability to form rosettes with
sheep red blood cells and for the presence of surface immunoglobulin.
In the entire series of 206 children, five patients were identified
whose blasts had surface immunoglobulin so the leukemia would be
classed as B cell. All of these patients had low values of terminal
transferase. However, the number of such children was small and

Table 2.   Characteristics of Four Patients with Terminal Transferase
           Activities Less Than 1 Unit/10^8 Bone Marrow Cells at
           Diagnosis

TDT Activity	Cells Assayed	Days Frozen	Marker	Age	Institution	Time To Relapse
0.5	3.6×10^7	2	Null	1 yr	3	154 days
0.7	12.0	5	Null	2	4	485
< 0.5	0.3	115	Null	4	2	>1915
< 0.5	1.1	52	Null	2	2	> 605

the majority of patients with low values of transferase had typical
null or T cell surface properties.

    We also examined the characteristics of patients with partic-
ularly high values of terminal transferase in their blast cells.
The characteristics of four such patients are listed in Table 3.
Transferase activities ranged from 563 to 876 units in bone marrow.
The ages of the patients were generally above the average for the
group of 206 children.   None of the blasts formed rosettes with
sheep red cells or had surface immunoglobulin.   Cells were obtained
from patients with a variety of white counts and the duration of
first remission appeared to be within the range expected.

    We analyzed survival of patients whose blasts contained high
levels of terminal transferase as opposed to patients whose blasts
contained low levels of terminal transferase.   High level of
terminal transferase was defined as greater than 20 units in bone

Table 3.   Characteristics of Four Patients with Terminal Transferase
           Activities Greater Than 500 Units/10^8 Bone Marrow Cells at
           Diagnosis

TDT Activity	Age	Sex	Peripheral Leukocyte Count	Marker	Time To Relapse
563	16 yrs	M	1,800	Null	733 days
695	9	M	179,000	Null	800
794	6	M	36,600	Null	>372
876	13	F	13,600	Null	145

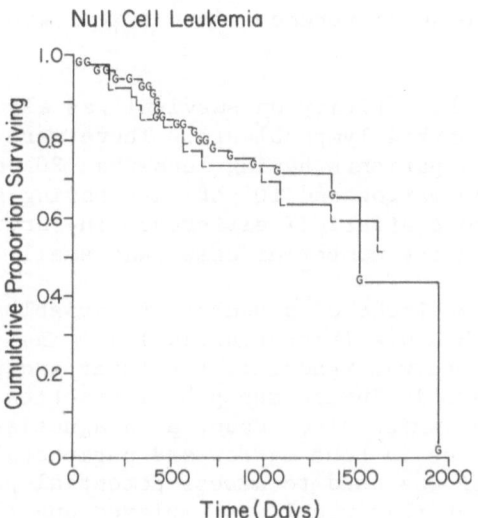

Fig. 3.   Comparative analysis of patients whose leukemic
           cells had low activities as opposed to high
           activities of TDT at diagnosis. All patients
           had "null cell" leukemia. "G" indicates patients
           with TDT activity in peripheral blood or bone
           marrow greater than or equal to 10 and 20
           units/10^8 cells, respectively. "L" indicates
           patients with TDT activity in peripheral blood
           and bone marrow less than 10 and 20 units,
           respectively.

marrow or greater than 10 units in peripheral blood. These figures
were originally chosen to distinguish activities characteristic of
acute myelogenous leukemia from those typical of acute lymphoblastic leukemia (Gordon et al, 1978). We analyzed survival
separately for children with T-marked lymphoblasts and those with
"null" lymphoblasts. There were 148 children whose "null" blasts
did not form rosettes with sheep red blood cells or have surface
immunoglobulin. Their survival is shown in Figure 3. The small "L"
marks children with low activity of transferase. The "G" marks

children with high transferase. There were 37 children in the low
transferase group and 111 children in the high transferase group.
Measurements were made at diagnosis and there is no significant
effect of transferase level on survival when these limits are used
(p > 0.1). The analysis was repeated using a cutoff of 10 units
of transferase in bone marrow to distinguish low from high values
and again there were no differences in survival within the "null"
cell group.

The effect of TDT activity on survival was also analyzed in
patients who had T-marked lymphoblasts. There were only 14 patients
in this group with 4 patients having less than 20 units of transferase in their bone marrow and 10 patients having greater than 20
units. There was no statistical difference in survival between the
two groups, although the number of cases was small.

We measured the effect of a number of variables on the prognosis of children with acute lymphoblastic leukemia. These included
institution where care was rendered, treatment group, mediastinal
mass, surface markers including sheep cell rosettes and surface
immunoglobulin, age, white blood count at diagnosis, and terminal
transferase activities in bone marrow and peripheral blood. The
Cox regression model was used to assess potential prognostic factors.
A stepwise version of this model was employed and the P-values corresponding to variables reflect only the presence of a relationship.
They do not measure the strength of the relationship or its predictive value. Because of the large number of patients analyzed,
variables may have statistically significant effects on prognosis,
but these effects may be so small that they are not clinically
significant.

We first analyzed factors associated with failure to achieve a
complete remission. Because 96% of the children achieved a complete
remission, we are looking at a very small subpopulation of 8 children.
Three factors were associated with failure to achieve a complete
remission. Of most importance, was that the child be diagnosed and
treated in institution 3. The second factor of importance was a low
value of terminal transferase in blasts at diagnosis. The third
factor was age above average for the group. When we looked at these
cases individually, it was apparent that this group of 8 patients
contained 5 children who were diagnosed as typical childhood acute
lymphoblastic leukemia, but had a B cell leukemia. All of the
children with B cell leukemia died within a year of diagnosis,
unlike patients with typical null or T cell disease. This poor
prognostic group of older children with low terminal transferase is
heterogeneous and contains children having lymphoma leukemia and
possibly non-lymphoid leukemias. They do not achieve complete
clinical remission with treatment appropriate for the usual childhood acute lymphoblastic leukemia.

Table 4.   Analysis of Prognostic Factors at Diagnosis

Predictive of Short Duration of First Remission

1.   Presence of Mediastinal Mass (p < 0.001)

Variables: Institution, Treatment Group, Mediasti-
             nal Mass, Surface Markers, Age, White
             Count, Terminal Transferase.

We next examined factors which predict a short remission (Table
4).   Relapse in childhood acute lymphoblastic leukemia is generally,
but not always, followed by death from the disease.   Factors which
predict short duration of remission are important in placing the
child in a high risk group that receives more intensive therapy.
Of the factors listed in Table 4, only the presence of a mediastinal
mass was associated with a short remission.   It is particularly
interesting that age and the presence of the T cell marker did not
predict short duration of remission in our series.

At the termination of our study, 59 patients had died, 4 were
lost to follow up, and 143 were alive.   Factors which predicted poor
survival include: the presence of a mediastinal mass, treatment in
institution 3, treatment group 2, and a high peripheral nucleated
cell count.   The analysis excluded the five patients with B marked
leukemia, since all of these patients died within a year of beginning
treatment and were not considered to have typical acute lymphoblastic
leukemia.

We next turned our attention to the question of whether quantitative terminal transferase measurements on cells from bone marrow
or peripheral blood obtained after induction are useful in predicting
relapse of the leukemia.   To illustrate the problems encountered,
Figure 4 shows the terminal transferase activity in cells from
peripheral blood over a period of 1600 days following the diagnosis

Table 5.   Analysis of Prognostic Factors at Diagnosis.

Predictive of Poor Survival

1.   Presence of Mediastinal Mass   (p < 0.001)
2.   Institution 3   (p = 0.01)
3.   Treatment Group 2   (p = 0.01)
4.   High Peripheral Nucleated Cell Count   (p = 0.12)

Variables: Institution, Treatment Group, Mediasti-
             nal Mass, Surface Markers, Age, White
             Count, Terminal Transferase.

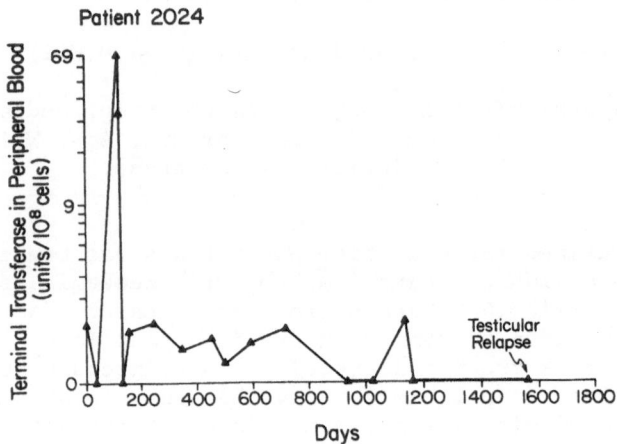

Fig. 4.   Terminal transferase activity in cells from the
peripheral blood of a child with acute lymphoblastic leukemia. The child was diagnosed on
day 0 and treatment was begun. He remained in
remission until testicular relapse occurred,
where indicated.

of acute lymphoblastic leukemia in a patient. At the time of initial
diagnosis, terminal transferase from peripheral blood was less than
1 unit, although it was elevated in bone marrow. During the first
year of therapy, the activity in peripheral blood was as high as
69 units per 10^8 nucleated cells. This high value was not associated with the appearance of blasts and treatment of the patient was
in the standard way according to protocol. During the five years
following initial diagnosis, except for the high values obtained on
two occasions during the first year, the terminal transferase value
in peripheral blood remain less than 1 unit. The patient eventually
relapsed in his testes. At the time of relapse, the terminal transferase value in peripheral blood was less than 0.1 units. Although
we did not analyze leukemic blasts from the testes of this particular patient, in other patients blasts from testis and spinal fluid
have contained transferase.

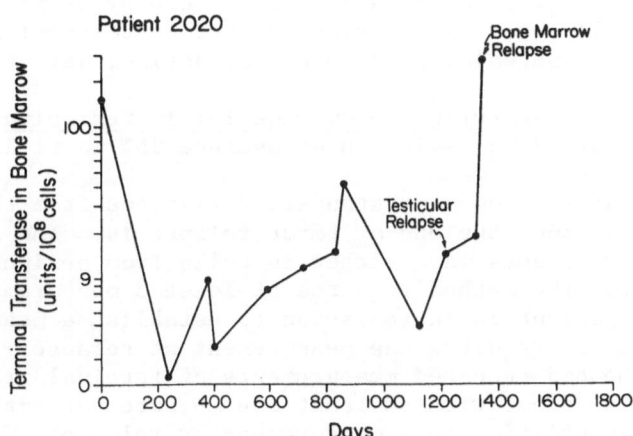

Fig. 5. Terminal transferase activity in cells from the
         bone marrow of a child with acute lymphoblastic
         leukemia. The child was diagnosed on day 0 and
         treatment was begun. He remained in remission
         until testicular relapse occurred, where marked.
         Hematologic relapse occurred some months later.

We have serially measured transferase in bone marrow from
several patients during treatment. In patient 2020, specimens were
obtained over a period of approximately 4 years (Figure 5). Bone
marrow transferase at diagnosis was greater than 100 units/10^8
cells. Over the next several years, this value fluctuated reaching
activities as high as 40 units/10^8 cells, but without morphological
relapse in bone marrow. Testicular relapse occurred on approximately day 1200. At this time, the bone marrow activity was within
the range typical of this patient. Bone marrow relapse occurred
around day 1400. This was not predicted by assays of TDT activity
in bone marrow obtained 2 months before relapse.

Statistical approaches to the question of whether increases in
terminal transferase occur before frank relapse are difficult to
design. If significant increases in TDT did occur, then we could

predict relapse of the leukemia and make some type of therapeutic intervention. A rough estimate of the potential usefulness of TDT measurements in predicting relapse can be made by examining the increases actually seen at the time of obvious clinical relapse. For this purpose, we calculated Z-scores, defined as:

$$\text{Z-score} = \frac{\text{TDT at relapse} - \text{average TDT in remission}}{\text{Standard deviation of average TDT in remission}}$$

The Z-score measures how many standard deviations from the mean of previous observations the one at first relapse is. The Z-score applies to measurements made either in cells from peripheral blood or bone marrow. The method requires at least 3 measurements of TDT while the patient is in remission to establish a mean and standard deviation. It requires one measurement at relapse. There were 19 children who had repeated measurements of terminal transferase in their peripheral blood with at least one measurement within 45 days of a relapse in addition to a measurement at relapse. We asked whether terminal transferase measurements in peripheral blood can predict the relapse. We used peripheral blood because most children did not have repetitive bone marrow specimens after initial induction and there were simply too few bone marrow specimens available for statistical analysis.

Table 6 shows the type of data used to calculate Z scores. Four patients are listed. For the first patient, the TDT in peripheral blood at relapse was 0. While the patient was in remission, the mean activity in nucleated cells from 10 separate specimens of peripheral blood had been 11 units with a standard deviation of 22 units. This means the value in peripheral blood had been fluctuating and was usually low, but had been high on several occasions. Since the TDT at relapse was 0, the Z score was negative. That is, the

Table 6.   Relative Increase in TDT Activity in Peripheral Blood at Relapse Compared to Average Activity While in Remission

| Patient | TDT at Relapse | TDT Remission | | | Z-score[a] |
		Mean	SD	N	
1	0	11	22	10	−0.5
2	37	3	6	10	5.7
3	0.9	10	13	7	−0.7
4	176	2	3	13	58

[a]The Z-score measures how many standard deviations from the mean of previous observations the one at first relapse is.

value at relapse was even lower than the average during remission. In patient 2, the terminal transferase value was elevated at relapse to 37 units compared to a mean of 3 units in 10 specimens of cells analyzed during remission. In this case the Z-score had a positive value of 5.7. This means the 37 units at relapse was 5.7 standard deviations above the mean TDT during remission. In a third patient, the TDT at relapse was 0.9 units while the mean during remission was 10 units, so again the Z-score was negative indicating no significant difference between TDT in remission and at relapse. In the fourth patient the TDT at relapse was 176 units, the mean during remission was 2 units with a standard deviation of 3. In this case the TDT at relapse was 58 standard deviations above the mean during remission.

We calculated Z scores for measurements made on cells from the peripheral blood of 23 patients and the bone marrow of 19 patients. Z was greater than 3 in 7 of the 23 patients with measurements in peripheral blood and in 11 of the 19 patients with measurements in bone marrow. In other words, in 30% of patients who have repeated measurements of TDT in peripheral blood and in 58% of patients who have repeated measurements of TDT in bone marrow, the activity of TDT at relapse is at least 3 standard deviations above the mean value during remission. On the other hand, in 42-70% of patients, TDT activity in cells at relapse is within 3 standard deviations of the mean value during remission. Because TDT activity in hemato-poietic cells fluctuates during remission of leukemia (and even in hematologically normal people), it is not possible to interpret fluctuations in terms of disease activity. In fact, in most patients TDT activity at relapse is within the ranges seen in remission. Conversely, we have observed very high transferase activity in cells from children who did not relapse during several years of follow-up after the high values were observed.

The essence of the problem of low TDT activities in marrow and blood at the time of relapse of leukemia is the high frequency of relapse outside the hematopoietic system, that is, in the testes and central nervous system. In nearly half of children who relapse with acute lymphoblastic leukemia, relapse occurs at a site other than bone marrow and peripheral blood. We had originally hoped that quantitative changes in transferase might occur in marrow or blood from these children. This does not seem to be the case. In four of the the 18 patients we studied most intensively, relapse was only in the testes. Four other relapses were limited to the central nervous system. Therefore, in 8 of 18 patients (44%) the relapse did not involve blood or marrow and could not be detected by quantitative measurements of TDT in cells from these sites. The bone marrow was involved in some fashion in the remaining 10 patients and elevated TDT activities were found.

CONCLUSIONS

We conclude that the quantitative activity of terminal transferase in blasts at diagnosis of children with typical acute
lymphoblastic leukemia does not convey prognostic information. The
exception to this statement would be low values that trigger further
investigation so that diagnostic errors are eliminated. Because our
study focused on young children, where acute lymphoblastic leukemia
is by far the predominant leukemia, we had relatively few diagnostic
errors. There was some difficulty distinguishing acute lymphoblastic
leukemia from Burkitt's lymphoma leukemia. The situation may be
different in adult patients where the classification of leukemia
is much more complex.

Repeated quantitative measurements of TDT activity in bone
marrow and peripheral blood from children with acute lymphoblastic
leukemia in remission have not provided clinically useful information. We have observed very high values of transferase activity in
children who did not relapse. We have also observed high values in
cells from the bone marrow of children without leukemia. For this
reason, high terminal transferase activity cannot be taken to
indicate impending relapse. In at least half of children who do
relapse, transferase is not notably elevated in bone marrow and
peripheral blood. We would direct our strategy for detection of
relapse toward improvement of multi-parameter analysis of cells from
bone marrow, peripheral blood, cerebrospinal fluid, and testis with
particular attention to methods permitting examination of individual
cells. This might help identify small numbers of leukemic blasts.
Most markers presently available are found on both leukemic blasts
and subsets of normal cells. Under these circumstances it is difficult to be certain that hematopoietic cells with "blastic" markers
are leukemic rather than normal.

REFERENCES

Coleman, M. S., Greenwood, M. F., Hutton, J. J., Bollum, F. J.,
    Lampkin, B., and Holland, P., 1976, Serial observations on
    terminal deoxynucleotidyl transferase activity and lympho-
    blast surface markers in acute lymphoblastic leukemia,
    Cancer Res., 36:120.
Coleman, M. S., and Hutton, J. J., 1981, Terminal transferase, in:
    "The Leukemic Cell," D. Catovsky, ed., Churchill Livingstone,
    Edinburgh.
Gordon, D. S., Hutton, J. J., Smalley, R. V., Meyer, L. M., and
    Vogler, W. R., 1978, Terminal deoxynucleotidyl transferase,
    cytochemistry, and membrane receptors in adult acute leukemia,
    Blood, 52:1079.

Greenwood, M. F., Coleman, M. S., Hutton, J. J., Lampkin, B., Krill,
    C., Bollum, F. J., and Holland, P., 1977, Terminal deoxynucleo-
    tidyl transferase distribution in neoplastic and hematopoietic
    cells, J. Clin. Invest., 59:889.
Hutton, J. J., Coleman, M. S., Keneklis, T. P., and Bollum, F. J.,
    1979, Terminal deoxynucleotidyl transferase as a tumor cell
    marker in leukemia and lymphoma: Results from 1000 patients,
    in: "Biological Basis For Cancer Diagnosis," M. Fox, ed.,
    Pergamon Press, Oxford.

THE PROGNOSTIC SIGNIFICANCE OF TERMINAL DEOXYNUCLEOTIDYL TRANSFERASE (TDT) IN PATIENTS WITH LEUKEMIAS AND MALIGNANT LYMPHOMAS

Roland Mertelsmann

Developmental Hematopoiesis Laboratory and
Hematology/Lymphoma Service, Department of Medicine
Memorial Sloan-Kettering Cancer Center, N.Y., N.Y.   10021

INTRODUCTION

Since the first report of elevated terminal deoxynucleotidyl transferase (TdT) activity in leukemic cells from a patient with acute lymphoblastic leukemia (ALL)[1], TdT has become one of the most widely studied phenotypic markers in hematopoietic neoplasias[2-8].

Starting in August of 1976, every patient with leukemia or malignant lymphoma seen at Memorial Sloan-Kettering Cancer Center (MSKCC) was evaluated for TdT activity and other phenotypic markers, if technically feasible. Approximately 6000 blood, bone marrow and lymph node samples from 2000 patients with hematopoietic neoplasias have been analyzed. In addition to confirming and extending reports by other investigators [2-6] regarding the diagnostic significance of TdT in a large, unselected, and consecutive patient population[7-11], these studies have documented for the first time biphenotypic acute leukemias[10-12] exhibiting features of both ALL and acute non-lymphoblastic leukemia (ANLL) and have provided evidence for the prognostic significance of elevated TdT activities in patients with preleukemic syndromes[13], chronic myelogenous leukemia (CML)[14], and malignant lymphomas[15]. Now adequate follow-up time has elapsed (up to 5 years) for patients treated on protocol allowing analysis of the prognostic significance of TdT in patients with hematopoietic tumors receiving standardized chemotherapy regimens.

This report will focus on the analysis of TdT activities and its relationship to remission frequency and duration as well as survival in patients with ALL, ANLL and malignant Non-Hodgkin's lymphomas (NHL) in adults. Observations made in blast crisis of

(CML) and other myeloproliferative (MPS) and myelodysplastic (MDS) syndromes will not be discussed in this report in view of previous reports by other investigators[3,6] as well as from this institution [10,13,14].

MATERIALS AND METHODS

All patients analyzed in this report were diagnosed and treated according to standardized chemotherapy protocols at Memorial Sloan-Kettering Cancer Center (MSKCC). Diagnoses of leukemia were made on bone marrow aspirates, evaluated with routine and cytochemical stains, and classified according to the FAB recommendations[15] as previously reported[16]. NHL diagnoses were made on lymph node biopsies and classified according to the Rappaport system as previously described [17,18].

TdT was determined biochemically in mononuclear cell fractions from anti-coagulated bone marrow or blood samples or in lymph node cell suspensions (kindly provided by Dr. Filippa, MSKCC) as previously described[8]. On more recent samples, immunofluorescence analysis using a commercial anti-calf thymus TdT antibody (BRL, Bethesda, Maryland) was performed in parallel with the biochemical assay.

The immunofluorescence technique used followed the procedure outlined by the manufacturer, with two exceptions: (1) A concentrated antibody solution was obtained and diluted 1:50 immediately before use. This was necessary since the 1:50 diluted antibody supplied in the manufacturer's kit was found to be unstable. (2) Slides were made from Ficoll-Hypaque separated mononuclear cell fractions rather than from unseparated blood and marrow, because of markedly lowered background and higher proportion of the cells of interest (mononuclear fraction). Slides were stained with the appropriate antibodies within 24 hours, since frequently weak, but definitely positive cells tended to become more and more difficult to recognize on slides, stored for only a few days.

Normal bone marrow samples obtained from volunteers yielded TdT activities <0.2 U/10^8 cells (1 U = 1 nmole (3H) dGMP incorporated per 10^8 cells in one hour under the conditions described[8]) in 99% of the samples. TdT activities were reported as positive, if dGMP incorporation was inhibited $>50\%$ by addition of ATP as inhibitor for TdT (cf.[8]) irrespective of the absolute level of activity. The specific activity expressed in U/10^8 cells was based on the difference in dGMP incorporation (Δcpm) in the absence and presence of ATP. If dGMP incorporation was inhibited $<50\%$ by addition of ATP, results were reported as TdT negative (TdT$^-$) and expressed in U/10^8 cells based on the Δcpm, to indicate the level above which TdT activity could be excluded.

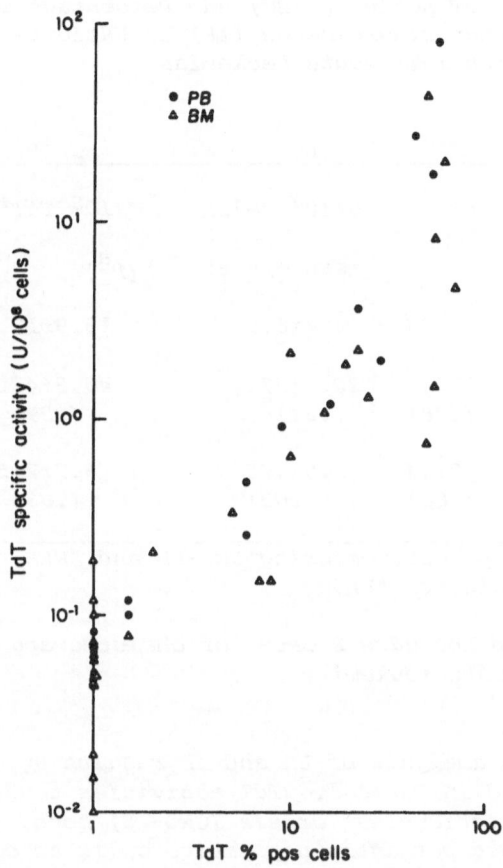

*Figure 1 Biochemical TdT Activities and Percentage of TdT⁺ Cells by
Immunofluorescence in Blood and Bone Marrow Samples from
Patients with Leukemias and Lymphomas and in Controls*

*RESULTS
Methodology*

*The previously described biochemical microassay[8] was found
highly reproducible and easy to use.*

*Biochemical (BC) and immunofluorescent (IF) results were
correlated (Fig. 1) However, biochemically determined TdT activities
expressed per TdT antigen positive cells varied by a factor of up to
100 (cf. Fig. 1), suggesting the possibility that TdT antigen containing cell phenotypes could be further subdivided according to
their different biochemical TdT activities.*

Table 1.

Biochemical TdT Activity (BC) and Percentage TdT$^+$
Cells by Immunofluorescence (IF) in Patients
with TdT$^+$ Acute Leukemias[a]

Diagnosis	n	IF	BC	BC/IF
		%+	U/10^8 cells	U/10^8 TdT$^+$ cells
			mean +/- SD	(P^a)
ALL	5	39±28	6.7±6.9	15.9±12.5
CML-LB	9	37±23 (.88)	20.2±27.4 (.31)	40.3±47.7 (.29)
ANLL[b]	11	17±14 (.06)	1.0±.75 (.02)	6.7±2.5 (.03)

a) Student's t test comparing CML-LB and ANLL,
   respectively, to ALL.

b) This group includes 2 cases of chemotherapy
   induced acute leukemia.

   A preliminary analysis of BC and IF results by clinical diagnosis is summarized in Table 1. TdT activities tended to be higher
in CML in lymphoblastic blast crisis (CML-LB) both, in absolute terms
as well as expressed per antigen positive cells as compared to ALL,
but these differences were not statistically significant in this
small sample size. Patients with TdT$^+$ ANLL, however expressed significantly lower percentages of TdT$^+$ cells, and lower biochemical
activities as well as lower TdT activities per TdT antigen positive
cells as compared to ALL (P=0.03, Tab. 1). While the lower proportion of TdT$^+$ cells in TdT$^+$ ANLL would have been expected in view of
the presence of TdT$^-$, non-lymphoid blasts ("biphenotypic leukemia")
the lower TdT activities per antigen positive cells could possibly
indicate a different TdT$^+$ cell type in ANLL than seen in ALL or
CML-LB.

TdT in ALL

   Of 113 adult patients treated on ALL protocols including 101
patients with ALL and 12 patients with lymphoblastic lymphoma (LBL),
44 cases were studied for TdT activity at diagnosis prior to treat-

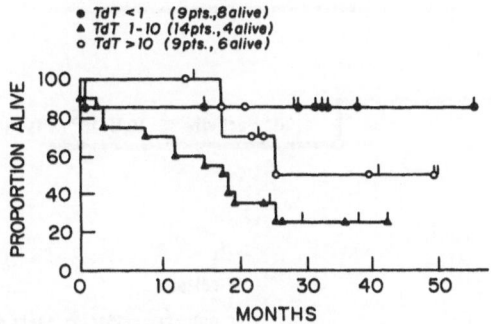

*Figure 2   Survival of Adult Patients with ALL by TdT activity
Determined in Bone Marrow Samples (P=0.01)*

ment and an additional 38 cases after initiation of treatment or at
relapse. Of the total number of 72 cases with evaluable TdT determinations only 2 with B cell ALL (excluded from survival analysis)
and 1 patient with T cell ALL failed to exhibit elevated TdT
activities.

Pre-treatment bone marrow TdT activities were found to be
significantly related to survival, with patients exhibiting TdT
activities <1 U/10^8 cells showing the longest survival (Fig. 2).
Identical results were obtained using TdT activities determined in
peripheral blood cells (data not shown). Since some patients had
TdT activities performed only in blood or lymph node specimens, while

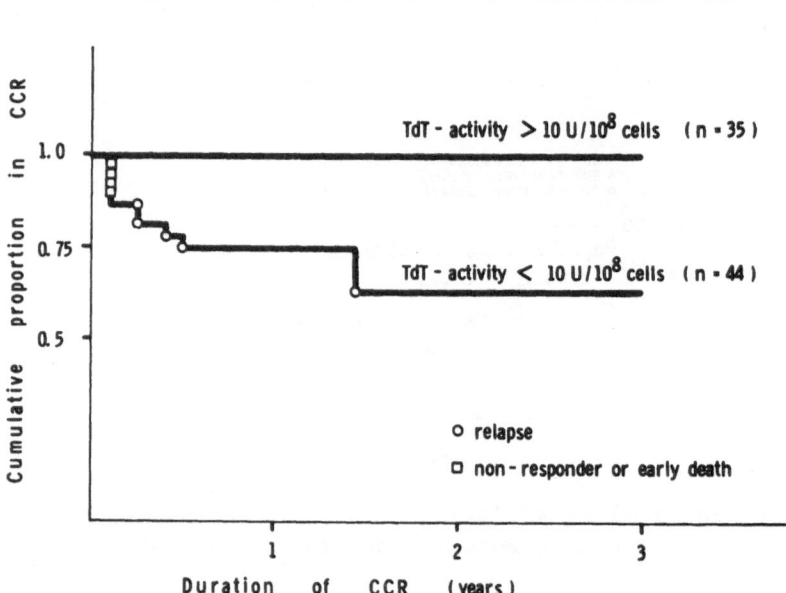

PROGNOSTIC VALUE OF TdT IN CHILDHOOD ALL

Figure 3 Survival of Children with All by TdT Activity (reprinted with kind permission of Dr. K. Welte).

some had determinations performed in more than one tissue, analysis was also performed on maximum TdT actitivies determined in any tissue of a given patient. This analysis showed 75% of patients exhibiting low TdT activities (<1 U/10^8 cells), 23% of cases with intermediate TdT activities (1.0-10.0 U/10^8 cells) and 58% of patients with high TdT activities (>10 U/10^8 cells) alive at 4 years confirming the prognostic significance of pre-treatment TdT levels in ALL of adults.

Similar observations demonstrating the prognostic significance of TdT in ALL were reported by Dr. K. Welte (Frankfurt, Germany, personal communication). In this study of 79 children with ALL treated on a cooperative study protocol[19], TdT values of <10 U/10^8 cells as determined in the standard assay described by Modak et al.[8] were associated with a worse prognosis compared to TdT activities >10 U/10^8 cells (Fig. 3).

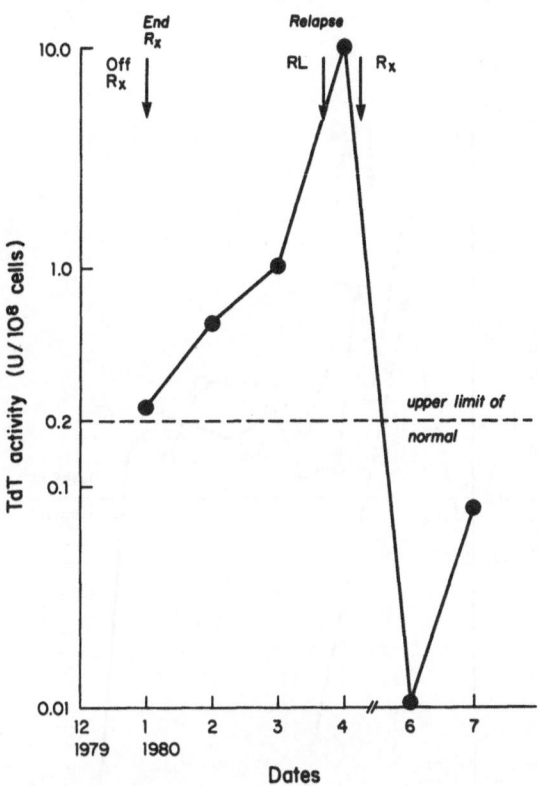

*Figure 4 Sequential Determination of Bone Marrow TdT Activities
in a Child with ALL in Complete Remission and in Relapse*

Further analysis of both the MSKCC adult population as well as the
children's group with ALL is required to determine whether similar
levels of TdT activity carry the same or different prognoses in the
pediatric and adult age groups and to assess the interrelationship
of TdT activities and other prognostic factors, before definitive
conclusions can be drawn.

With the intensive treatment protocols used in both, the
pediatric[19] and adult[20] group, T and null surface marker phenotypes
(with the exception of B cell types) were no longer of prognostic
significance (unpublished), suggesting that TdT activities might be
a more sensitive phenotypic parameter, recognizing prognostically
important cell types in ALL even in patients treated on the more
recent intensive chemotherapy regimens.

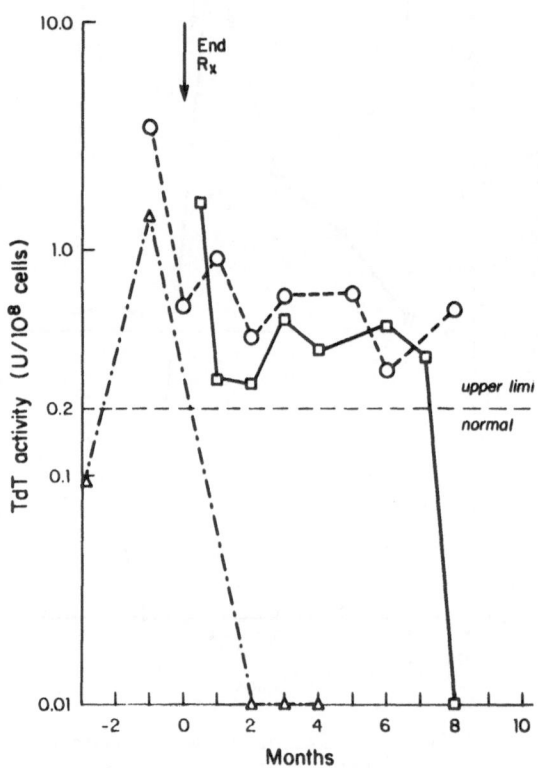

Figure 5   Sequential Determination of TdT Activities in Three Child-
ren with ALL in complete Remission Prior to and After
Completion of Chemotherapy

     Consecutive TdT determinations in patients with ALL in complete
remission have yielded intermittently elevated TdT activities in a
high proportion of patients[7]. While in some patients clinical
relapse was preceded by rising TdT values in marrow samples obtained
at monthly intervals (Fig. 4), other patients staying in clinical
remission have not progressed and in some cases reverted to normal
values even without further chemotherapy (Fig. 5).

     Although elevated TdT activities as a result of rebounding
normal TdT+ progenitor cells following chemotherapy could explain
this phenomenon, further analysis of TdT activities in ALL in remi-
ssion on and off chemotherapy and comparison to NHL receiving a very
similar treatment protocol (Fig. 6),

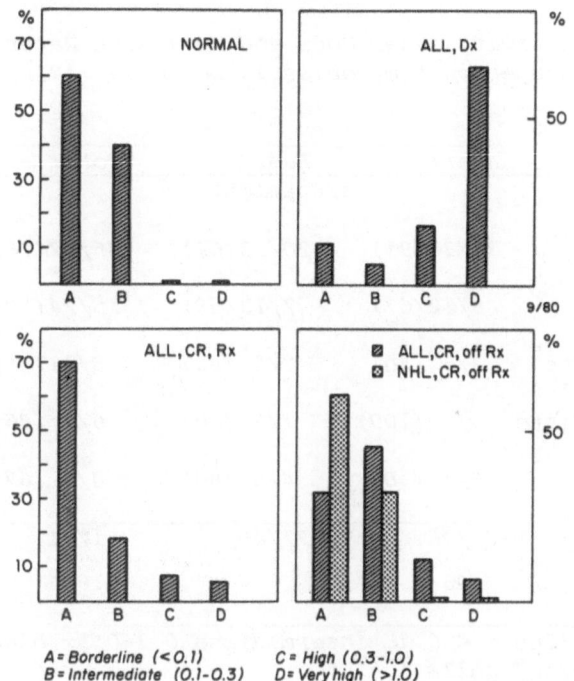

Figure 6   Distribution of TdT Activities in Controls and Patients
with ALL at Diagnosis, and in Remission on and off Chemotherapy and in Patients with NHL after Completion of Chemo-
Therapy

suggest that the observed pattern of elevated marrow TdT activities
in unique to ALL and cannot be explained by chemotherapy alone.
Further clinical follow-up and more detailed studies employing other
marker techniques are required to determine the pathophysiological
and clinical significance of these observations.

TdT in ANLL

Ninety-one patients with ANLL receiving protocol chemotherapy
were analyzed for TdT activity. Thirteen (14%) of these exhibited
elevated TdT activities as seen in ALL. In addition, 11 patients

Table 2.

TdT Activity, Auer Rods and Remission Rates
in Acute Non-Lymphoblastic Leukemia (ANLL)

TdT	Auer+	Auer-	Total
		n/total(%)	
negative	31/34(91)	20/33(61)	46/67(76)
positive	8/12(67)	7/12(58)	15/24(63)
borderline[a]	3/6 (50)	2/5 (40)	5/11(45)
intermediate[a]	5/5 (100)	1/2 (50)	6/7 (86)
high[a]	0/1 (0)	4/5 (80)	4/6 (67)
Total	29/46	27/45	61/91
P[b]	.064		.804

a borderline = < 0.1, intermediate 0.1-0.5, high
>0.5 $U/10^8$ cells.

b P value based on Fisher's exact test comparing
TdT^+ and TdT^- groups

(12%) exhibited low (<0.1 $U/10^8$ cells) but unequivocal TdT activity.
Although these low TdT values were within the range of activities
seen in normal marrows, the presence of TdT activity was still considered abnormal in view of the replacement of normal marrow cells
by leukemic cells which would have been expected to dilute out any
residual normal TdT^+ marrow progenitor cells to non-detectable levels.

Remission rates were lower in the TdT^+ group (15/24, 63%) compared to the TdT^- group (46/67, 66%), but not at a statistically
significant level (Tab. 2). Analysis of clinical and morphological
features failed to reveal any recognizable relationship of elevated
TdT activities to age, sex, FAB classification (data not shown) or
Auer rods (Tab. 2).

We have previously reported that the presence of Auer rods in
patients with ANLL is associated with a better prognosis. Twelve
of 24 patients (50%) in the TdT^+ group and 34 of 67 (51%) in the
TdT^- group exhibited Auer rods. Of the Auer rod positive (A^+), TdT^-

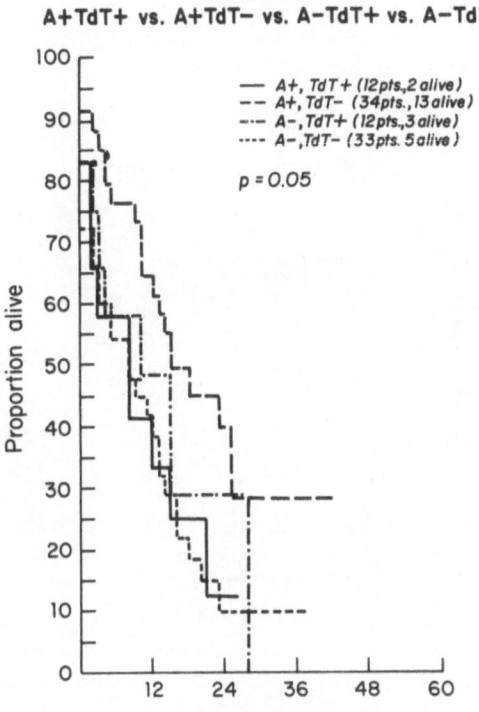

Figure 7 *Survival of Patients with ANLL by Auer Rods and TdT*

Patients, 31/34 or 91% achieved a complete remission compared to 8/12 or 67% in the A^+, TdT^+ group (P=0.06, Tab. 2). Remission durations were also considerably shorter in the TdT^+ patients, with only 1 of 15 patients (6%) continuing in CR as of last follow-up, compared to 13 of 51 patients (26%) in the TdT^- group (P=0.1 by survival curve analysis). These shorter remissions were observed in all TdT^+ patients irrespective of their level of TdT activity (data not shown), although small sample sizes are involved.

Analysis of survival of patients by Auer rods and TdT activities (Fig. 7), revealed a significant difference between TdT^+ and TdT^- patients in the A^+ group (P=0.03), while no difference was observed between TdT^+ and TdT^- cases in the A^- patients (overall P=0.05). Determination of

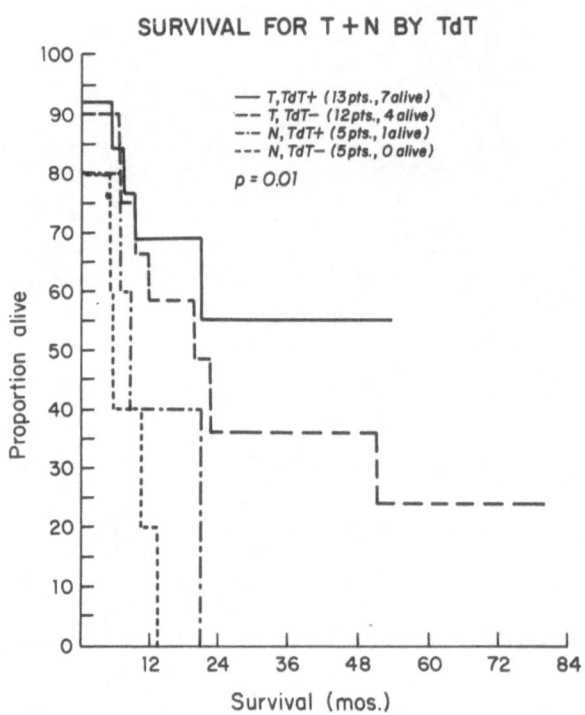

SURVIVAL FOR T + N BY TdT

Figure 8   Survival of Patients with NHL of Diffuse Histological Types Exhibiting T and Null Cell Markers by TdT.

both, Auer rods and TdT activities in patients with ANLL therefore appears to recognize a subgroup of patients with an exceptionally good prognosis (~30% long term survival for A$^+$, TdT$^-$), not recognized by any other clinical or phenotypic feature.

TdT in NHL

While nodular lymphomas exhibit B cell markers in the majority of cases[18], diffuse NHL comprise a broad spectrum of cell phenotypes including TdT$^+$ and TdT$^-$ T and null cell types, B cell as well as histiocytic or macrophage phenotypes. In a recent analysis of 52 patients with diffuse NHL treated with protocol chemotherapy at MSKCC[21], 50% of null cell lymphomas and 52% of T cell lymphomas exhibited high levels of TdT activity.

One of 26 B cell cases in this group, showed 35% monoclonal B cells (Ig MAD, ϰ) and 65% null cells and also exhibited elevated TdT activity (1.65 U/10^8 cells). Unfortunately, no further studies (cell separations, staining for cytogplasmic IgM) could be done in this case to further investigate the nature of the TdT⁺ cells. All other B cell lymphomas studied (total n=81) did not show detectable levels of TdT activity.

Analysis of survival patterns showed a significantly worse prognosis for the TdT⁻ T and null phenotypes as compared to the TdT⁺ T and null phenotypes (Fig. 8). The majority of TdT⁺ cases carried a histological diagnosis of lymphoblastic lymphoma (LBL), which by cell markers as well as in its clinical course closely resembles ALL[22]. The case numbers in each TdT category in the LBL group were relatively small, however, precluding further analysis by levels of TdT activity in this group of patients.

Overall, patients with TdT⁺ NHL tended to be younger (median age 25) compared to the TdT⁻ group (median age 46; P=0.001). Although males and stages IV, which are generally associated with a poorer prognosis in lymphomas[23], were more frequent in the TdT⁺ group, the CR rate was significantly higher in TdT⁺ cases with 13/18 (72%) achieving complete remission compared to the TdT⁻ group with 17/44 (39%) (P=.046).

DISCUSSION

The present analysis of 215 cases of ALL, ANLL, and diffuse NHL with long-term follow-up treated at a single institution on standardized chemotherapy protocols documents for the first time the prognostic significance regarding survival of levels of TdT activity in these disease entities.

The ongoing comprehensive analysis of other prognostic factors in ALL and ANLL including additional, more recently studied patients, suggests that TdT is one of the single most important prognostic parameters in adult patients with acute leukemia. TdT activities appear to be independent of other factors of prognostic significance such as morphological variants (e.g. Auer rods), cytogenetics[24] and clinical variables (unpublished). The observations presented here, if verified in a larger patient population, will be of great help in stratifying patients to different treatment protocols in view of the reproducibility of the TdT assay and its quantitative results.

The biological significance of elevated TdT activities in

patients with ANLL has to be further analyzed. In a recent study of
TdT distribution in ANLL by Bradstock et al.[4], 13 of 63 patients
studied showed a high proportion of leukemic cells (7-90%) staining
for TdT by immunofluorescence. In three cases TdT appeared to be
present in cells distinct from the predominating ANLL cells supporting the concept of biphenotypic leukmeias[12]. In two cases TdT
appeared to be expressed in myeloid cells, suggesting an aberrant
expression of TdT in myeloid cells[4]. In the remaining cases no
further analysis of the distribution of TdT in morphologically
identifiable cells was possible. Patients with ANLL expressing TdT
had a significantly lower CR rate in this study[4], in contrast to our
own observations. Our analysis, however, confirms a poor prognosis
for TdT^+ ANLL but with respect to remission duration and survival,
while no significant difference between CR rates of TdT^+ and TdT^-
cases was seen (see Results). It is unclear at this point whether
differences in technique (biochemical vs. immunofluorescence assay),
treatment protocols or patient population are responsible for this
difference.

    The demonstrated prognostic significance of TdT levels in
patients with ALL, will most likely be highly dependent on the
treatment protocol used. While most other institutions fail to
achieve any long-term remissions in patients with ALL (e.g. [25]), our
protocols for adult ALL have reproducibly achieved long-term remissions (follow-up >10 years in some patients), with >50% of patients
in continuous first remission after 5 years on the more recent
protocols[26].

    After recognizing the clinical and phenotypic similarity between
ALL and LBL[22], all patients with LBL at MSKCC were treated with ALL
protocols, achieving >50% 4-year survival in the TdT^+ T cell group
(Fig 8, cf. [21]), in this, until recently, universally fatal disease
(cf. [27]).

    The observations in diffuse NHL demonstrating the prognostic
significance of TdT in T and null cell types, have to be confirmed
in a larger patient population. It is of interest that null cell
NHL carry a significantly worse prognosis than T cell NHL, while no
difference between the T and null cell variants of ALL was observed
in patients treated with the same protocol[26]. Recent studies using
monoclonal anti T cell antibodies would support the concept that LBL
and ALL are caused by different T cell phenotypes not recognized by
conventional marker techniques[28]. Further study of TdT by biochemical and, in parallel, by immunofluorescence assay are required to
determine whether these cell types also express different levels
of TdT.

    The previously reported observation of intermittently elevated
levels of TdT in marrow samples of patients with ALL in CR, was

further documented in the present study, demonstrating that this phenomenon is only seen in ALL and not in similarly treated patients with NHL. This persistent abnormality of hematopoiesis in patients with ALL in CR even off all chemotherapy might be similar in its pathogenesis to the persisting derrangements of myelopoiesis[29] and of regulator molecule expression[30] persisting in ANLL in CR. So far we have not been able to demonstrate a statistically significant relationship between elevated TdT values in ALL in CR and subsequent clinical course.

The clinical significance of TdT+ phenotypes in the blast crisis of CML, of other MPS and of MDS with a therapeutic response to vincristine and prednisone, has been well documented in many studies[3,5,13,14]. When treated with an intensive ALL-regimen, even long remission can be achieved[14]. Apparently, all chronic clonal disorders of hematopoiesis, i.e. MPS and MDS, that are caused by an early stem cell defect, can develop blast crisis syndromes involving any of the cell lineages of hematopoiesis[13]. As a result, all patients with MDS or MPS in the accelerated or blastic phase of their disease should be evaluated for TdT prior to selection of a chemo-therapeutic regimen.

The indications for performing TdT determinations in patients with leukemias and lymphomas at this point include all acute leukemias, MPS and MDS in accelerated phase or blast crisis and biopsy specimens from patients with lymphomas of diffuse histological types.

In MDS and MPS in blast crisis the selection of treatment will largely depend on the presence or absence of significant numbers of TdT+ cells or of elevated TdT activity, respectively. Further analysis of the prognostic significance of TdT activities in patients with ALL and ANLL with long-term follow-up is needed, before therapeutic stratification can be advocated, because no superior treatment to currently used regimens can be recommended as yet for the poor prognosis groups. Although combinations of ALL and AML-type regimens might be superior for TdT+ ANLL, initial observations do not show a benefit of this approach for most patients[4] (unpublished observations) while occasional patients will respond to such a combination regimen[12].

Based on own observations and published reports by other investigators, TdT continues to be one of the most important phenotypic markers in the clinical evaluation of patients with jematopoietic neoplasias. In addition, analysis of TdT is expected to continue to provide insight into the underlying pathogenetic and pathophysiological mechanisms (cf. [31,32]) of hematopoietic neoplasias.

ACKNOWLEDGEMENTS

I am grateful to Dr. B. Clarkson and Dr. M.A.S. Moore for many

valuable discussions and to Drs. Z. Arlin, T. Gee, S. Kempin, B.
Koziner, L. Murphy, D. Straus, and N. Wollner for allowing me to
study their patients and to Drs. D. Filippa and P.H. Lieberman for
providing lymphnode specimens for TdT analysis. I would like to
thank Ms. Lorna Barnett and Ms. Sa Kurosawa for their expert technical assistance and Ms. Cynthia Garcia for the typing of the manu-
script. This work was supported in part by NCI grant PO1-CA-20194
and the Gar Reichman Foundation.

The kind permission of Thieme Verlag, Stuttgart to reprint
Figures 1, 3-8 from a recently published monograph[11] is appreciated.

REFERENCES

1. R. McCaffrey, C. F. Smoler, and D. Baltimore, Terminal deoxy-
    nucleotidyl transferase in a case of childhood acute lympho-
    blastic leukemia, Proc. Nat. Acad. Sci. USA. 70:521 (1973).

2. F. J. Bollum, Terminal deoxynucleotidyl transferase as a hema-
    topoietic cell marker, Blood 54:1203 (1979).

3. S. M. Marks, D. Baltimore, and R. P. McCaffrey, Terminal trans-
    ferase as predictor of initial responsiveness to vincristine
    and prednisone in blastic chronic myelogenous leukemia, N. Engl.
    J. Med. 298:812 (1978).

4. G. Janossy, A. V. Hoffbrand, M. F. Greaves, K. Ganeshaguru, C.
    Pain, K. F. Bradstock, H. G. Prentice, H. E. M. Kzy and T. A.
    Lister, Terminal transferase enzyme assay and immunological
    membrane markers in the diagnosis of leukemia: a multipara-
    meter analysis of 300 cases, Brit. J. Hemat. 44:221 (1980).

5. K. F. Bradstock, A. V. Hoffbrand, K. Ganeshaguru, P. Llewellin,
    K. Patterson, B. Wonke, A. G. Prentice, M. Bennett, G. Pizzolo,
    F. J. Bollum, and G. Janossy, Terminal deoxynucleotidyl trans-
    ferase expression in acute non-lymphoid leukemia: an analysis
    by immunofluorescence, Brit. J. Hemat. 47:133 (1981).

6. G. Janossy, R. K. Woodruff, M. J. Pippart, G. Prentice, A. V.
    Hoffbrand, A. Paxton, T. A. Lister, C. Bunch, and M. F. Greaves,
    Relation of "lymphoid" phenotype and response to chemotherapy
    incorporating vincristine - prednisone in the acute phase of
    Ph' positive leukemia. Cancer 43:426 (1979).

7. R. Mertelsmann, I. Mertelsmann, B. Koziner, M. A. S. Moore, and
    B. D. Clarkson, Improved biochemical assay for terminal deoxy-
    nucleotidyl transferase in human blood cells: Results in 89
    adult patients with lymphoid leukemias and malignant lymphomas
    in leukemic phase, Leuk. Res. 2:57 (1978).

8. M. J. Modak, R. Mertelsmann, B. Koziner, R. Pahwa, M. A. S. Moore, B. D. Clarkson, and R. A. Good, A micromethod for determination of terminal deoxynucleotidyl transferase (TdT) in the diagnostic evaluation of acute leukemias, J. Can. Res. Clin. Oncol. 98:91 (1980).

9. R. Mertelsmann, D. A. Filippa, B. Koziner, E. Grossbard, J. Beck, S. Gupta, R. A. Good, M. A. S. Moore, P. H. Lieberman, and B. D. Clarkson, Correlation of biochemical and immunological markers with conventional morphology and clinical featues in 120 patients with malignant lymphomas, Adv. Exp. Med. Biol. 114:553 (1979).

10. R. Mertelsmann, B. Koziner, D. A. Filippa, E. Grossbard, G. Incefy, M. A. S. Moore, and B. D. Clarkson, Clinical significance of TdT, surface markers and CFU-c in 297 patients with hematopoietic neoplasias, in: "Modern Trends in Human Leukemia III", R. Neth, R. C. Gallo, P. H. Hofschneider, and K. Mannweiler eds., Springer-Verlag, Berlin-Heidelberg-New York, (1979).

11. R. Mertelsmann, Leukaemien und maligne Lymphome. Phaenotypische und pathophysiologische Untersuchungen und ihre klinische Bedeutung, Georg Thieme Verlag, Stuttgart (1981).

12. R. Mertelsmann, B. Koziner, P. Ralph, D. Filippa, S. McKenzie, Z. A. Arlin, T. S. Gee, M. A. S. Moore, and B. D. Clarkson, Evidence for distinct lymphocytic and monocytic populations in a patient with terminal transferase positive acute leukemia, Blood 51:1051 (1978).

13. R. Mertelsmann, M. A. S. Moore, and B. D. Clarkson, Sequential marrow culture studies and terminal deoxynucleotidyl transferase activities in myelodysplastic syndromes, in: "Preleukemia", F. Schmalzl, and K. P. Hellreigel eds., Springer Verlag, Berlin--Heidelberg-New York, (1979).

14. Z. A. Arlin, K. K. Jain, T. S. Gee, S. J. Kempin, R. S. K. Chaganti, and B. D. Clarkson, Improved response in terminal deoxynuclcotidyl transferase (TdT) positive blastic phase of chronic myelogenous leukemia (CML) treated with an intensive regimen for acute lymphoblastic leukemia (ALL), ASCO Abstract #690, 509 (1981).

15. B. Koziner, D. A. Filippa, R. Mertelsmann, S. Gupta, B. Clarkson, R. A. Good, and F. P. Siegal, Characterization of malignant lymphomas in leukemic phase by multiple differentiation markers of mononuclear cells, Am. J. Med. 62:556 (1977).

16. J. M. Bennett, D. Catovsky, M. T. Daniel, G. Flandrin, D. A. G.
     Galton, H. R. Gralnick and C. Sultan, Proposals for the
     Classification of the acute leukaemias, Brit. J. Hemat.
     33:451 (1976).

17. R. Mertelsmann, H. T. Thaler, L. To, T. S. Gee, S. McKenzie,
     P. Schauer, A. Friedman, Z. Arlin, C. Cirrincione, and B.
     Clarkson, Morphological classification, response to therapy and
     survival in 263 adult patients with acute non-lymphoblastic
     leukemia, Blood 56:773 (1980).

18. D. A. Filippa, P. H. Lieberman, R. A. Erlandson, B. Koziner,
     F. P. Siegal, A. Turnbull, A. Zimring, and R. A. Good, A study
     of malignant lymphomas using light and ultramicroscopic, cyto-
     chemical, and immunological techniques. Correlation with cli-
     nical features, Am J. Med. 64:259 (1978).

19. H. Riehm, H. Gadner, G. Henze, H. J. Langermann, and E. Oden-
     wald, The Berlin childhood acute lymphoblastic leukemia therapy
     study, Am. J. Pediatr. Hematol. Oncol. 2:299 (1980).

20. B. Clarkson, P. Schauer, R. Mertelsmann, T. S. Gee, Z. Arlin,
     S. Kempin, M. Dowling, P. Dufour, C. Cirrincione and J.
     Burchenal, Results of intensive treatment of acute lymphoblastic
     leukemia in adults, in: "Cancer: Achievements, challenges and
     prospects for the 1980s", J. H. Burchenal, and H. F. Oettgen,
     eds., Grune & Stratton, New York, 2:317 (1981).

21. E. Gold, R. Mertelsmann, D. Filippa, B. Koziner, T. H.
     Szatrowski, P. H. Lieberman, T. S. Gee, and B. Clarkson,
     Prognostic significance of surface markers, TdT, clinical para-
     meters and three histopathological classifications in diffuse
     lymphomas, Am. Soc. Hematol. (Abstract) (1981).

22. B. Koziner, R. Mertelsmann, and D. A. Filippa, Adult lymphoid
     neoplasias of T and null cell types, in: "Differentiation of
     normal and neoplastic hematopoietic cells", B. Clarkson P. A.
     Marks, J. E. Till, eds., Cold Spring Harbor Conferences on
     Cell Proliferation, 5:843 (1978).

23. R. I. Fisher, S. M. Hubbard, V. T. DeVita, C. W. Berard, R.
     Wesley, J. Cossman and R. C. Yound, Factors predicting long-
     term survival in diffuse mixed, histiocytic, or undifferentiated
     lymphoma, Blood 58:45 (1981).

24. M. G. Kris, R. Mertelsmann, S. Jhanwar, R. Chaganti, T. H.
     Szatrowski, T. S. Gee, S. Kempin, Z. Arlin, and B. Clarkson,
     Prognostic significance of pretreatment marrow cytogenetic stu-
     dies in acute non-lymphocytic leukemia (ANLL), Am. Soc. Hemat.
     (Abstract), (1981).

25. *E. J. Freireich, M. J. Keating E. A. Gehan, K. B. McCredie, G. P. Bodey, and T. Smith, Therapy of acute myelogenous leukemia, Cancer 42:874 (1978).*

26. *P. Schauer, Z. Arlin, R. Mertelsmann, C. Cirrincione, A. Friedman, T. S. Gee, M. Dowling, S. Kempin, D. J. Straus, B. Koziner, S. Mackenzie, H. T. Thaler, P. Dufour, C. Little, C. Dellaquilla, S. Ellis and B. Clarkson, Treatment of acute lymphoblastic leukemia in adults. Results of the L-10 and L-10M protocols, submitted for publication.*

27. *C. N. Coleman, J. R. Cohen, J. S. Burke and S. A. Rosenberg, Lymphoblastic lymphoma in adults: Results of a pilot protocol, Blood 57:679 (1981).*

28. *A. Bernard, L. Boumsell, E. L. Reinkerz, L. M. Nadler, J. Ritz, H. Coppin, Y. Richard, F. Valensi, J. Dausset, G. Flandrin, J. Lemerle, and S. F. Schlossman, Cell surface characterization of malignant T cells from lymphoblastic lymphoma using monoclonal antibodies: Evidence for phenotypic differences between malignant T cells from patients with acute lymphoblastic leukemia and lymphoblastic lymphoma, Blood 57:1105 (1981).*

29. *M. A. S. Moore, Prediction of relapse and remission in AML by marrow culture criteria, Blood Cells 2:109 (1976).*

30. *H. E. Broxmeyer, E. Grossbard, N. Jacobsen and M. A. S. Moore, Persistence of inhibitory activity against normal bone-marrow cells during remission of acute leukemia, N. Engl. J. Med. 301:346 (1979).*

31. *R. Mertelsmann, S. Gillis, B. Koziner, and M. A. S. Moore, Abnormal T cell growth factor (Interleukin 2) response pattern in human leukemias exhibiting high terminal transferase activity, in: "Differentiation factors in cancer", M. A. S. Moore, eds., Raven Press, New York, (1981) in press.*

32. *R. Mertelsmann, S. Gillis, G. Steinmann, P. Ralph, M. Stiehm, B. Koziner, and M. A. S. Moore, T-cell growth factor (Interleukin-2) and terminal transferase in human leukemias and lymphoblastic cell lines, Blut, (1981) in press.*

CLINICAL RELEVANCE OF TERMINAL TRANSFERASE AND ADENOSINE DEAMINASE IN LEUKEMIA

E. Brusamolino[1], U. Bertazzoni[2], P. Isernia[1], E. Ginelli[3], A. I. Scovassi[2], M.G. Zurlo[4], P. Plevani[3], N. Sacchi[3] and C. Bernasconi[1]

[1]Divisione di Ematologia, Policlinico S. Matteo, Pavia
[2]Istituto CNR Genetica Biochimica Evoluzionistica, Pavia
[3]Istituto di Biologia Generale, Università di Milano
[4]Clinica Pediatrica De Marchi, Università di Milano

SUMMARY

Terminal Transferase (TdT), Adenosine Deaminase (ADA), immuno-logical membrane markers, cytochemical reactivity and cytogenetics were analyzed in 226 patients with ALL, AUL and AML, in 70 patients with CML and in 3 cases of Ph' positive acute leukemia presenting as ALL.

TdT was tested in peripheral blood and bone marrow with both the biochemical and immunofluorescence (IF) methods, and ADA was determined biochemically only in peripheral blood cells.

By using conventional cytochemistry, cell surface markers de-terminations, TdT and ADA analysis, three distinct groups are recog-nized in ALL at presentation: T-ALL with TdT+ and very high ADA values; non-T, non-B ALL with TdT+ and intermediate levels of ADA; B-ALL with TdT absence and low levels of ADA.

Clinical presentation and response to therapy in adult and chil-dren ALL were correlated to TdT determinations. The median survivals in adults, calculated for TdT+ and TdT- groups, were 14.2 and 5.6 months, respectively.

TdT and ADA were determined in ALL during remission. The wide fluctuation observed for TdT IF and ADA values prevented a reliable monitoring of remissions. At relapse, TdT and ADA values were similar to those found for ALL at presentation; TdT IF determinations were diagnostic in cases showing CNS involvement as the only localiza-tion.

279

Forty per cent of AUL and 11% of AML cases were positive for TdT; the medians of ADA values of the TdT+ cases in both AML and AUL were several times higher than those obtained in the TdT- group. While TdT positivity and high ADA had a favorable prognostic value in AUL, similar conclusions can not be drawn at the moment for AML.

In chronic phase of CML, TdT was strictly negative and ADA values were increased over the control line only in cases showing initial signs of transformation. In acute phase, the cases positive for TdT (32%) presented a significantly higher ADA activity than the TdT negative ones. The actuarial survival curves for the TdT+ and TdT- groups differ significantly, presenting median survivals from onset of acute phase of 11 and 4.8 months, respectively. The three cases of Ph' positive ALL were all TdT+, presented high ADA values and entered chronic phase of CML after therapy.

INTRODUCTION

Enzyme assays have recently become available as useful aids in the biochemical characterization of leukemic blasts. Among these, Terminal Transferase (TdT) is now considered a very specific marker for immature lymphoid cells (review by Bollum, 1979).

Purified antibodies to TdT have recently become available for the analysis of individual cells in bone marrow (Bollum, 1975) and have proved to be particularly valuable for the identification of neoplastic cells in leukemic sanctuaries (Janossy et al., 1980).

High levels of the enzyme have been found in blast cells of ALL (McCaffrey et al., 1973; Coleman et al., 1976; Hoffbrand et al., 1977), in approximately 30% of patients with Chronic Myelogenous Leukemia (CML) in blastic transformation (Sarin et al., 1976; Marks et al., 1978). The presence of TdT has also been described in a very low percentage of adult Acute Myeloid Leukemia (Srivastawa et al., 1976), in one case of Ph' negative acute leukemia with mixed lympho-myelomonoblastic morphology (Mertelsmann et al., 1978). More recently, a new phenotype of ALL, called pre-B, was found to be positive for Terminal Transferase (Vogler et al., 1978; Brouet et al., 1979).

Adenosine Deaminase (ADA), which catalyzes the deamination of adenosine to inosine, is found in most animal tissues. The congenital absence of ADA results in a hereditary disease called Severe Combined Immunodeficiency. A subnormal level of the enzyme has been reported in Chronic Lymphocytic Leukemia (Tung et al., 1976; Ramot et al., 1977) whereas in blasts of adult and childhood acute leukemia the levels of ADA were raised with few exceptions (Smyth and Harrap, 1975; Meier et al., 1976).

In the present work, the correlation between TdT and ADA in

acute leukemias has been investigated. This analysis should permit
a more reliable distinction between lymphoid and myeloid leukemias,
allow the attribution of the unclassifiable blasts to their lymphoid
or myeloid origin, and contribute to the clinical follow-up of complete remissions.

We have also analyzed TdT and ADA in adult Ph'+ leukemias for
monitoring the chronic phase of CML, obtaining prognostic information in CML blastic phase and characterizing the cases of Ph'+ acute
leukemias presenting as ALL. TdT and ADA analyses were evaluated in
connection with conventional cytology, cytogenetics, response to
therapy and overall survival.

MATERIALS AND METHODS

Patients Selection

The study population was as follows: 119 adult patients affected
with acute leukemia at presentation; 63 children affected with acute
leukemia at presentation; 40 cases with ALL at the time of clinical
relapse; 90 cases with acute leukemia in clinical remission; 42
adult patients with Ph'+ CML in chronic phase; 28 adult patients
with Ph'+ CML in blastic phase and 3 cases of Ph'+ acute leukemia
presenting as ALL; 15 non-neoplastic human controls.

All adult patients involved in this study were admitted to the
Division of Hematology of Policlinico San Matteo in Pavia between
January 1979 and March 1981. All children were admitted to the De
Marchi Pediatric Clinic of the University of Milan between January
1979 and March 1981.

Analysis of results was blind; clinical features, cytochemistry,
surface markers and enzyme assay were analyzed separately by different investigators.

The classification of acute leukemia was based on conventional
morphology and cytochemical findings, as stated in 1976 by an international panel (FAB classification, Bennett et al., 1976).

Membrane Markers

Surface Immunoglobulins (SIg) and E-rosettes were usually evaluated in peripheral blood and occasionally in bone marrow. For the
determination of surface immunoglobulin positive cells, 5×10^6
cells were suspended in 100 µl of a 1:8 dilution of FITC-conjugated
goat antihuman antiserum and incubated for 15 min at room temperature
and 30 min at 4°C. The cells were washed three times with PBS and
fluorescent cells counted with an epi-illuminated Leitz Orthoplan
microscope, equipped with an HBO 50 W mercury lamp, a 100x oil ob-

jective and a selective filter for FITC.

For the detection of E-rosette positive cells (E+ blasts) 5 x 10^6 cells, suspended in 0.4 ml of adsorbed fetal calf serum, were added to 0.4 ml of 0.5% washed sheep red blood cells in PBS. The suspension was incubated at 37°C for 30 min, spun into a pellet and kept at 4°C for 3 hours. The cells were then resuspended by gently rolling the tube and the percent of E-rosette forming cells was calculated by counting only the cells with three or more adherent erythrocytes.

Reagents

Monospecific rabbit antibodies to homogeneous calf TdT (R-anti-TdT) and purified goat anti-rabbit IgG antibody coupled to fluorescein isothiocyanate (FITC) were donated by Prof. F. Bollum. Goat anti-human immunoglobulins conjugated to FITC were obtained from Behring-werke. Poly d(pA)$_{50}$ was a gift of P. Plevani. ^{3}H dGTP was purchased from Amersham, and GF/C and DE 81 papers were obtained from Whatman. Nucleosides and nucleotides were from Boehringer.

TdT Biochemical Assay

Mononuclear cells, separated on Phicoll-Hypaque gradient, were washed with PBS, counted and frozen at -20°C. The extraction was made by resuspending the cells at a density of 1-2 x 10^8 in 0.25 M K phosphate, 1 mM mercaptoethanol and sonicating 2-4 times for 15 sec with an MSE microtip sonicator set at 10-20 mA. The cellular debris was removed by centrifugation at 40,000 RPM for 60 min at 2°C and the supernatant was used as enzymatic extract.

The reaction mixture contained in a vol. of 0.25 ml: 0.2 M K cacodylate pH 7.5, 0.5 mM ^{3}H dGTP (100 cmp/pmole), 4 mM MgCl$_2$, 0.01 mM poly d(pA)$_{50}$, 1 mM mercaptoethanol, and enzyme extract. At different time intervals, 50 µl aliquots of the reaction were spotted on GF/C and treated for acid insoluble radioactivity. One unit of TdT activity catalyzes the incorporation of 1 nmole of nucleotide in 1 hour.

ADA Biochemical Assay

The enzymatic extract was the same as that prepared for TdT from peripheral blood (see above). The assay measures the conversion of adenosine to inosine and is essentially as described by Coleman and Hutton (1975). The reaction mixture contained in a vol. of 0.1 ml: 0.05 M K phosphate pH 7.5, 0.6 mM ^{14}C adenosine (6000 cpm/nmole), 1 mM 2-mercaptoethanol and enzyme extract. At 3, 6, 9, 15 min, aliquots of the reaction were placed on strips of DE 81 chromatography paper previously spotted at the origins with 5 µl of 100 mM unlabeled inosine. The spots were immediately dried and developed by ascending liquid chromatography on 1 mM ammonium formate for about 90 min. After careful drying, the inosine spot was localized with the aid of an UV lamp, cut out and counted. One unit of enzyme activity corresponds to 1 nmole of inosine produced in 1 min. The mean value obtained from the mononucleated cells of the peripheral blood of 10 adult controls was 113 ± 16 U/10^8cells.

TdT Immunofluorescence (IF)

The assay was performed essentially as described by Stass et al. (1979). Testing was made on smears of bone marrow aspirates, bone marrow touch preparations, cytocentrifugal spread of cerebrospinal fluid, which were air-dried and kept at room temperature for no longer than 10 days. The slides were fixed in methanol at 4°C for 30 min. After drying, primary antibody (R-anti-TdT) was placed on a selected area and the slide incubated for 30 min at room temperature in a humidified chamber. The slides were washed 3 times for 5 min in PBS and reincubated for 30 min in a humid chamber after the addition of the secondary antibody (FITC goat anti-rabbit IgG). After washing three times in PBS for 5 min, the slides were mounted in PBS containing 10% glycerol. The specimens were examined with a Leitz Orthoplan microscope equipped with a 40 Phaco objective and IR filter for FITC. The field was observed under phase microscopy for counting the total number of white cells and epifluorescent light for scoring the percentage of cells with fluorescent nuclei.

Cytogenetics

Chromosomal analyses were performed using Q-, GT- or RB-banding techniques either on direct preparations of bone marrow and/or on blood cultures with and without phytohemagglutinin (Pasquali et al., 1979). Analyses were performed at the Istituto di Biologia Generale and Genetica Medica, University of Pavia.

Statistics

The actuarial survival curves were calculated according to Berkson and Gage (1950). Means are given with standard errors. Student's t was used to test the significance of the differences between means after log transformation of the variables. Significance of the differences between actuarial survival curves was calculated by the method of F-distribution.

RESULTS

TdT and ADA in Acute Leukemias

The 119 cases of adults with acute leukemia at presentation were defined by conventional methods as Acute Lymphoblastic Leukemia

(ALL, 45 cases), Acute Myeloid Leukemia (AML, 59 cases) and Acute Unclassifiable Leukemia (AUL, 15 cases). The 63 cases of childhood acute leukemia presented as ALL (55 cases), AML (6 cases) and AUL (2 cases).

In Table 1 we have summarized the results of surface markers analysis and its correlation to FAB classification in ALL.

Table 1. Immunological Classification in Childhood and Adult ALL and Correlation with FAB Classification.

	Total (n=100)	Children (n=55)	Adults (n=45)	FAB (adults+children) L_1	L_2	L_3
ALL subgroups						
T-ALL	18	11(20%)	7(15%)	6	12	
B-ALL	4	1(1.5%)	3(6%)			4
non-T, non-B ALL	78	43(78.5%)	35(79%)	43	33	2

In Table 2 the cytochemical reactivity is correlated to immunological subgroups of ALL and to the presence of TdT. Of interest is the observed difference in the pattern of positivity for Acid Phosphatase and the absence of acid esterase in all cases negative for TdT.

The values of TdT and ADA enzymatic activities obtained in acute lymphoid leukemias are reported in Fig. 1, and the results of TdT immunofluorescence (IF) are shown in Fig. 2. TdT was tested in peripheral blood and bone marrow with both the biochemical and/or the IF method in adults and mainly by the IF test in children. ADA was determined only in peripheral blood cells. In adults, all T-ALL (7 cases) were positive for TdT whereas the three B-ALL cases were TdT negative; the group of non-T, non-B ALL (28 cases) was TdT positive with the exception of 2 cases in which no fluorescent cells were found in bone marrow. In children, TdT determinations were positive in 8 of the 9 cases with T-ALL and in 35 of the 41 cases with non-T, non B ALL; the single B-ALL case was TdT negative.

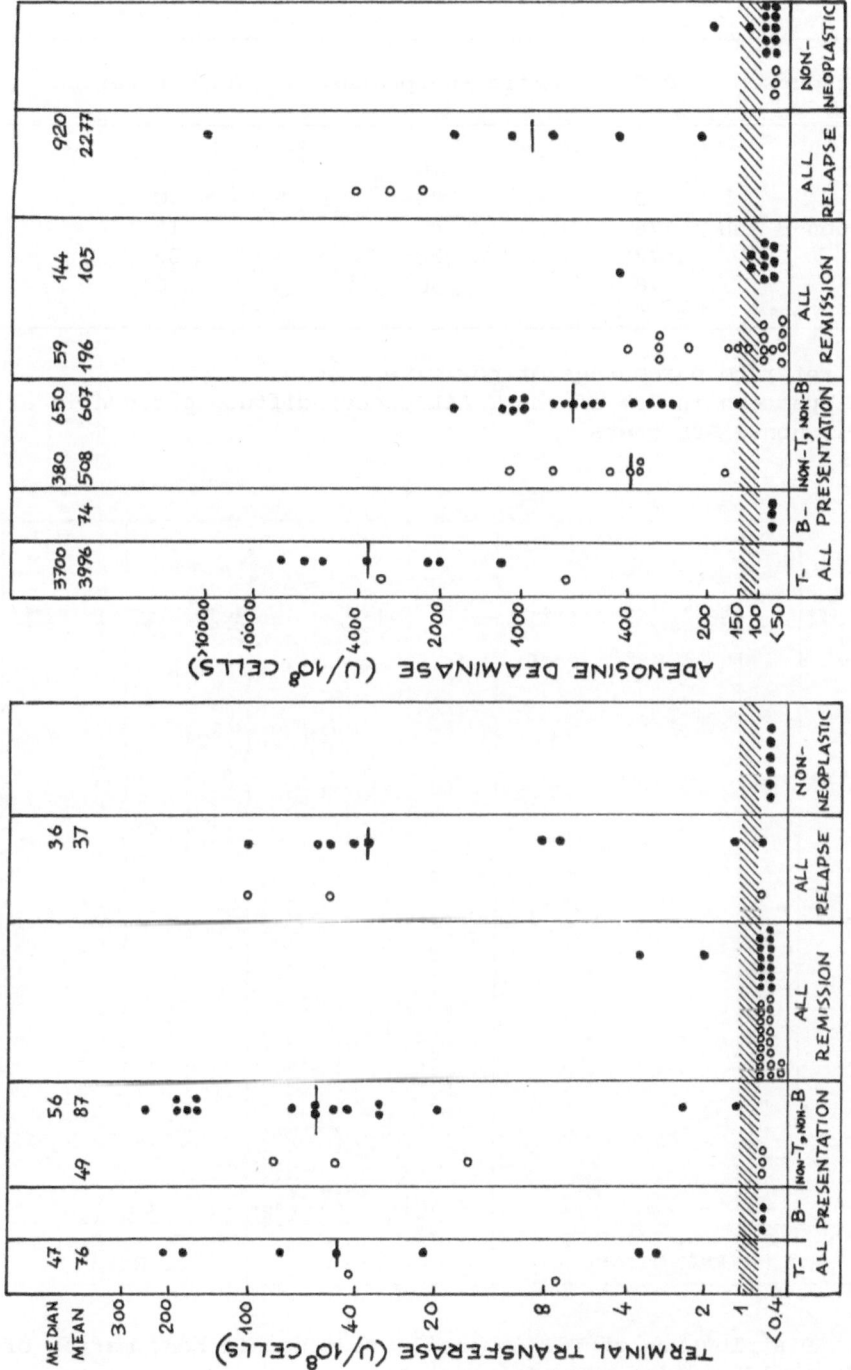

Fig. 1. Distribution of TdT and ADA enzymatic activities in childhood (o) and adult (•) ALL.

Table 2.   Cytochemical Reactivity in Immunological ALL Subgroups
           (Adults and Children)

ALL Subgroups	PAS	Acid Phosphatase	Acid Esterase
T-ALL	60	70	50
B-ALL	20	25	0
non-T, non-B ALL	78	47	15
TdT+	72	58	32
TdT-	76	30	0

Data refer to percentage of positive cases.
Focal pattern in 75% of the T-ALL cases; diffuse pattern in all
non-T, non-B ALL cases

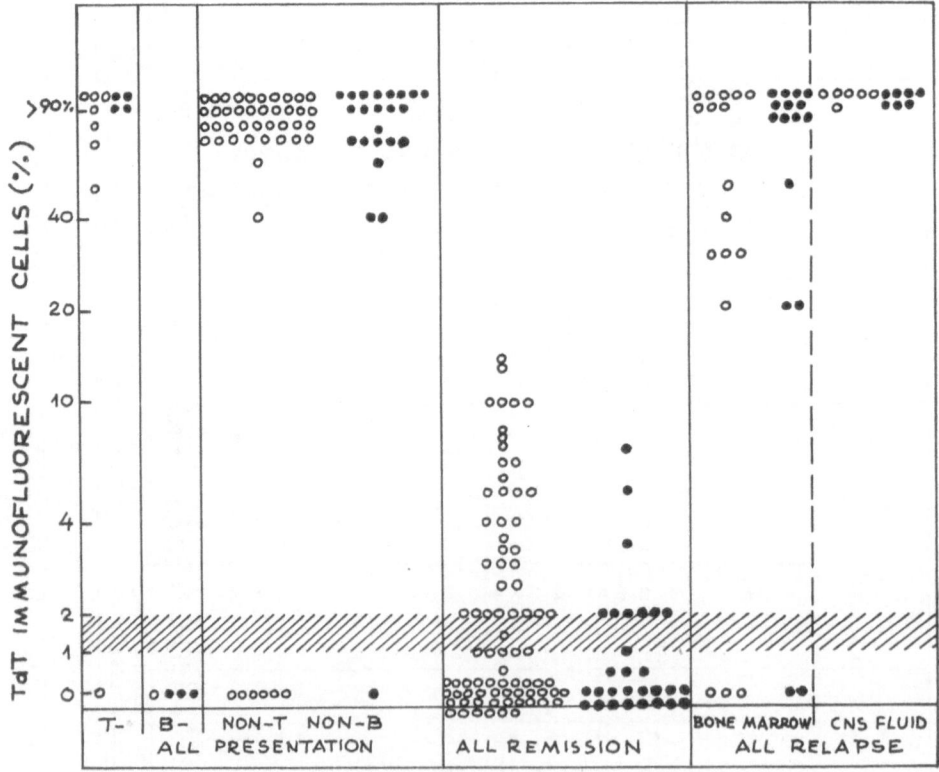

Fig. 2.   Distribution of TdT immunofluorescence in bone marrow of
          childhood (o) and adult (•) ALL.

A close correlation between enzymatic and IF assays was found. In the positive cases the percent of fluorescent cells ranged between 40 and 100% of total nucleated cells.

The ADA enzymatic assay was performed in 27 adults and 9 children with ALL (see Fig. 1). The range of ADA varied between 20 and 7580 U/10^8 cells, with a mean of 1458±393 U/10^8 cells in adults and 745±334 U/10^8 cells in children. These values were significantly higher (P<0.001) than those found in non-neoplastic controls (113±16; see Methods and Fig. 1).

The mean value of ADA in adult T-ALL (3996±931 U/10^8 cells) is significantly higher (P<0.001) than the mean value of 607±78 U/10^8

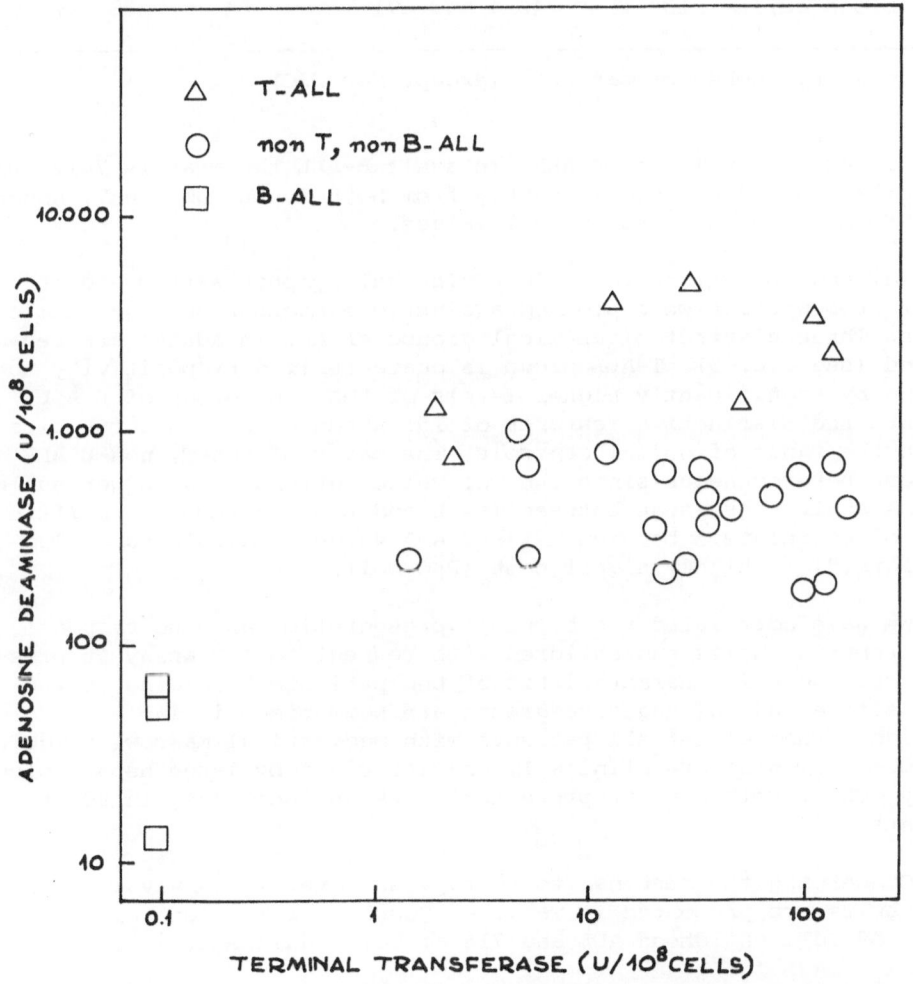

Fig. 3. Correlation between TdT and ADA in adult ALL at presen-

Table 3.   Relationships Between TdT and Clinical Presentation in
           Childhood and Adult ALL

Features	Children		Adults	
	TdT+	TdT-	TdT+	TdT-
	46(84%)	9(16%)	41(91%)	4(9%)
Mediastinal enlargement	10(22%)	1(11%)	7(17%)	O
Lymph node enlargement	22(48%)	3(33%)	19(46%)	2(50%)
Liver and/or spleen enlarg.	25(54%)	8(88%)	30(73%)	3(75%)
CNS involvement	O	O	5(12%)	1(25%)
White blood cells ($\times 10^9$/l)	42	71	64	38

Data refer to number of patients (except for WBC)

cells in adult non-T, non-B ALL. In adult B-ALL the mean is 74 ± 27 U/
10^8 cells and differs significantly from both T-ALL and non-T, non-B
ALL (P<0.001) but not from normal values.

     TdT and ADA values found in peripheral lymphoblasts of adult
ALL at presentation were plotted against one another on logarithmic
scales. Three distinct biochemical groups of ALL in adults are recognized (see Fig. 3). T-ALL group is characterized by positivity to
TdT and by significantly higher levels of ADA; the group of B-ALL
presents the distinctive features of TdT absence and ADA activities
within the range of normal controls. The cases of non-T, non-B ALL
are more heterogeneous since the TdT values extend on a larger scale
and ADA activities range between T-ALL and B-ALL cases. The coefficient of correlation between TdT and ADA values, calculated on log
transformed, is highly significant (P<0.001).

     We have correlated the clinical presentation and the response
to therapy in adults and children with respect to TdT assay at presentation. The main characteristic of the patients belonging to the
TdT positive and TdT negative groups are summarized in Table 3. It
is evident that almost all patients with mediastinal mass were TdT+.
The TdT- patients were clinically characterized by large hepatospleno-
megaly. CNS involvement at presentation was present only in adult
patients.

     Concerning the response to therapy and overall survival in ALL,
the results are presented in Table 4. Complete remission was achieved
in 93% of TdT+ childhood ALL and 71% of TdT+ adult ALL.

     No difference in Complete Remission (CR) rate and overall sur-

Table 4. Response to Therapy in Childhood and Adult ALL. Evaluation
          by TdT

| | Children | | Adults | |
	TdT+	TdT-	TdT+	TdT-
n. of patients	46	9	41	4
Median follow-up (mos)	15	14	16	12
Range (mos)	2-24	2-22	2-28	2-20
Number of complete remissions	43(93%)	8(88%)	29(71%)	1(25%)
Number of partial remissions	3(7%)		7(17%)	0
Remission death or no response	0	1	5(12%)	3
Number of B.M. relapses	3(7%)	1	17(58%)	0
Number of CNS relapses	7(16%)	0	3(10%)	0

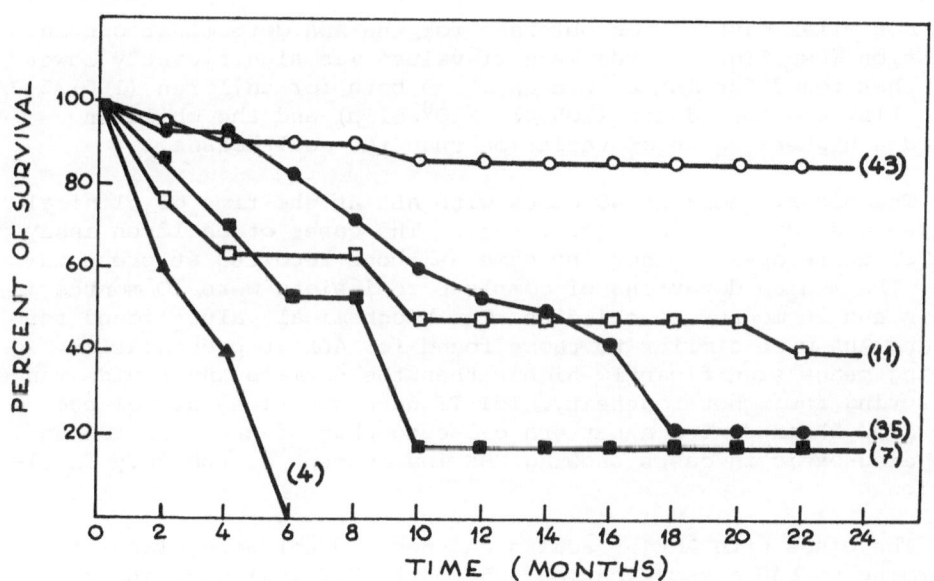

Fig. 4. Actuarial survival curves for adults and children with
         ALL of different immunological subtypes.
         Onon-T, non-B ALL, children; ●non-T, non-B ALL, adults
         ◘T-ALL, children; ■ T-ALL, adults; ▲B-ALL
         Number of patients is given in parenthesis

vival was observed between TdT+ and TdT- childhood leukemias.

We also calculated the survival curves for adults and children with ALL according to their immunological subclassification (see Fig. 4): significant differences were seen between adult and childhood ALL and between non-T, non-B and T-ALL subgroups.

TdT and ADA were determined in ALL at the time of the first hematological remission and during remission, while continuing maintenance chemotherapy. The TdT biochemical activity in peripheral blood was non-detectable and in bone marrow fell within normal control values (see Fig. 1); in 2 adults showing only partial remission the TdT activity presented values of 2 and 3.5 U/10^8 cells. As shown in Fig. 2 in all cases of adults with apparent complete remission, the percent of TdT fluorescent cells was below 2% with few exceptions: one case with 7% TdT+ cells overtly relapsed four months after the test; two cases showing 3-5% TdT+ cells during a phase of chemotherapy induced aplasia have not yet relapsed at the present time.

In children the fluctuation was much wider, ranging from 0 to 13%; however no direct correlation was found between cases showing TdT positivity higher than 2% and the observed relapses. The 10 cases of children off therapy showed TdT IF values of 1-2%.

A similar pattern was obtained for the ADA determinations in remission (see Fig. 1): the mean of values was significantly lower than that found for ALL at presentation both for children (196±36 U/10^8 cells) and for adults (105±41 U/10^8cells) and the children showed a higher degree of variation than the adult cases.

The observations in 40 cases with ALL at the time of clinical relapse are reported in Figs. 1 and 2. The cases of children assayed for TdT at relapse are not the same as those reported at presentation. The median durations of complete remissions were 10 months in adults and 26 months in children. The biochemical values found for TdT and ADA were similar to those found for ALL at presentation, showing means significantly higher than the normals and a wide range reflecting their heterogeneity. TdT IF determinations at relapse (see Fig. 2) presented a pattern close to that of presentation and were diagnostic in cases showing CNS involvement as the only localization.

The cases with AML(59 adults and 6 children) were classified according to FAB classification and to the TdT analysis, as summarized in Table 5.

Chromosomal analysis was carried out in 35 patients and anomalies were observed in 6 patients. Two cases presented loss of Y and translocations 8, 21 and four had more complex anomalies including monosomy X, loos of 2, trisomy 7 and variable structural re-

Table 5. TdT in Adult and Childhood AML

FAB classif.	% distribution	TdT+ cases	TdT- cases
M_1	14	2	7
M_2	38	1	24
M_3	8		5
M_4	26	3	14
M_5	11	1	6
M_6	3		2
Total	100	7 (11%)	58 (89%)

arrangements.

TdT analysis was performed by using the biochemical assay on peripheral blood and bone marrow of 35 patients, the IF test on 44 patients and both tests on 17 patients (see Fig. 5). Five adults and two children were found to be positive for TdT, representing 11% of the tested population. Immunofluorescence analysis was strictly negative in all other patients studied, except in a single case of childhood AML in relapse showing 5% fluorescent cell in bone marrow. The IF test, performed in 12 cases of AML in complete remission gave less than 2% fluorescent cells in bone marrow.

The main characteristics of TdT+ AML at presentation are summarized in Table 6. The TdT+ adult patients were treated according to a protocol for Acute Myelo-Monocytic Leukemias with a four-drug combination chemotherapy consisting of daunorubicin, cytosine arabinoside, 6-thioguanine and VP-16213 (Bernasconi et al., 1976). One case went into complete remission lasting 8 months, then relapsed and lost TdT positivity.

Fifteen cases of Acute Leukemias unclassifiable at presentation with conventional methods (AUL) were also tested for TdT. Five cases found TdT positive (see Fig. 5) were classified as lymphoid and treated with ALL therapy; 10 TdT negative cases were classified as myeloid and treated with AML therapy.

The analysis of ADA in AML and AUL at presentation is reported in Fig. 5 by subdividing ADA values into two different lanes according to their positivity or negativity to TdT. Although the number of TdT+ cases is low, it appears that the medians of ADA activity of the TdT+ cases in both AML and AUL are about three times higher than those obtained in the TdT- cases.

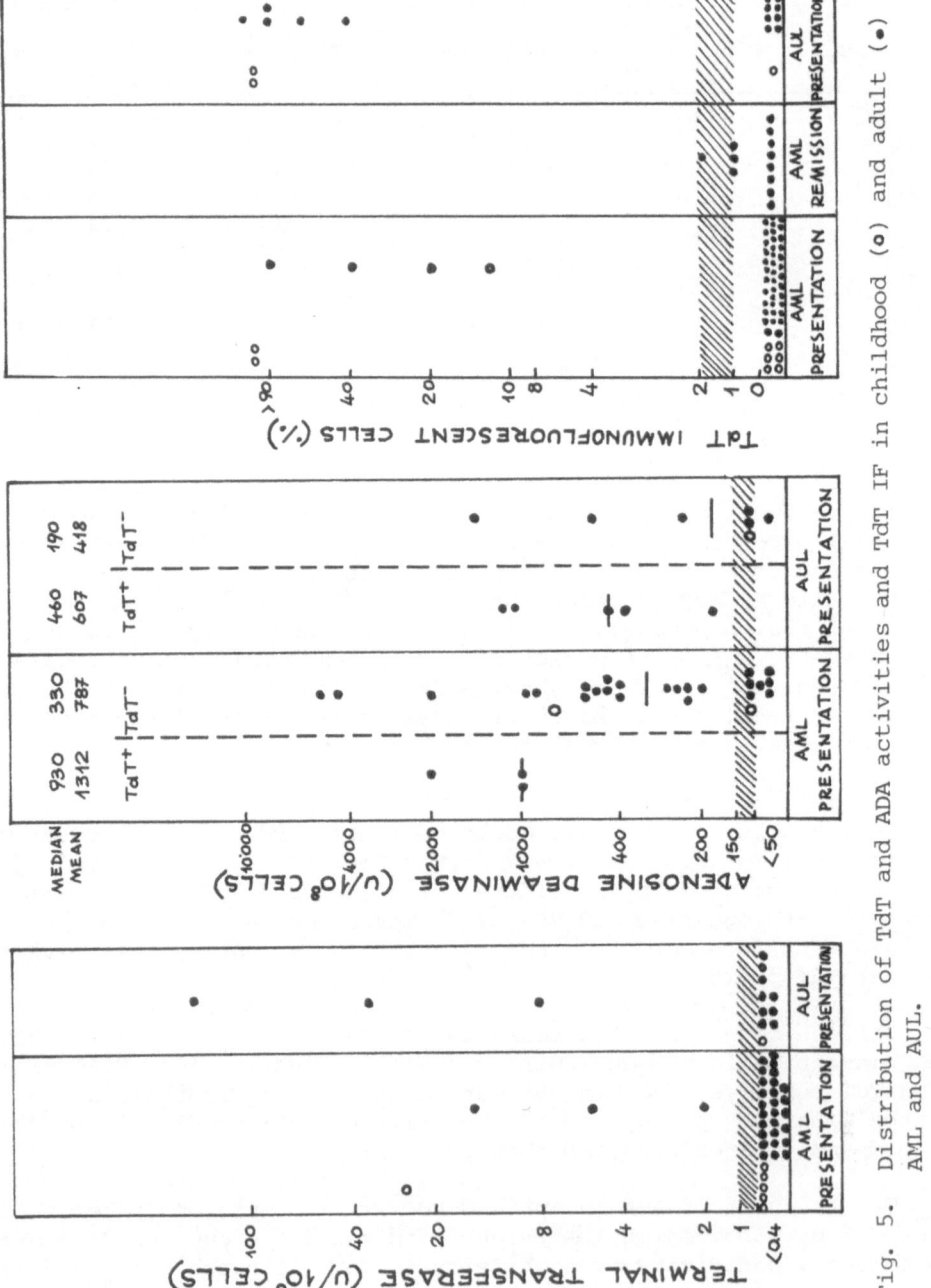

Fig. 5. Distribution of TdT and ADA activities and TdT IF in childhood (o) and adult (•)
AML and AUL.

Table 6. Characteristics of TdT Positive AML at Presentation.

Case	Sex	Age (yrs)	FAB class.	Auer rods%	Cytogenetics	TdT ($U/10^8$)	TdT IF	ADA ($U/10^8$)	Therapy	Response to ther.	Survival (mos)
ADULTS
1	M	33	M_4	3	46xy	5.2	20	927	DR,CA, TG,VP	CR	11
2	M	77	M_4	2	46xy	1.8	12	2100	CA,TG,VP	NR	1
3	M	62	M_4	0	46xy	14.5	0	910	CA,TG,VP	NR	1
4	M	29	M_2	10	46xy		40		DR,CA,TG	CR	5+
5	M	50	M_5	0	46xy		60		DR, CA, TG,VP	Induct. therapy	1+
CHILDREN
1	M	9	M_1	0	ND		50			NR	4+
2	F	9	M_1	0	ND		80			CR	2+

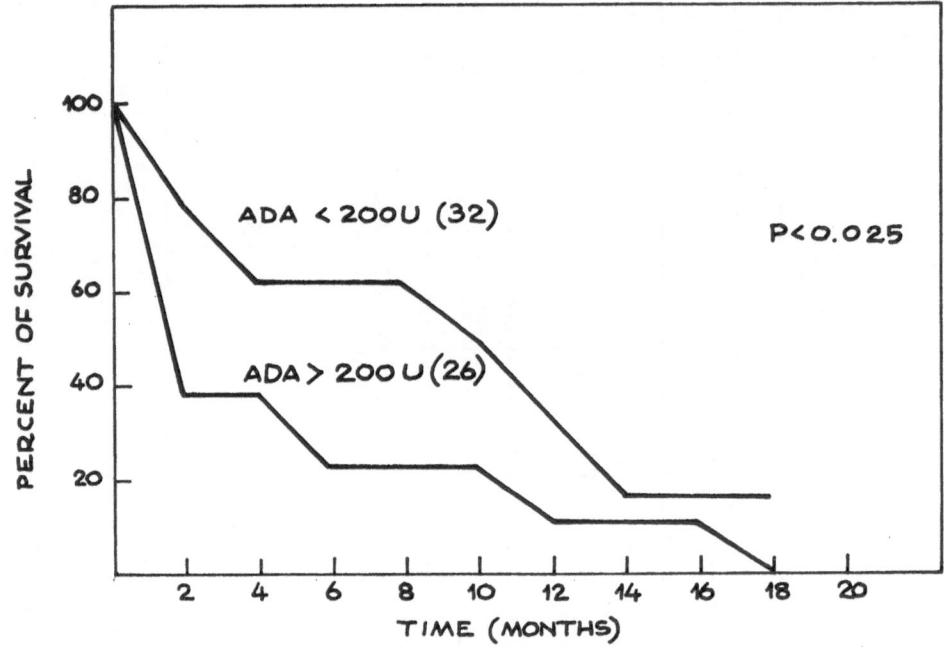

Fig. 6. Actuarial survival curves for AML patients according to
ADA. Number of patients is given in parenthesis.

We have also calculated the survivals for the two groups of
AML subdivided according to the levels of ADA activity: higher or
lower than 200 U/10^8 cells, respectively. The actuarial survival
curves given for the two groups (see Fig. 6) indicate that the patients with low levels of ADA present a significantly longer survival than the patients with higher levels of ADA.

TdT and ADA in Ph' Positive Leukemias

Morphological and cytogenetical analyses were performed in 42
cases of CML in stable phase showing the presence in all cases of
Philadelphia chromosome and in 3 cases of additional chromosomal a-
nomalies: three chromosome rearrangements involving, in addition to
9 and 22, chromosomes 2 and 6 (2 cases) and loss of Y (one case).

TdT and ADA were analyzed biochemically (see Fig. 7) in the
peripheral blood of 18 and 25 cases, respectively; the TdT IF test
was performed in 22 cases. The TdT enzymatic assay was strictly
negative, giving values below 0.4 U/10^8 cells and the IF values in
bone marrow were below 1%, with the exception of 3 cases ranging
between 2 and 4%.

The average value of ADA activity in stable phase was 103±15 U/
10^8 cells, which is very close to that obtained for 10 non-neoplastic

Table 7.   Acute Phase of Ph' Positive Leukemia. Evaluation by TdT

Clinical Features	TdT+ Group	TdT- Group	Sign. Diff.
n. of patients	10(32%)	21(68%)	
Lymphoid morphology	7	2	
Myeloid morphology	0	13	
Mixed morphology	3	6	
Additional chromos. anomalies	3	6	
ADA activity (mean U/10^8 cells)	1753	274	P<0.001
Response to V and P	9	0	
Duration of chronic phase (mos)	31	34	
Median survival from onset (mos)	11	4.8	P<0.01

controls (113±16 U/10^8 cells). The values of ADA were increased over
the control line only in 2 cases showing the early signs of transformation.

The clinical features of 31 adult patients in acute phase of
Ph'+ leukemia are given in Table 7. The morphology was lymphoid in
9 cases, myeloid in 13 cases; in 9 cases it presented a mixed type
since in peripheral blood and bone marrow granulated blasts with
clear lymphoid morphology and agranulated blasts with lymphoid appearance were present.

Cytogenetical analyses showed that nine cases had, in addition
to the standard 9/22 translocation, other single or multiple anomalies
in various combinations, including: three-chromosome rearrangements
involving chromosomes 2, 3 and 6 (3 cases); duplication of Ph' (2
cases); isochromosome for the long arm of 17 (1 case); monosomy 7
(1 case); loss of Y (1 case) and a complex rearrangement consisting
in the presence of Ph' chromosome with translocation (5;22), trisomy
8 and isochromosome for the long arm of 17 (1 case).

Twenty-three patients were analyzed for TdT by the enzymatic
assay, 21 patients by the IF test and 10 by both methods. Ten of 31
patients (32%) were found positive for TdT. These included 7 patients
developing blastic transformation following CML and 3 patients with
no history of chronic phase and presenting as Ph+ ALL. The mean value
of TdT enzymatic activity obtained in bone marrow of TdT+ cases was
82±34 U/10^8 cells. This value is close to the mean value found in a-
dult ALL at presentation. The percent of bone marrow cells with nuclear fluorescence ranged between 15 and 40% of total nucleated cells.
In the TdT positive cases, the correlation between morphology by
standard methods and TdT assay shows that 7 of 10 cases were clearly

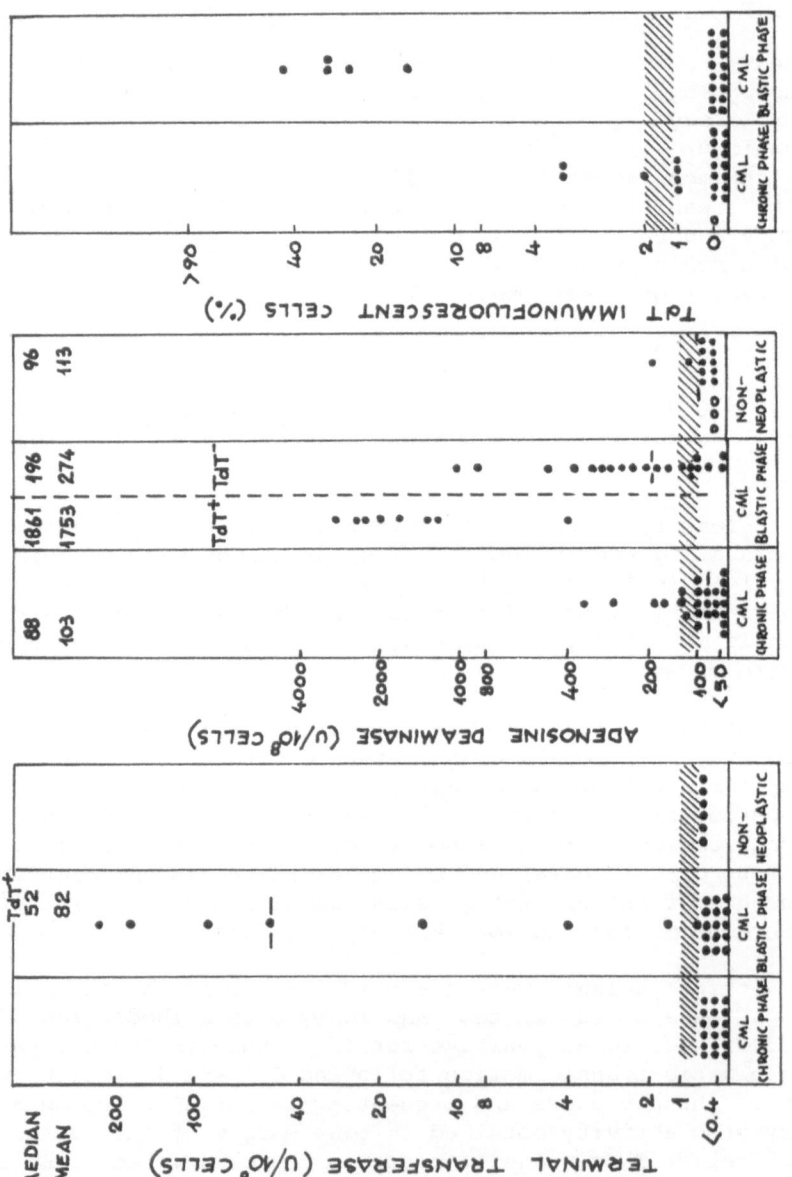

Fig. 7. Distribution of TdT and ADA in Ph' positive leukemias.

lymphoid. The three other positive cases had a mixed type morphology and lower levels of TdT activity. The TdT negative cases (in which the enzyme was not detectable and with no fluorescent cells in bone marrow) included 20 patients in blastic transformation of Ph' positive CML and a single case presenting as Ph'+ AML without an apparent chronic phase. The morphology was purely myeloid in 11 cases, monocytic in 2 cases and mixed in 6 cases. Two cases showed a "lymphoid" appearance, suggesting a possible micromegakaryocytic origin.

In cases developing blastic transformation, no difference was observed between the TdT+ and TdT- groups for what concerns the duration of chronic phases, which had medians of 31 and 34 months, respectively.

The ADA determinations are of particular interest (see Fig. 7) since the mean value of the TdT+ group (1753 ± 275 U/10^8 cells) is significantly higher (P<0.001) than that found in the TdT- group (274 ± 58 U/10^8 cells). An additional observation concerns the distribution of ADA values in the TdT- group. It seems that a positive trend exists between low ADA values and extent of survival. In fact, the median survivals for the cases with ADA higher than 200 U/10^8 cells and for those with ADA lower than 200 U/10^8 cells were 4 and 8 months, respectively.

No reversions to chronic phase were observed in TdT negative cases; response to chemotherapy was poor in all cases with only partial and short-lasting reduction of blast cells. TdT positive patients temporarily responded to chemotherapy with Vincristine (V) and Prednisone (P).

The actuarial survival curves calculated for the TdT positive (CML in blastic phase and Ph'+ ALL) and TdT negative patients are shown in Fig. 8. The two curves differ significantly (P<0.01) and the median survival for the two groups is 11 and 4.8 months, respectively. Only one patient was surviving at 16 months in each group.

The clinical features and response to therapy of the cases presenting as Ph'+ acute leukemias are given in Table 8. All three cases entered a chronic phase of CML after treatment with V and P. Of these latter, the first case reverted to CML after a prolonged period of severe aplasia and subsequently developed a CNS relapse with TdT positive blasts in CNS fluid; the second case after a stable phase lasting four months had a second acute phase but turned TdT negative; the third case with additional monosomy 7 had a stable phase lasting 12 months, then transformed into a TdT positive crisis (TdT activity in peripheral blood: 52 U/10^8 cells) which did not respond to therapy with vincristine and prednisone.

Table 8.   Patients Presenting in Ph' Positive Acute Leukemia

Sex	Age (yrs)	Cytology	Cytogenetics	TdT (U/10^8)	TdT IF%	ADA (U/10^8)	Response to V,P therapy	Surviv. (mos)
TdT+								
F	17	Lymphoid	Ph'+		30	1170	CP	13+
F	35	Lymphoid	Ph'+		40		CP	8+
M	44	Lymphoid	Ph'+; -7	52.0		2380	CP	12-
TdT-								
F	40	Myeloid	Ph'+,t(6;9;22)		O		NR	3-

CP = reversion to chronic phase;   NR = no response

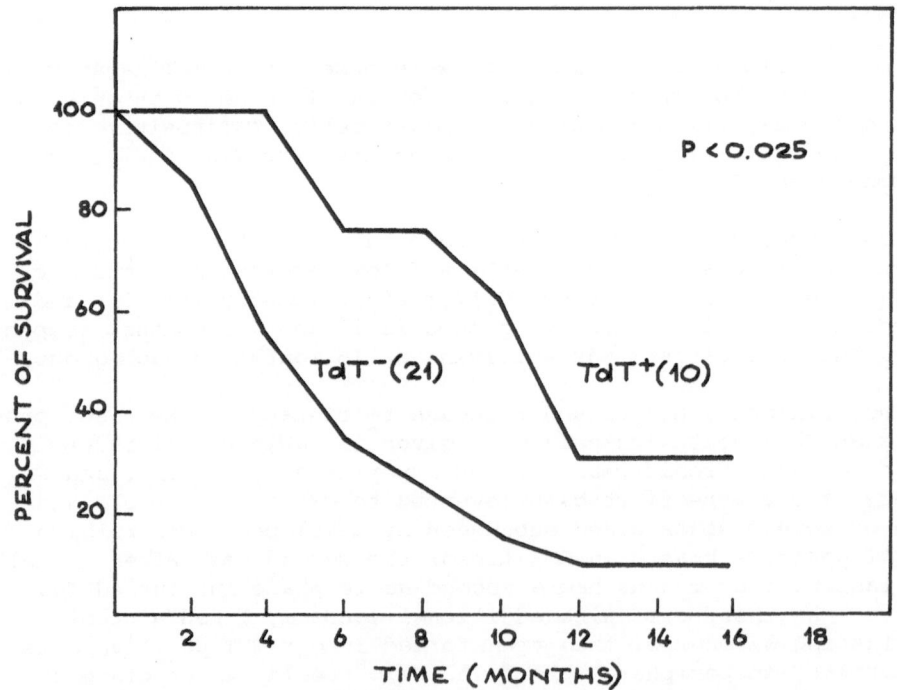

Fig. 8. Actuarial survival curves for adult patients in acute phase
of Ph' positive leukemia according to TdT. Number of patients
is given in parenthesis.

DISCUSSION

A good correlation was observed between enzymatic and IF assays for TdT. This permitted the IF test only to be used in the majority of childhood ALL at presentation and in particular situations such as bone marrow hypocellularity or bone marrow touch preparations in adult ALL.

The diagnostic relevance of TDT and ADA correlation in adult ALL was evaluated by plotting on logarithmic scales the values obtained for each patient. The subtypes defined by membrane markers fell into three distinct biochemical groups: the group of T-ALL with positivity to TdT and very high levels of ADA; the group of B-ALL which is TdT negative and presents low levels of ADA; the cases of non-T, non-B ALL with positivity to TdT and intermediate values of ADA. A similar pattern was obtained by Coleman et al. (1978) in childhood ALL. Very high levels of ADA could be of diagnostic interest in the identification of T-ALL devoid of E^+ blasts, with Acid Phosphatase reactivity and TdT positivity.

The possibility of detecting, with the TdT IF assay, residual or relapsing leukemic cells after therapy has been exploited for the diagnosis of ALL diseases in the CNS fluid, whereas the observed wide fluctuation of TdT IF values prevented a reliable monitoring of remissions as well as the detection of residual disease in bone marrow. In this respect, the determination of ADA in remission is recommended, since the level of this enzyme is well correlated to the rate of cellular proliferation. In fact, the mean ADA value was raised by at least 10 times in our cases of ALL at relapse.

Most cases of acute leukemias classified as myeloid by conventional methods were TdT negative. However, in 7 (11%) AML cases the presence of TdT was ascertained by biochemical and IF assays. All positive cases had a high percent of blast cells in bone marrow, presented cytochemical features of myeloid origin and were Philadelphia chromosome negative. The activity of TdT in the positive cases was significantly lower than in ALL at presentation.

The finding of TdT in AML would indicate the presence of a mixed lymphoid and myeloid cell population, as recently reported in human Ph' negative acute leukemias other than CML in blast crisis (Mertelsmann et al., 1978). This could result from the evolution of the original leukemic cell into clones presenting different phenotypes. In our cases, the TdT positive cells appeared as agranulated blasts and had, as seen by phase contrast, a larger size than usually observed in typical lymphoblasts. It seems therefore that the expression of TdT in non-lymphoid cells can not be ruled out, as also suggested by Bradstock et al(1981), who found TdT fluorescence and myelopreroxidase positivity in the same cell.

The comparison of the ADA values found in ALL and in AML indicates that this marker is not useful in distinguishing lymphoid and myeloid leukemias. However, when the ADA values found in AML and AUL are divided into the two TdT+ and TdT- subgroups, it is evident that in both cases the medians of the TdT+ groups are several times higher than those of the TdT- groups.

At present, the prognostic significance of TdT positivity in AML can not be assessed, since more observations are needed. An interesting observation concerns the significantly longer survival of the AML patients with low levels (<200 U/10^8 cells) of ADA in peripheral blasts.

In Ph'+ leukemias TdT and ADA analyses provided new elements for monitoring the chronic phase of CML, by recognizing the appearance of early transformation and yielding prognostic information. The stable phase of CML was characterized by TdT absence (Brusamolino et al., 1981) and ADA ranging within normal controls in the majority of cases studied. The ADA values were increased over the control line only in cases showing the early signs of transformation.

In the acute phase of Ph'+ leukemias, the TdT assay was positive in 32% of the cases, a value similar to that reported by other authors (Kung et al., 1978; Marks et al., 1978; Hoffbrand et al., 1979). Responsiveness to treatment with Vincristine and Prednisone resulted in a significantly longer survival time for the TdT+ group. Of particular interest is the observation of a significant increase of ADA values in the TdT+ patients in acute phase, with respect to the TdT- transformations. This represents an additional useful tool in the definition of the two, TdT+ and TdT-, groups which are known to be treated with different protocols (Marks et al., 1978).

The recent report on the utilization of 2'-deoxycoformycin, a specific inhibitor of ADA, in the treatment of ALL (Mitchell et al., 1980) would suggest that this drug could be tested in all cases of acute leukemias (including the acute phase of Ph'+ leukemias) which show high levels of this enzyme and which have become resistant to all conventional therapies.

The characterization of Ph'+ leukemia presenting as ALL is of particular interest to possibly enlighten the relationship between Ph'+ ALL and Ph'+ CML. Our three cases of Ph'+ ALL were all TdT positive, had high values of ADA, responded to ALL therapy reverting to a stable phase of CML; the Ph' chromosome was present in all the metaphases at onset and did not disappear in chronic phase after therapy. This altogether suggests that these two forms of leukemia are strictly related, arising from a common "target" cell.

ACKNOWLEDGEMENTS

We are indebted to Prof. F.J. Bollum for providing us with the antibody against the terminal transferase. We thank S. Jayakar for useful suggestions in statistical calculations and L. Riva for editorial assistance.

This work was supported in part by Grant 80-01561-96 of the Programma Finalizzato "Controllo della crescita neoplastica" of the Italian National Research Council and by EURATOM Contract 152-76IBlOl.

This publication is contribution n. 1748 of the Radiation Protection Programme of the European Community Commission.

E. Brusamolino was granted a fellowship by the Associazione Italiana per la Ricerca sul Cancro.

REFERENCES

Bennett, J.M., Catovsky, D., Daniel, M.T., Flandrin, G., Galton, D.
    A.G., Gralnick, H.R. and Sultan, C., 1976, Proposal for the
    classification of the acute leukemias. French-American-
    British (FAB) Co-operative group, Brit. J. Haem., 33:451.
Bernasconi, C., Lazzarino, M., Morra, E., 1976, Erkrankungen der
    Myelopoese (A. Stacher and P. Höcker, eds.) pp. 224-227,
    Urban and Schwarzenberg, München, Berlin, Wien.
Berkson, J. and Gage, R.P., 1950, Calculation of survival rates for
    cancer, Proc. Staff Meeting Mayo Clinic, 25:270.
Bollum, F.J., 1975, Antibody to terminal deoxynucleotidyl transferase,
    Proc. Natl. Acad. Sci. U.S.A, 72:4119.
Bollum, F.J., 1979, Terminal deoxynucleotidyl Transferase as a hemato-
    poietic cell marker, Blood, 54:1203.
Bradstock, K.F., Hoffbrand, A.V., Ganeshaguru, K., Llewellin, P.,
    Patterson, K., Wonke, B., Pizzolo, G., Prentice, A.G., Bennett,
    M., Bollum, F.J., Janossy, G., 1981, Terminal deoxynucleotidyl
    Transferase expression in acute non-lymphoid leukemia. An
    analysis by immunofluorescence, Brit. J. Haem., 47:133.
Brouet, J.C., Preud'homme, J.L., Penit, C., Valensi, F., Rouget, P.,
    and Seligmann, M., 1979, Acute lymphoblastic leukemia with
    pre-B cell characteristics, Blood, 54:269.
Brusamolino, E., Bernasconi, C., Isernia, P., Scovassi, A.I., and
    Bertazzoni, U., 1981, Terminal Transferase and Adenosine De-
    aminase in adult Philadelphia positive leukemias, in "Leukemia
    Markers",(W. Knapp, ed.), Academic Press, London (in press).
Coleman, M.S. and Hutton, J.J., 1975, Micromethod for quantitation
    of adenosine deaminase activity in cells from human peripheral
    blood, Bioch. Med., 13:46.
Coleman, M.S., Greenwood, M.F., Hutton, J.J., Bollum, F.J., Lampkin,
    B., and Holland, Ph., 1976, Serial observations on terminal

deoxynucleotidyl transferase activity and lymphoblast surface
markers in acute lymphoblastic leukemia, Cancer Res., 36:120.

Coleman, M.S., Greenwood, M.F., Hutton, J.J., Holland, Ph., Lampkin,
B., Krill, C., and Kastelic, J.E., 1978, Adenosine Deaminase,
Terminal Deoxynucleotidyl Transferase (TdT) and Cell Surface
Markers in Childhood Acute Leukemias, Blood, 52:1125.

Hoffbrand, A.V., Ganeshaguru, K., Janossy, G., Greaves, M.F., Catovsky,
D., Woodruff, R.K., 1977, Terminal deoxynucleotidyl Transferase.
Levels and membrane phenotypes in diagnosis of acute leukemia,
Lancet, II:520.

Hoffbrand, A.V., Ganeshaguru, H., Llewelin, P., and Janossy, G., 1979,
Biochemical Markers in Leukaemia and Lymphoma, in "Recent Results in Cancer Research" (R. Gross and K.-P. Hellriegel, eds.,)
69:25-41, Springer-Verlag, Berlin, Heidelberg, New York.

Janossy, G., Bollum, F.J., Bradstock, K.F., and Ashley, J., 1980,
Cellular phenotypes of normal and leukemic hemopoietic cells
determined by analysis with selected antibody combinations,
Blood, 56: 430.

Kung, P., Long, J.C., McCaffrey, R.P., Ratfliff, R.L., Harrison, T.
S., and Baltimore, D., 1978, Terminal deoxynucleotidyl Trans-
ferase in the diagnosis of leukemia and malignant lymphoma, Am.
J. Med., 64:788.

Marks, S.M., Baltimore, D., McCaffrey, R., 1978, Terminal Transferase
as a predictor of initial responsiveness to vincristine and
prednisone in blastic chronic myelogenous leukemia, N. Engl.
J. Med., 298:812.

McCaffrey, R., Smolder, D., and Baltimore, D., 1973, Terminal deoxynucleotidyl Transferase in a case of childhood acute lymphoblastic leukemia, Proc. Natl. Acad. Sci. USA, 70:521.

Meier, J., Coleman, M.S., and Hutton, J.J., 1976, Adenosine deaminase
activity in peripheral blood cells of patients with haematological malignancies, Brit. J. Cancer, 33:312.

Mertelsmann, R., Koziner, B., Ralph, P., Filippa, D., McKenzie, S.,
Arlin, Z.A., Gee, T.S., Moore, M.A.S. and Clarkson, B.D.,
1978, Evidence for distinct lymphocytic and monocytic populations in a patient with Terminal Transferase positive acute
leukemia, Blood, 51:1051.

Mitchell, B.S., Koller, C.A., Heyn, R., 1980, Inhibition of adenosine
deaminase activity results in cytotoxicity to T lymphoblasts
in vivo, Blood, 56:556.

Pasquali, F., Casalone, R., Francesconi, D., Peretti, D., Fraccaro,
M., Bernasconi, C. and Lazzarino, M., 1979, Transposition of
9q34 and 22 (q11 qter) regions has a specific role in chronic
myelocytic leukemia, Hum. Gen., 52:55.

Ramot, B., Brok-Simoni, F., Barnea, N., Bank, I. and Holtzmann, F.,
1977, Adenosine Deaminase (ADA) activity in lymphocytes of
normal individual and patients with chronic lymphoblastic
leukemia, Brit. J. Haem., 36:67.

Sarin, P.S., Anderson, P.N. and Gallo, R.C., 1976, Terminal deoxynucleotidyl transferase activities in human blood leukocytes

and lymphoblast cell lines: high levels in lymphoblast cell lines and in blast cells of some patients with chronic myelogenous leukemia in acute phase, Blood, 47:11.

Smyth, J.F. and Harrap, K.R., 1975, Adenosine Deaminase activity in leukaemia, Brit. J. Cancer, 31:544.

Srivastava, B.I.S., Khan, S.A. and Henderson, E.S., 1976, High terminal deoxynucleotidyl transferase activity in acute myelogenous leukemia, Cancer Res., 36:3847.

Stass, S.A., Schumacher, A.R., Keneklis, T.P. and Bollum, F.J., 1979, Terminal Deoxynucleotidyl Transferase immunofluorescence on bone marrow smears: experience in 156 cases, Am. J. Clin. Path., 72:898.

Tung, R., Silver, R., Quagliata, F., Conklyn, M., Gottesman, J. and Hirschhorn, R., 1976, Adenosine Deaminase activity in chronic lymphocytic leukemia. Relationship to B- and T-cell subpopulations, J. Clin. Invest., 57:756.

Vogler, L.B., Crist, W.M., Bockman, D.E., Pearl, E.R., Lawton, A.R. and Cooper, M.D., 1978, Pre-B-cell Leukemia. A new phenotype of childhood lymphoblastic leukemia, N. Engl. J. Med., 298: 872.

STUDIES OF TERMINAL DEOXYNUCLEOTIDYL TRANSFERASE IN NORMAL AND NEOPLASTIC HUMAN CELLS*

Francesco Campagnari,[1] Libero Clerici,[1] Emilio Bombardieri,[2] Paolo Vezzoni,[2] Giovanni Di Fronzo,[2] Maria Luisa Villa,[2] and Gian Luigi Buraggi[2]

[1]Laboratory of Biochemistry, Biology Group D.G. XII, C.E.C. Joint Research Centre, 21020 Ispra (Va) Italy
[2]Istituto Nazionale per lo Studio e la Cura dei Tumori via Venezian 1, 20133 Milano, Italy

INTRODUCTION

Terminal deoxynucleotidyl transferase, TdT, is an unusual DNA polymerase that does not require a template and catalyzes the random addition of deoxynucleotidyl units to the 3'-OH ends of single DNA chains and of oligodeoxynucleotides. The enzyme was first identified as a distinct entity by Krakow et al. (1962), but it was purified from calf thymus and extensively characterized in the laboratory of F.J. Bollum (Yoneda and Bollum, 1965; Kato et al., 1967; Chang and Bollum, 1971a). For several years TdT was considered almost exclusively as a convenient tool for synthesizing DNA-like molecules (Bollum, 1966; Ratliff et al., 1967; Chang and Bollum, 1971b) and also we made large use of it in our Euratom program of preparing tailored polydeoxynucleotide substrates for the enzymes of DNA repair (Campagnari et al., 1973).

TdT was overwhelmingly reproposed to the general attention when it was detected in the blasts cells of a child with acute lymphoblastic leukemia, ALL (McCaffrey et al., 1973). This finding was

---

* This publication is, in part, contribution n. 1786 of the Programme Biology, Radiation Protection and Medical Research, Directorate General XII of the Commission of the European Communities.

soon complemented by other pathological observations and new basic
data. In mammals and very likely in man, the enzyme is associated
with subsets of very immature hematopoietic cells that, after a
short period of the neonatal life, are located only in the thymus
cortex and in the active bone marrow (see review by Bollum, 1978).
These cells are classified as undifferentiated blasts or scarcely
differentiated lymphoblasts and their neoplastic transformation gives
rise to hematological malignancies with enhanced levels of TdT.
The enzyme is present in the tumoral cells of: most patients with
ALL (McCaffrey et al., 1975; Coleman et al., 1976; Hoffbrand et al.,
1977); lymphoblastic lymphomas, LL (Donlon et al., 1975; Koziner et
al., 1977; Habeshaw et al., 1979); numerous cases of acute undifferentiated leukemias, AUL, and a third of the blastic crisis in chronic myelogenous leukemias, CML (Hoffbrand et al., 1977; Kung et al.,
1978; Hutton et al., 1979). TdT has rarely been found in other leukemic disorders (Bollum, 1979; Hoffbrand et al., 1979).

The detection of this marker enzyme in human cells from proliferative diseases of the hematopoietic system is of practical inte-
rests since the TdT positive blasts are generally sensitive to the
hormonal therapy with glucocorticoids (Kung et al., 1975; Marks et
al., 1978). Optimized measurements of the enzyme activity in crude
extracts from lymphoid tissues and cells (Coleman et al., 1974; Coleman et al., 1977a,b; Vezzoni et al., 1981) and immunofluorescence
methods for cells containing TdT (Goldschneider et al., 1977; Kung
et al., 1978) have been established. We have standardized procedures for these two techniques and the present paper reports data.
about their clinical application. The responses obtained in the biochemical assays correlated with the results of the cytoimmunological
tests carried out with monospecific antibodies against purified TdT.

METHODS

TdT was isolated from frozen calf thymuses by a modification
of the method proposed by Chang and Bollum (1971a). In fact, the
step corresponding to the pH 4.5 treatment was omitted and the active fractions recovered after the final chromatography on hydroxyapa-
tite were reprocessed for a second gel filtration on Sephadex G-100.

The purified TdT was assayed for polymerization of Mg-dATP on
a $p(dA)_8$ initiator in conformity with the requirements for optimal
catalysis. The reaction mixture comprised: 200 mM K-cacodylate, pH
7.0; 8 mM $MgCl_2$; 2 mM 2-mercaptoethanol; 0.3 mM $ZnSO_4$; 2 mM (^{14}C)

dATP; 20 μM 3'-OH termini of the p(dA)$_8$ primer; 1.5 mg of bovine serum albumin, BSA, per ml; and 15 to 25 μg of enzyme protein per ml.
The incubation was at 37 °C for 15 min. At intervals of 5 min, 20 μl
of the assay solution were spotted on disks of DEAE-81 cellulose from
Whatman. The unpolymerized dATP substrate was removed from the DEAE
paper by washing 5 times with 0.15 M Na$_2$HPO$_4$, once with distilled
water and 5 times with ethanol. The disks were dried under an infra-
red lamp and counted for the retained radioactivity in a liquid scintillation spectrometer with a combination of 0.4 % 2,5-diphenyloxa-
zole, PPO, and 0.008 % p-bis-(o-methylstyryl)-benzene, bis-MSB, in
toluene (w/v) as a fluor.

Under the conditions of the assay, the isolated TdT had the
same specific activity as that reported for the homogeneous enzyme by
Chang and Bollum (1971a). In fact, it catalyzed the incorporation of
102,000 nmol of radioactive dA nucleotides into the initiator per hr
and per mg of protein.

TdT was crosslinked with 0.2 % glutaraldehyde (w/v), diluted to
200 μg per ml in 100 mM NaCl, 50 mM K-phosphate, pH 7.2, and injected
subcutaneously to rabbits for production of monospecific antibodies.
About 0.5 ml of the enzyme preparation and a dose of complete Freund's
adjuvant were administered to each of 5 rabbits.During the following
month, 3 injections of TdT supplemented with incomplete Freund's
adjuvant, were given to the animals at regular intervals. A booster
dose of the enzyme was administered at the end of the second month.
As estimated by immunodiffusion tests in soft agar, the sera from
4 rabbits contained appreciable amounts of antibodies against TdT.

The immunologically active sera were pooled and processed for
isolation of the type-G immunoglobulins, IgG. The proteins in the
sera were salted out with (NH$_4$)$_2$SO$_4$ at 50 % saturation and separated by centrifugation. The sedimented material was redissolved in a
small volume of 50 mM K-phosphate, pH 7.0, and fractionated by gel
filtration on a Sephadex G-100 column in equilibrium with the same
buffer. The portions of the effluent recovered at the position corresponding to the void volume displayed antibody properties versus
TdT and were loaded onto a 5 ml column of Protein A-Sepharose CL-4B
from Pharmacia. Most proteins were not retained by the IgG specific
adsorbent which was washed by percolation with 20 ml of 100 mM K-
phosphate, pH 7.0. The gel was eluted with 0.1 M acetic acid and
1 ml fractions were collected. The active IgG emerged from the column
as a sharp peak of u.v. absorbing material. The proteins in the
fractionated effluent were salted out by adding solid (NH$_4$)$_2$SO$_4$ up

to 75% saturation, sedimented by centrifugation, redissolved in and extensively dialyzed against 100 mM NaCl, 25 mM K-phosphate, pH 7.2 The final preparation contained 4 mg of IgG per ml and, at equal protein concentration, had a titer of TdT-antibodies which was at least 200-fold higher than the one of the starting antisera. The isolated IgG served to standardize immunofluorescence tests for TdT positive cells by the indirect method using a second fluorescent antibody against the rabbit IgG.

Samples of normal and neoplastic lymphoid tissue from patients with lymphomas and of thymus glands from children with thymomas were obtained in the operating room during surgery. The specimens were either immediately frozen to $-70\,^{\circ}C$ for the preparation of tissue extracts or finely minced with scissors in saline solution for obtaining suspensions of single cells by filtering through 4 layers of gauze. The latter procedure was applied to isolate cells from fresh thymuses of laboratory animals and from clinical specimens of human bone marrow. Mononuclear cells from heparinized blood of leukemic individuals were separated by centrifugation on Ficoll gradients according to Boyum (1968), washed with 100 mM NaCl, 25 mM K-phosphate, pH 7.2, and sedimented by centrifugation.

The frozen tissues and the packed mononuclear leukocyte were suspended in 250 mM K-phosphate, pH 7.0, 2 mM 2-mercaptoethanol at concentrations of 0.2 g of tissue or 2×10^{8} cells per ml, respectively, and were disrupted in small conical homogenizers with ground glass mortars and pestles. The homogenates were spun in ultracentrifuge at 100,000 x g for 30 min. The recovered supernatants were considered as the crude extracts of lymphoid tissues and cells.

The determination of TdT activity in these extracts was carried out with a reaction mixture that much resembled the one previously adopted for crude preparations (Vezzoni et al., 1981). The assay solution comprised: 200 mM K-cacodylate, pH 7.0; 0.5 mM $MnCl_2$; 0.3 mM $ZnSO_4$; 2 mM 2-mercaptoethanol; 0.5 mM (^{3}H)dGTP or (^{14}C)dGTP substrates; 6 µM 3'-OH termini of a $p(dA)_{55-95}$ initiator, 1.5 mg of BSA per ml; and crude extracts at a final dilution of 10 to 20 times. Incubation was at 37°C for 10 min and 20 µl of the reaction mixture were spotted on disks of glass-fiber paper GF/C from Whatman at various intervals. The disks were processed for counting the radioactivity of the acid insoluble material, as previously outlined (Bekkering-Kuylaars and Campagnari, 1972). When less crude preparations of TdT were used, the polymerization rates in this assay were 3-fold lower than in the assay standardized for the pure enzyme (see above). However, a unit of TdT in the crude extracts was defined as the amount that changed 1 nmol of dGTP substrate into acid insoluble products under the stated conditions.

For immunofluorescent staining of TdT, the isolated cells were resuspended in the buffered saline at an approximate concentration of 10^6 cells per ml and cytocentrifuge slides were prepared from 2-3 drops of the suspensions. The slides were dried in air, fixed with methanol at 2°C and rinsed with saline. Then, the cells were covered with the buffer containing the rabbit antibodies against TdT at a dilution of 50 µg of protein per ml and incubated for 1 hr in a moist chamber at 25°C. The slides were rinsed with saline and a fluorescein conjugated antibody against the rabbit IgG was applied to the cells. This second layer antibody was a commercial preparation of fluorescent anti-rabbit γ-globulins from Behringwerke AG (Marburg, West Germany) that were derived from sera of goats immunized against the $F(ab)_2$ fragment of rabbit IgG. The goat γ-globulins were used as a 50-fold dilution of the reconstituted 1 ml sample and the staining reaction was carried out for 45 min at 25°C in the moist chamber. After washing with the neutral saline solution the slides were mounted for cytological analyses in an Olympus fluorescence microscope equipped also for phase-contrast illumination and photography.

RESULTS

TdT Immunogen and its Antibodies

The catalytic potency and the properties of our TdT preparation from calf thymus were similar to those described for the highly purified enzyme of Chang and Bollum (1971a). In fact, zone centrifugation on sucrose gradients of as many as 400 µg of the isolated enzyme yielded a single peak of protein coinciding with the profile of TdT activity and having a sedimentation coefficient of 3.5 S (Fig. 1). Although a 40% enzyme inactivation occurred during the long ultracentrifugation, the ratio of catalytic units to protein concentration remained practically constant for all the active fractions recovered in an approximately normal distribution along the gradient. This was to be expected for an enzyme preparation of homogeneous protein composition. In electrophoretic analyses carried out at nondenaturing conditions, the isolated protein behaved as a single major species with basic properties and a high isoelectric point. It migrated very slightly toward the anode during gel electrophoresis

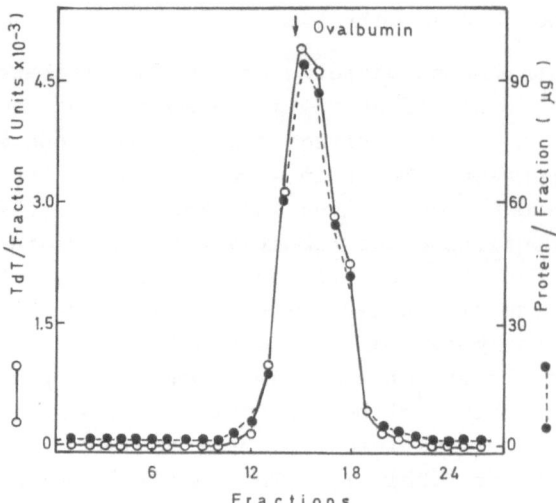

Fig. 1   Sedimentation analysis of purified bovine TdT in a linear
         sucrose gradient. 400 µg of active proteins were spun over
         a 4.4 ml gradient for 23 hr at 47,000 r.p.m. in the SW 60
         Ti rotor of a Beckman ultracentrifuge. See also the text.

buffered with Tris-barbituric acid at pH 8.7, whereas it traveled
toward the cathode in the runs with a mixture of β alanine and acetic acid at pH 4.5 as a buffer. The disclosed properties conform to
the features of the small size TdT with a molecular weight of 32,000
and two distinct component polypeptides (Bollum, 1974). Up-to-date
however, we failed to obtain a net separation of the subunits by
SDS-polyacrylamide gel electrophoresis where additional protein
bands with estimated molecular weights exceeding those of the single
component polypeptides and of the native enzyme were noted. Very
likely, the reported proneness of the TdT polypeptides to reag-
gregate at random (Chang and Bollum, 1971a) was enhanced somehow in

our experiments and became one of the causes of the results. The reasons for this are not understood and the matter requires further investigation.

As already anticipated, the rabbit antibodies against bovine TdT corresponded to an IgG enrichment by more than two orders of magniture over the original antisera. Thus, specific precipitation reactions in the immunodiffusion tests were observed also when the isolated antibodies were used at the same range of protein concentrations as the diluted antigen (Fig. 2).

The purified antibodies represented the gross IgG fraction of the original antisera and reflected its composition. Therefore, our preparation comprised a collection of monospecific antibodies against the purified TdT and the other unknown immunoreactive IgG molecules present in the peripheral blood of the rabbit at the same time. It should then be expected that the isolated IgG could also react with cellular components differing from TdT. Actually, this was not a problem in our cytoimmunofluorescence tests.In fact, the adopted dilution of the rabbit IgG allowed us to detect preferentially the immunological reactions due to the prevailing anti-TdT species. For maximal selectivity versus the enzyme protein, it would have been necessary to separate the monospecific TdT antibodies from the

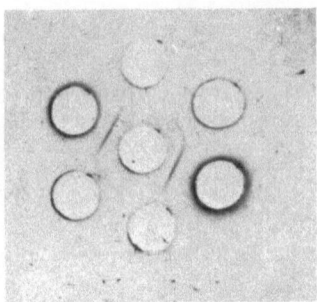

Fig. 2. Microdiffusion of rabbit antibodies against TdT antigen. Center well contained purified bovine TdT (1mg/ml). Wells in the clockwise positions contained the isolated IgG fraction (4mg/ml) serially diluted as follows: 1 hr, 2-fold; 7 hr, 4-fold; 11 hr, 8-fold; 5 hr, 16-fold; 9 hr, 32-fold; 3 hr, 64-fold.

other immunoreactive IgG molecules.

   Although produced by immunization versus TdT from bovine thymus,
the rabbit antibodies reacted efficiently with thymocytes from the
mouse, the rat and man and with other cell populations from which
detectable amounts of TdT could be extracted. Such intraspecific
immunological crossreactivity was to be expected for antibodies
against an intracellular mammalian protein. Actually, this was in-
strumental for establishing the immunofluorescence tests for TdT-
containing cells from man. Fig. 3 shows photographs of cells from
human bone marrow with positive and negative immunofluorescence
staining for TdT. The fluorescence was restricted to the nucleus
and this appeared as the predominant pattern also in the tests carried out in malignant lymphoid cells and discussed below.

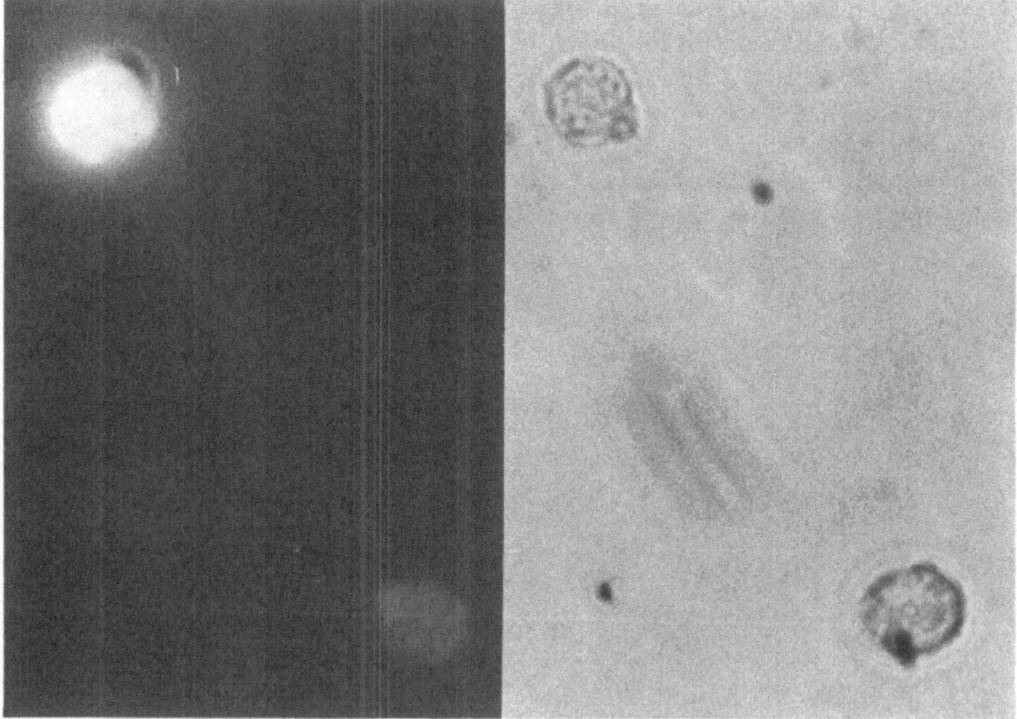

Fig. 3.   Immunofluorescence staining for TdT in human cells from bo-
          ne marrow. View for fluorescent light is on the left; view
          for phase contrast light is on the right. The upper cell is
          TdT-positive; the lower cell is negative.

Tests for TdT in Leukemic Cells and Lymphomas of Man

Mononuclear cells of peripheral blood from 36 leukemic individuals and specimens from 75 malignant lymphomas and from 12 normal
lymph nodes were examined for their TdT content. The analyses were
performed by determination of the enzyme activity in the crude extracts and/or an immunofluorescence staining of TdT in the isolated cells. All patients were not receiving chemotherapy.

Diagnosed leukemias were classified on the basis of routine
morphological and cytochemical data. Moreover, the acute forms, negative for peroxidase and unspecific esterase, were grouped according to the distribution of immunologically reacting markers in
their cell populations as proposed by Thiel et al. (1980). The cell
markers taken into consideration were: common non-T, non-B ALL antigen, cALL; T-cell antigens, HuTLA; sheep erythrocyte receptor, E-R;
cytoplasmic type M immunoglobulin, CyIgM; and surface membrane immunoglobulin, SmIg. For the present investigation, the subgroup phenotypes were identified as the following combinations: AUL with no marker; common ALL with cALL only; Thymic ALL, Thy ALL, with HuTLA and
E-R; Pre-B ALL with CyIgM and cALL; B-ALL with SmIg only. No determination of these markers was performed in the solid lymphoid tumors
that were simply classified histologically with adoption of the criteria suggested by Nathwani, Kim and Rappaport for diagnosis of lymphoblastic lymphoma, LL, as reported by Vezzoni et al. (1981).

Measurements of TdT units in the extracts and data from the immunofluorescence staining of TdT-positive cells were combined in a
same graph. Separate graphs, however, will be presented for leukemias
and lymphomas. TdT was almost undetectable in the preparations from
white blood cells of chronic lymphocytic leukemias, CLL, and from normal lymph nodes. In such cases, the enzyme levels were well below the
values of 10 units, U, per 10^8 cells or per 0.1 g of lymphoid tissue,
that were established as limits for discriminating between the trivial and the pathologically significant findings. Moreover the cells
from both sources did not fluoresce in the immunoassay for TdT. Any-
way, we have cautiously regarded as lymphoid cellular populations with
enhanced levels of TdT only those having more than 2% of cells stain-
able by immunofluorescence.

Fig. 4 shows the quantitative distribution of TdT in the white
blood cells from leukemic patients. As anticipated, the data are expressed as concentrations of enzyme units and of fluorescent TdT-po-

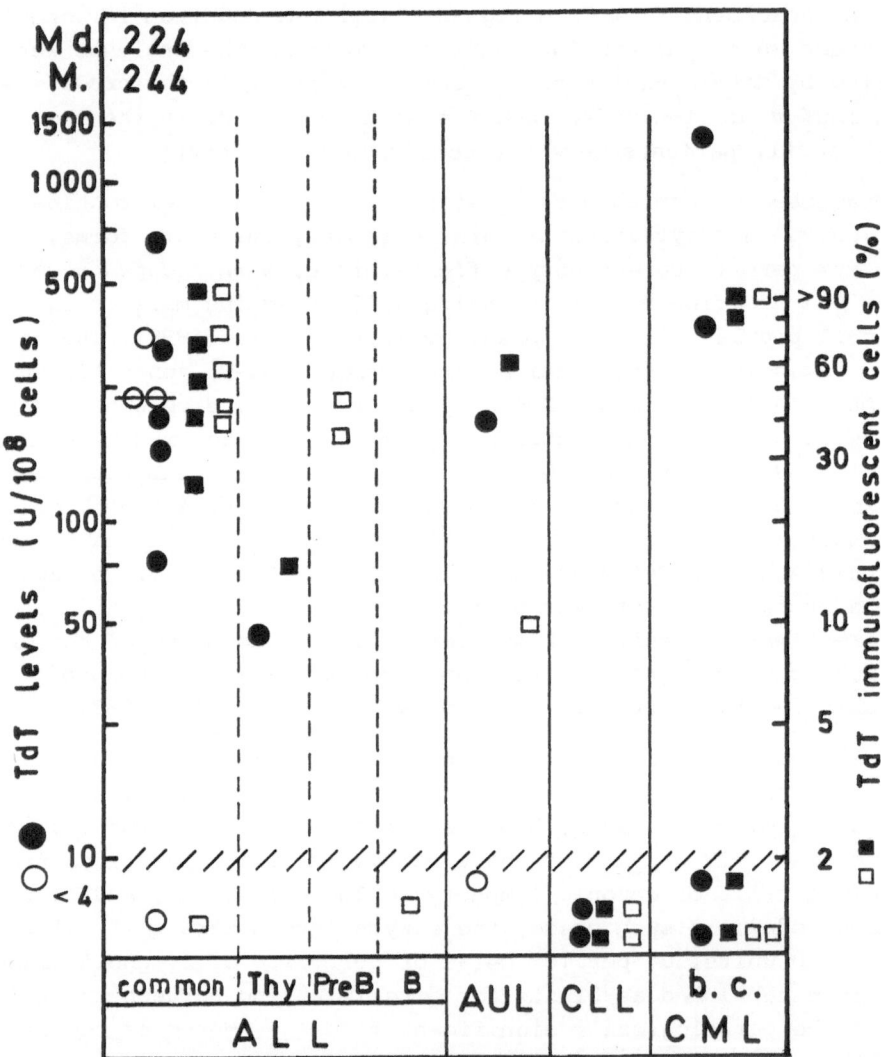

Fig. 4   Distribution of TdT in leukemias. Biochemical assays, circles:
immunofluorescence tests, squares. Open symbols, cases analyzed by one method; closed symbols, cases analyzed by two methods. M, mean value; Md, median value in ——.

sitive cells for the biochemical and the immunological assay, respectively. The two arbitrary scales of concentration values in the or-

dinates are consensual and suitable to reveal correlations between
responses by the different tests. In terms of gross quantitation of
the results, a net concordance of data was always evident in any
single case analyzed by the two methods and recorded with closed
symbols in Fig. 4. Chronic lymphocytic leukemia, CLL, is for control.

The overall pattern of the presence and levels of TdT in human
leukemia resembled closely the one described under INTRODUCTION and
derived from evidence firmly established by a number of investigators. In fact, TdT was elevated in the neoplastic cells from 16 out
of 19 patients with ALL, being practically undetectable only in 2 cases of common ALL and the single case of B ALL. In the group of 8
common ALL whose blasts were analyzed biochemically and found to contain appreciable amounts of enzyme, the concentration of TdT per
10^8 cells ranged from 78 to 648 U with a mean value of 244 U. Our
set of data was too small for drawing conclusions about the modulation of the intracellular enzyme levels in the different subgroups
of ALL. The isolated observations concerning Thy- and preB-ALL conform to expectations according to the recent survey by Hoffbrand and
Janossy (1980) and by Thiel et al. (1981). The findings of high TdT
levels in mononuclear cells from 2 out of 3 individuals with ALL and
from 4 out of 7 patients with CML in blastic cryses, b.c. CML, confirm the pattern described as distintive for the distribution of
TdT in the malignant cells of these hematological disorders by previous reports (Hoffbrand et al., 1977; Kung et al., 1978; Hutton et
al., 1979). It should be noted that TdT-positive cell populations
from b.c. CML displayed maximal levels of enzymatic activity and percentage of fluorescence stained elements even above the corresponding values obtainable in mononuclear blood cells from most cases of
ALL.

Fig. 5 summarizes the results of the analyses for TdT that were
performed in solid tumors from lymphoid tissues. For a most convenient graphical presentation, these human malignancies were grouped
simply as LL, Hodgkin's and non-Hodgkin's lymphomas. TdT was restricted to the cells from LL, as indicated by preceding papers (Donlon
et al., 1975; Koziner et al., 1978; Habeshaw et al., 1979). The enzyme was found in all 10 cases of LL that were examined, its concentration ranging from 289 to 1301 U per 0.1 g of tissue with a mean
value of 630 U. This supports the indications of our preliminary report that optimized methods of extraction and assay for TdT in crude tissue preparations led to detect very high levels of enzymatic
activity in LL neoplasms (Vezzoni et al., 1981). These levels are

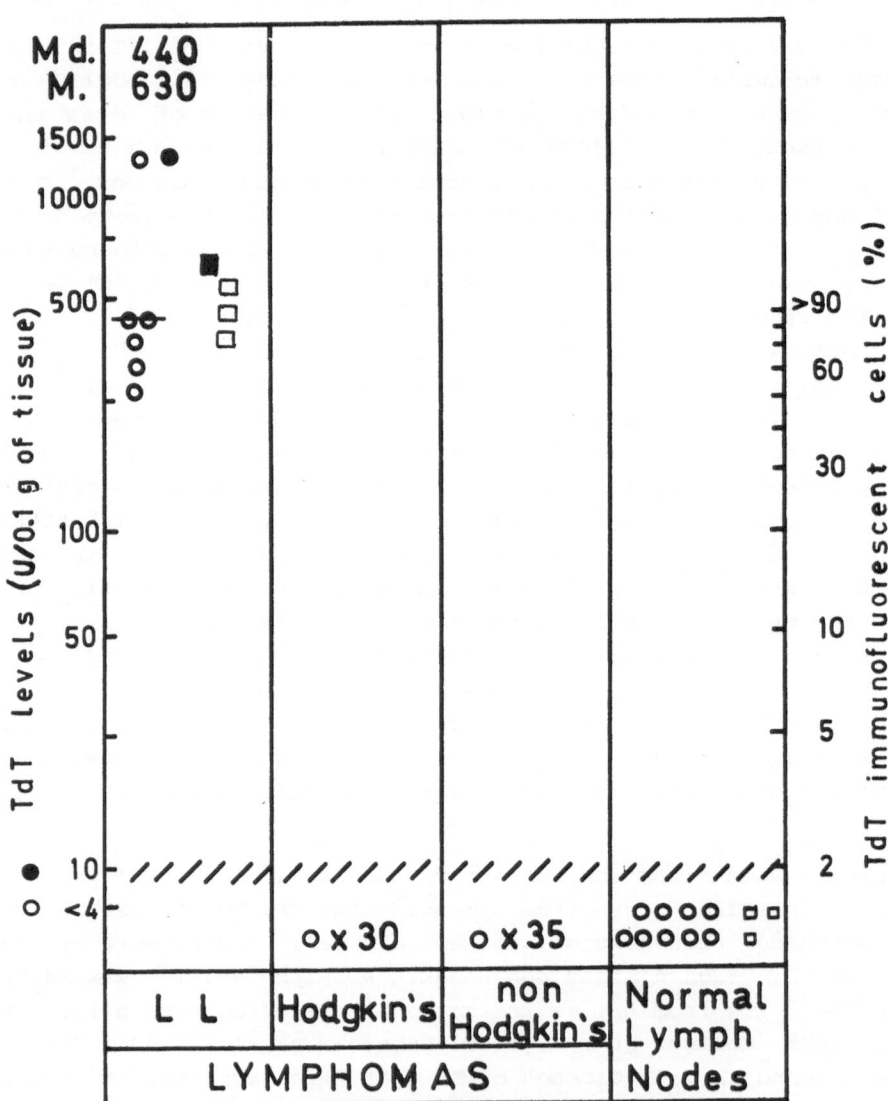

Fig. 5.   Distribution of TdT in lymphomas. Biochemical assays, cir-
          cles; immunofluorescence tests, squares. Open symbols, ca-
          ses analyzed by one method; closed symbols, cases analyzed
          by two methods. M, mean value; Md, median value in ——

comparable with those observed in the circulating blasts from ALL and other leukemic diseases expressing the TdT marker.

In most malignant lymphoblasts the TdT immunofluorescence was restricted to the nuclei like it was noted for the TdT-positive bone marrow cells (see Fig. 3). A few cases with cytoplasmic fluorescent material were observed. The nuclear location of TdT in human leukemic lymphoblasts was detected (Bollum, 1978; Hutton et al. 1979) and soon related to an enzyme form of high molecular weight (Bollum and Brown, 1979; Deibel and Coleman, 1979). Then the large TdT from the nuclei of bone marrow cells and of immature lymphoblasts would be the functional form undergoing peptide chain cleavage and degradative change to the low molecular weight enzyme of the thymic form (Bollum and Brown, 1979). However, the real significance of the limited proteolysis affecting native TdT remains to be ascertained. Thus characterization of subgroups of leukemic lymphoblasts with respect to constitutive physyological levels, molecular form and intracellular location of the TdT marker may well provide dubious data.

SUMMARY

Optimized biochemical assays and cytoimmunofluorescence tests were used to detect terminal deoxynucleotidyl transferase, TdT, in malignant cells of 36 leukemias and 75 lymphomas from patients not receiving chemotherapy.

TdT was virtually absent from normal lymph nodes and from leukocytes of chronic lymphocytic leukemia, CLL, taken as controls. Its quantitative distribution in the neoplasms matched the current knowledge. Appreciable amounts of TdT were found in all the 10 lymphomas of lymphoblastic type, LL, and in the white blood cells of: 16 out of 19 acute lymphoblastic leukemia, AAL, perhaps with modulation in the various phenotypes; 2 out of 3 acute undifferentiated leukemias, AUL; and 3 out of 7 blastic crises in chronic myelogenous leukemia, b.c. CML. Biochemical and cytoimmunologycal analyses yielded concordant responses and even roughly comparable estimates in the same patients. TdT immunofluorescence was clearly nuclear in most cells and was cytoplasmic occasionally. Definite correlations between concentrations of enzymatic activity and percentage of immunofluorescent cells could not be established. Further detailed work will be required to identify putative subgroups in TdT-positive blast populations.

REFERENCES

Bekkering-Kuylaars, S. A. M., and Campagnari F., 1972, Purification
    of a DNA polymerase from calf Thymus nuclei,
    Biochem. Biophys. Acta, 272: 526.

Bollum, F. J., 1966, Biosynthetic polydeoxynucleotides, in: "Proce-
    dures in Nucleic Acid Research," vol. 1, p. 577, G. L. Canto-
    ni and D.R. Davies, eds., Harper and Row, New York.

Bollum, F. J., 1974, Terminal deoxynucleotidyl transferase, in:"En-
    zymes," vol. 10, p. 145, P.D. Boyer, ed., Academic Press, New
    York.

Bollum, F. J., 1978, Terminal deoxynucleotidyl transferase: Biolo-
    gical studies, in: "Advances in Enzymology," vol. 47, p.347 ,
    A. Meister, ed., John Wiley and Son, New York.

Bollum, F. J., and Brown, M., 1979, A high molecular weight form of
    terminal deoxynucleotidyl transferase,
    Nature, 278: 191.

Boyum, A., 1968, Isolation of mononuclear cells and granulocytes
    from human blood,
    Scand. J. Lab. Inv., 21 (Suppl. 97): 77.

Campagnari, F., Clerici, L., and Bertazzoni, U., 1973, Studies in
    mammalian genetical biochemistry.
    Annual Rep. 1972 Joint Research Center, Ispra, EUR 5060: 403.

Chang, L. M. S. and Bollum, F. J., 1971, Deoxynucleotide-polymeri-
    zing enzymes of calf thymus gland. V. Homogeneous terminal deo-
    xynucleotidyl transferase,
    J. Biol. Chem., 246: 909.

Chang, L. M. S. and Bollum F. J., 1971, Enzymatic synthesis of oli-
    godeoxynucleotides,
    Biochemistry, 10: 536.

Coleman, M. S., Hutton, J. J., De Simone, P., and Bollum, F. J., 1974,
    Terminal deoxyribonucleotidyl transferase in human leukemia,
    Proc. Natl. Acad. Sci. 71: 4404.

Coleman, M. S., Greenwood, M. F., Hutton, J. J., Bollum, F. J., 1976,
    Lampkin, B., and Holland, P., 1976, Serial observations on termi-
    nal deoxynucleotidyl transferase activity and lymphoblasts sur-
    face markers in acute lymphoblastic leukemia.
    Cancer Res., 36: 120.

Coleman, M. S., 1977, Terminal deoxynucleotidyl transferase: charac-
    terization of extraction and assay conditions from human and
    calf tissue,
    Arch. Biochem. Biophys. , 182: 525.

Coleman, M. S., 1977, A critical comparison of commonly used proce-
    dures for the assay of terminal deoxynucleotidyl transferase
    in crude tissue extracts,
    Nucleic Acids Res., 4: 4305.
Deibel, M. R. Jr., and Coleman, M. S., 1979, Purification of a high
    molecular weight human terminal deoxynucleotidyl transferase,
    J. Biol. Chem., 254: 8634.
Donlon, J. A., Jaffe, E. S., and Braylan, R. C., 1977, Terminal de-
    oxynucleotidyl transferase activity in malignant lymphomas,
    New Engl. J. Med., 297: 461.
Goldschneider, J., Gregoire, K. E., Barton, R. W., and Bollum, F. J.,
    1977, Demonstration of terminal deoxynucleotidyl transferase
    in thymocytes by immunofluorescence,
    Proc. Natl. Acad. Sci., 74: 738.
Habeshaw, J. A., Catley, P. F., Stansfield, A. G., Ganeshaguru, K.,
    and Hoffbrand, A. V., 1979, Terminal deoxynucleotidyl transfe-
    rase activity in lymphoma,
    Br. J. Cancer, 39: 566.
Hoffbrand, A. V., Ganeshaguru, K., Janossy, G., Greaves, M. F., Ca-
    towsky, D., and Woodruff, R. K., 1977, Terminal deoxynucleoti-
    dyl transferase levels and membrane phenotypes in diagnosis of
    acute leukemia,
    Lancet, 2: 520.
Hoffbrand, A. V., Ganeshaguru, K., Llewelin, P., and Janossy, G.,
    1979, in: "Biochemical markers in leukemia and lymphoma," p. 25,
    R. Gross and Hellriegel K. P., eds., Springer Verlag, Berlin.
Hoffbrand, A. V., and Janossy, G., 1981, Enzyme and membrane markers
    in leukemia: recent developments,
    J. Clin. Pathol., 34: 254.
Hutton, J. J., Coleman, M. S., Keneklis, T. P., and Bollum, F. J.,
    1979, Terminal deoxynucleotidyl transferase as a tumor cell
    marker in leukemia and lymphoma: results from 1000 patients,
    Adv. Med. Oncol. Res. and Educ., 4: 165.
Kato, K., Goncalves, J. M., Houts, G. E., and Bollum, F. J., 1967,
    Deoxynucleotide-polymerizing enzymes of calf thymus gland. II.
    Properties of the terminal deoxynucleotidyl transferase,
    J. Biol. Chem., 242, 2780.
Koziner, B., Filippa, D. A., Mertelsmann, R., Gupta, S., Clarkson,
    B., Good, R. A., and Siegal, F. P., 1977, Characterization of
    malignant lymphomas in leukemic phase by multiple differentia-
    tion markers of mononuclear cells. Correlation with clinical
    features and conventional morphology,
    Am. J. Med., 63: 556.

Krakow, J. S., Coutsogeorgopoulos, C., and Canellakis, E. S., 1962, Studies on the incorporation of deoxyribonucleotides and ribonucleotides into deoxyribonucleic acid, Biochim. Biophys. Acta, 55: 639.

Kung, P. C., Silverstone, A. E., McCaffrey, R. P., and Baltimore, D., 1975, Murine terminal deoxynucleotidyl transferase: cellular distribution and response to cortisone, J. Exp. Med., 141: 855.

Kung, P. C., Long, J. C., McCaffrey, R. P., Ratliff, R. L., Harrison, T., A., and Baltimore, D., 1978, Terminal deoxynucleotidyl transferase in the diagnosis of leukemia and malignant limphoma, Am. J. Med., 64: 788.

Marks, S. M., Baltimore, D., and McCaffrey, R. P., 1978, Terminal deoxynucleotidyl transferase as a predictor of initial responsiveness to vincristine and prednisone in blastic chronic myelogenous leukemia, New Engl. J. Med., 298: 812.

McCaffrey, R. P., Smoler, D. F., and Baltimore, D., 1973, Terminal deoxynucleotidyl transferase in a case of childhood acute lymphoblastic leukemia, Proc. Natl. Acad. Sci., 70: 521.

McCaffrey, R. P., Harrison, T. A., Parkman, R., and Baltimore, D., 1975, Terminal deoxynucleotidyl transferase activity in human leukemic cells and in normal human thymocytes, New Engl. J. Med., 292: 775.

Ratliff, R. L., Hoard, D. E., Ott, D. G., and Hayes, F. N., 1967, Heteropolynucleotide synthesis with terminal deoxynucleotidyl transferase, Biochemistry, 6: 851.

Thiel, E., Rodt, H., Huhn, D., Netzel, B., Grosse-Wilde, H., Ganeshaguru, K., and Thierfelder, S., 1981, Multimarker classification of acute lymphoblastic leukemia: evidence for further T subgroups and evaluation of their clinical significance, Blood, 56: 759.

Vezzoni, P., Campagnari, F., Di Fronzo, G., and Clerici, L., 1981, Terminal deoxynucleotidyl transferase in human lymphomas: possible existence of forms with high and low molecular weights, Br. J. Cancer, 43: 312.

Yoneda, M., and Bollum, F. J., 1965, Deoxynucleotide-polymerizing enzymes of calf thymus gland. I. Large scale purification of terminal and replicative deoxynucleotidyl transferases, J. Biol. Chem., 240: 3385.

NUCLEAR TERMINAL DEOXYNUCLEOTIDYL TRANSFERASE (TdT) IN LEUKEMIC INFILTRATES OF TESTICULAR TISSUE

G. Janossy*, J.A. Thomas*, O.B. Eden[+], F.J. Bollum[§]

*Department of Immunology, Royal Free Hospital
London, U.K., [+]Bristol Children's Hospital & S.W.
Regional Blood Transfusion Centre, Southmead
Bristol U.K. and[§]Uniformed Services University of the
Health Sciences, Bethesda, Maryland

INTRODUCTION

Carefully monitored chemotherapy regimes and prophylactic control of meningeal disease[1,2]have contributed to prolonged hematological remissions in childhood acute lymphoblastic leukemia (ALL). With this lengthened remission duration leukemic infiltration of the testis has become an increasingly recognised complication in boys especially after cessation of maintenance therapy[3]. Many centers have therefore instituted the policy of routine testicular biopsy before cessation of therapy in order to detect residual disease in this site [4,5].

Leukemic infiltration of the testis is nevertheless difficult to diagnose by routine histological methods, particularly if the tissue is only minimally affected and distorted by the scarring effects of chemotherapy [6]. In recent years membrane and enzyme markers (such as nuclear terminal deoxynucleotidyl transferase; TdT) have been used to identify leukemic blasts [7-9]. In this study indirect immunofluorescence (IF) and immunoperoxidase (IP) techniques were used in order to determine the optimal method for identifying TdT positive leukemic blasts in testicular tissue.

MATERIALS AND METHODS

Preparation of tissue biopsies

Samples of normal thymus were obtained from children undergoing cardiac surgery. Tonsils were collected after elective surgical removal. A total of 40 testicular biopsies were examined.

Some of these had been processed in other laboratories for conventional histology. The tissues were prepared in the following manner:

(i)   <u>Cytocentrifuge preparations</u> (cytospin): Bone marrow and blood leucocytes were separated on Ficoll-Triosil, resuspended and washed in phosphate buffered saline pH 7.2 (PBS). Smears were made in a cytocentrifuge and dried rapidly at RT. These cytospin preparations were fixed in cold methanol at 4°C (30 min) unless otherwise stated. Thymocytes were obtained by teasing out finely cut infant thymus tissues and processed similarly.

(ii)   Fixed and paraffin embedded material: was processed as shown in Table 1.

Table 1

Processing of testicular biopsies

---

fixed in 6% formol-sucrose (6hrs, 20°C)
    or 10% buffered formalin
    alcohol (3 x 2 hrs)
    chloroform (2 x 2 hrs)
    Vacuum and wax embedding

---

dewaxing in xylene (20 mins)
dehydration in methanol (20 mins)
blocking endogenous peroxidase 0.3% H_2O_2 (30 mins)

---

enzyme digestion
    DNA-ase 0.1% for 30 min at RT[*]

---

immunoperoxidase
    normal swine serum (1:30, 30 mins, 20°C)
    rabbit anti-TdT (1µg/10-15µl) 4 hrs 20°C
                or overnight, 4°C) .
    wash (20-30 mins, 20°C)
    swine anti-rabbit IgG (1:50; 30 mins, 20°C)
    wash (30 mins; 20°C)
    rabbit peroxidase anti-peroxidase (1:50; 30 mins, 20°C)
    wash (30 mins; 20°C)
    1 mg/ml diamino-benzidine; 5 mins.
    1 drop of H_2O_2 (5 mins; 20°C)
    wash in tap water
    mounting

---

[*]When tested with Feulgen staining only pale magenta nuclear stains remain. Trypsin is less effective.

Enzyme digestion

0.1% trypsin (BDH) in 0.1% calcium chloride pH 8.6 and different batches of deoxyribonuclease (DNA-se) in 0.1M sodium acetate buffer with 0.005 $MgCl_2$, pH 7.4 were tested. The conditions required for optium digestion of protein and of DNA was assessed by methyl green pyronin and Feulgen staining, respectively (see below). Both enzyme solutions were freshly prepared and used immediately; the magnesium buffer was, however, stored at 4°C.

Antisera

All antisera were used in sufficient amounts to cover the cell preparations on slides (10µl on cytocentrifuge smears and 30-40µl on sections). The incubation was carried out in moist chambers in order to prevent the evaporation or reagents. Rabbit anti-calf TdT (R-anti-TdT) antibody was purified on a TdT immuno-adsorbent column[10] and used in 0.1 µg/µl concentration. The reactivity pattern of this reagent has been extensively character-ised in normal and leukemic tissues[8,9,11]

In the IF test a purified goat anti-rabbit IgG (Fab$_2$) antibody coupled to fluorescein isothiocyanate (G-anti-R-Ig-FITC; 1 µg/1 µl) was used. In the IP test a swine anti-rabbit IgG (Sw-anti-R-IgG: Dako) second layer was used at a 1:50 dilution.

Staining procedure

IP staining was carried out as shown in Table 1. Some sections were lightly counterstained with light green (0.05%, 30 sec) and examined under a light microscope. Other sections were studied without counterstaining under phase contrast.

Parallel samples were processed similarly with IF staining, except that the blocking of endogeneous peroxidase was omitted. After incubation with anti-TdT the preparations were washed for 30 minutes, incubated with G-anti-R-IgG-FITC and washed again (30 min). The mounted samples were examined under a Zeiss fluorescence microscope.

Table 2

Effects of fixation on the preservation and demonstration of TdT in cytocentrifuge preparations of human thymocytes

	6% Formol Sucrose		10% Buffered Formalin		Bouins		Carnoys	
	20°C	4°C	20°C	4°C	20°C	4°C	20°C	4°C
30 mins	N.T.	N.T.	N.T.	N.T.	-	±	-	-
1 hr	+	+	+	+	-	-	-	-
2 hrs	+	+	+*	+	-	-	-	-
3 hrs	+*	+	±	+*	-	-	-	-
4 hrs	+	+*	±	+	-	-	-	-
16 hrs	+	+	-	±	-	-	-	-

* + good staining (see Fig.1)   - staining destroyed

± weak staining   N.T. Not tested

Controls

      To evaluate the staining specificity the test layer (R-anti-
TdT antibody) was replaced by normal swine serum (absorbed with
human tonsil, three times; 1:50 dilution). Additional controls in
the IP test involved omission of the second antibody and sections
stained with DAB only with and without hydrogen peroxide. The
latter was to test the efficacy of blocking endogeneous peroxidase.

RESULTS

Effects of fixation on TdT staining of isolated thymocytes

      Cytospin preparations of thymocyte suspensions were fixed in
various fixatives for different periods of time (Table 2). Excellent
labelling of nuclear TdT was obtained when the smears were fixed in
6% formol sucrose for up to 4 hours at RT and up to 16 hours at
4°C. Similarly, good staining was seen after fixing in 10% formalin
for up to 2 hours at RT or to 4 hours at 4°C. The intensity of
staining was comparable to the strong labelling seen on the smears
fixed in cold methanol by the conventional method. In contrast,
Bouin and Carnoys fixatives quickly destroyed the antigenicity of TdT.

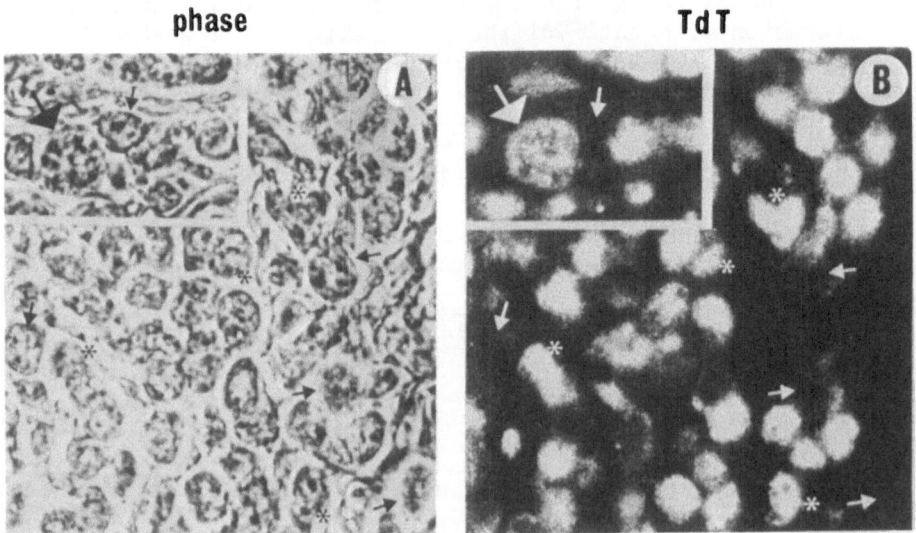

Fig.1   Analysis of TdT with immunofluorescence. The
sample of involved testis was photographed with phase (A)
and fluorescein filter (FITC;B). Some blasts have characteristically shaped nuclei. These are clearly stained for TdT
(asterisk). Note that the nucleoli remain unstained ('holes'
on TdT stain). The blasts are clearly heterogeneous, some
are TdT negative (small arrows). Dividing cells release the
TdT into the cytoplasm (large arrow) from ref.14 with permission.

Effects of enzyme treatment on TdT staining in tissue sections

The experiments above suggested that 10% formalin and 6% formol sucrose may be suitable fixatives. It was soon observed however that the blocks of infant thymus had to be treated with DNA-ase to obtain satisfactory staining for nuclear TdT. Only 2 of 6 batches of DNA-ase were found to be suitable. The other 4 batches of DNA-ase either did not remove the Feulgen-positivity or damaged the sections.

Investigation of leukemic infiltrate in testis

These samples were fixed in formal and routinely processed using optimal DNA-ase treatment. The sections were stained for TdT with both the IP and IF methods.

In 11 cases the routine histology of testicular biopsies (taken as routine procedure or for clinical suspicion) showed heavy leukemic infiltrates. (Table 3) . In 9 of these cases strong nuclear TdT staining was observed in the infiltrating lymphoblasts (Fig.1) while in two cases blast cells were present but were TdT negative. In some cases the leukemic blasts were heterogeneous and some contained no identifiable TdT enzyme in the vicinity of strongly positive blasts.

Table 3

Histological and immunohistological (TdT) analysis of paraffin embedded testicular biopsies taken from 26 cases of ALL in remission

Cases	Histology	Staining for TdT IF or IP*
1-8	+	+
9	+	(+)
10	+	-
11-12	?	-
13-17	-	(+)
18-19	-	+
20-25	-	-

* Immunofluorescence or immunoperoxidase. Cases 6-8 contained blasts heterogeneous in respect of TdT staining (Fig.3)

+ unequivocal leukaemic infiltrate

(+) minimal leukaemic infiltrate

- no evidence of leukaemia

? equivocal histology

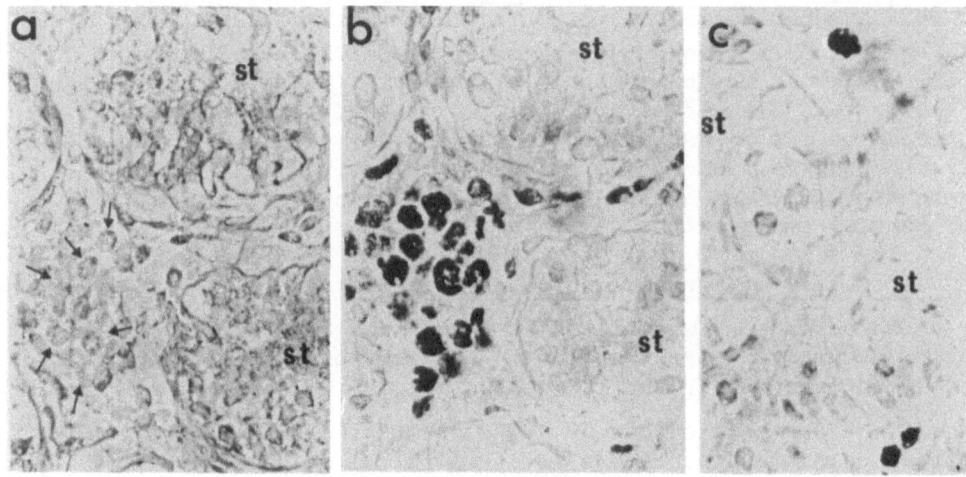

Fig.2   Analysis of testicular biopsies with immuno-peroxidase
a: negative control preparations were incubated with second
layers only but not with anti-TdT antibody. The lymphoblasts
are peroxidase negative (arrows). b. Adjacent section to 'a'
stained for TdT. Infiltrating blasts are heavily labelled.
c: in this testicular sample the anti-TdT antibody detects
scattered TdT positive blasts which could not be identified
with conventional morphology. St: seminiferous tubules.

    The morphological details obtained were also remarkable. The
TdT positivity corresponded to the nucleus and clear cytoplasmic
staining was seen only in the dividing cells where TdT is known to
be released from the disrupted nucleus. (Fig.1). Some positive
cells had cigar-shaped nuclei and represented elongated migrating
forms (Fig.2) which were easily overlooked on the conventional
preparations. Other cells contained small TdT negative 'holes'
within the nucleus, which corresponded to the nucleolus. These
patterns in some blasts gave the appearance of a pseudo-lobulated
nucleus (Fig.2)

    The quality of TdT staining was poorer when involved testicular
biopsies were frozen and the sections cut in a cryostat, and were
transferred onto 'warm' slides. When the tissue thaws up on the
warm slide the TdT molecules diffuse out from the nucleus and the
staining looks cytoplasmic. This technique can be recommended.

    It is important to point out, however, that the sections of
frozen biopsies can be fixed within the cryostat (in either
methanol or in 10% buffered formalin at -20° for 15 minutes), and
then transferred onto albumin-gelatin coated cold dry slides. These
fixed sections stick to the slides well and do not require DNA-ase
treatment for TdT staining. We have shown that the intensity of

TdT staining and the histological preservation of the slides prepared this way is superior to the paraffin embedded material where
the TdT staining is weak and the DNA-ase treatment may slightly
damage the morphology.

Analysis of minimal testicular involvement and negative controls

These were paraffin embedded blocks treated with DNA-ase.   In
7 biopsies from patients with treated ALL (nos 13-19) the normal
structure of seminiferous tubules was preserved and there was no
histological evidence of increased lymphoid or lymphoblastic infiltration (Table 3).   Large areas of the tissue were void of TdT
positive cells.   Interestingly, however, in three cases (Nos 13-
15) scanty and widely dissmeninated cells in some areas of interstitial tissue contained TdT.   In one case (No 15) a few TdT
positive cells were seen in the subcapsular area (Fig.2).   When
these cells were reinvestigated with a 100x oil objective,
characteristic patchy IF staining and IP-PAP complexes deposited
within the 'crevasses 'in between the ridges of nuclear chromatin
bands could be seen.   This pattern is typical of TdT staining.   It
is most markedly seen in thymocytes, where the amount of detectable
TdT is moderate or low.

No identifiable TdT positive cells were present in another
seven testicular biopsies (Nos.20-26) from patients with treated
ALL.   With one exception (see above) all cases were regarded at
the histological investigation as apparently non-involved samples.
The negative controls studied included two samples of normal
autopsy material from non-leukemic individuals and five samples of
human tonsil tissue.   No identifiable TdT positive cells were seen
in any of these biopsies.   A further negative control was the
staining with second layer antibodies only omitting the R-anti-TdT
antibody.   Sections adjacent to those used for staining with R-anti-
TdT were studied.   The TdT positive lymphoblasts were totally unstained in the negative controls. (Fig.2).

Comparison of IP-PAP and IF analysis

Using optimal methods both peroxidase and fluorescence
analysis gave reliable observations.   The advantages of the IP
technique were two-fold.   The permanent preparations allowed the
careful snlysis of nuclear details with high manification on
positively labelled cells.   The counterstaining with methyl green,
although not essential with phase contrast microscopes, helped
the comparison with conventional haematoxylin-eosin staining. The
IP method was nevertheless very difficult to standardize.

The IF staining was the quicker and easier test but it had one
disadvantage .   The red cells showed autofluorescence which has to
be distinguished from TdT staining.   This was done by demonstrating
red cell fluorescence on both the fluorescein (green) and rhodamine
(red) fluorescence channel, as opposed to the genuine TdT staining
which was visible only on the fluorescein (FITC,green) channel.

DISCUSSION

        In normal human tissues TdT positive cells can only be
found within the bone marrow and thymus. At any other site TdT
positive cells suggest residual ALL [8,9]. This finding has
recently been exploited to detect ALL cells in the cerebrospinal
fluid [12]. In this study we standardized the method to investigate
residual testicular ALL.

        Three factors have proved to be important. First, biopsy
material fixed in formalin-sucrose was superior to other fixatives
such as Bouin or Carnoy, and gave better nuclear TdT localization
than conventional sections made from frozen biopsies. Second, the
sections obtained from formalin-fixed paraffin embedded blocks had
to be digested with DNA-ase. Only selected batches worked in our
hands and it was very difficult to find the optimal conditions for
TdT staining. It seems therefore that the TdT staining of
testicular biopsies may remain, at least until the development of
a standard kit, in the realm of the specialized immunohistological
laboratories. Third, we have developed a simple alternative for
the difficult method of using paraffin-embedded blocks with DNA-
ase treatment. This modification uses frozen biopsies which are
<u>cut and fixed</u> in the cryostat at -20° thus preventing the diffusion
of TdT from the nucleus (for details see ref. 14).

        The advantage of establishing the TdT method in paraffin embedded
tissue is that it is now possible to perform retrospective studies
on samples fixed and embedded by routine histological methods. The
staining for TdT can clearly detect even scanty ALL blasts which
are scattered in the interstitial testicular tissue (Fig.2) and
do not form lymphoid cell clusters. It is therefore possible that
the TdT staining is a more sensitive method than conventional
histology in this specialized tissue.

        The significance of this approach is as follows. In one series
of 170 biopsies performed during or at the end of maintenance ALL
therapy 18 samples were frankly positive for leukemic cells in
routine histology[5]. The observations on a further 17 samples (10%)
were difficult to interpret because of the presence of inflammatory
and large mesenchymal cells or due to inadequate 'crushed' material.
In addition, 13 samples (7%) were reported as histologically egative
but subsequently relapsed in the testis. The median time between
these biopsies and the relapses was 12 months. These 'false negative'
samples might result from a missed focal deposit. We have already
observed one such case where TdT$^+$ cells were detected in the subcapsular space (Table 3) and the patient relapsed four months later.
Alternatively, relapse may derive from a diffuse cellular infiltrate
which, when scanty, might be difficult to notice without a specific
marker. In this respect it is important to point out that even
where a definite infiltration is diagnosed histologically it is
most frequently diffusely distributed and not focal [13].

One cannot help feeling that once the histopathologists realise the importance of TdT test in testicular ALL they might start storing frozen biopsies for analysis. This would make the tests easier and more reliable, and it would be in the patients' interest. The modification of this test for obtaining optimal TdT staining is simple (see above).

ACKNOWLEDGEMENTS

We thank Miss Wendy Daniels for technical help and Dr.J. Pincott, Histopathology Department, Institute of Child Health, Drs. K. Tiedemann and A. Smith, Medical Oncology Group, St Bartholomew's Hospital for allowing us to analyse their patients.   We thank Debra Warren for her excellent secretarial assistance.

REFERENCES

1.   D. Pinkel: Treatment of acute leukemia, Ped.Clin. North Am. 23: 117 (1976).
2.   D.R. Miller, Acute Lymphoblastic Leukemia, Ped.Clin. North Am. 27: 269 (1980)
3.   O.B. Eden, R.M. Hardisty, E.M. Innes, H.E.M Kay and J. Peto: Testicular disease in acute lymphoblastic leukemia in childhood, Brit.Med.J. i: 334 (1978)
4.   T.H. Kim, V.K. Liu, R.D. Woodruff, A.H. Ragab: Testicular biopsy prior to termination of leukemic therapy. J.Paediatr. 94: 95 (1979)
5.   O.B. Eden, A. Rankin: Testicular biopsies in childhood lymphoblastic leukemia. Presented 12th meeting International Society of Pediatric Oncology, Budapest (1980)
6.   M. Lenden, I.M. Hann, M.K. Palmer, S.M. Shalet, and P.H Morris-Jones: Testicular histology after combination chemotherapy in childhood for acute lymphoblastic luekemia. Lancet ii: 439 (1978)
7.   R.P. McCaffrey, A. Harrison, B.S.Parkman, D. Baltimore: Terminal deoxynucleotidyl transferase activity in human leukemic cells and normal thymocytes. N.Eng.J.Med. 292: 775 (1975)
8.   F.J. Bollum: terminal deoxynucleotidyl-transferase as a hemopoietic cell marker: Blood 54: 1203 (1979)
9.   G. Janossy, F.J. Bollum, K.F. Bradstrock, and J. Ashley: Cellular phenotyes of normal and leukemic hemopoietic cells determined by selected antibody combinations:Blood 56: 430 (1980)
10.   F.J. Bollum: Antibody to terminal deoxynucleotidyl transferase: Proc.Nat.Acad.Sci USA 72: 4119 (1975)
11.   K.F. Bradstock, G. Janossy, G. Pizzolo, A.V. Hoffbrand, A. McMichael, J.R. Pilch, C. Milstein, P. Beverley,and F.J Bollum. Subpopulations of normal and leukemic human thymocytes: An analysis with the use of monoclonal antibodies. J.Nat. Cancer.Inst 65 : 33 (1980)

12.   K.F. Bradstock, E.S. Papageorgiou, and G. Janossy: Diagnosis
of meningeal involvement in patients with acute lymphoblastic
leukemia using immunofluorescence for terminal transferase:
Cancer 47: 2471 (1981)
13.   T. Kuo, T-P. Tschang, and J-Y. Chu: Testicular relapse in
childhood during bone marrow remission: Cancer 38: 2604 (1976).
14.   J.A. Thomas, G. Janossy, O.B. Eden, F.J. Bollum: Immunological
identification of leukemic infiltrates in testes: Blood (in press)

PHENOTYPIC CHANGES DURING BLAST CRISIS OF CML CHARACTERIZED BY OCCURRENCE AND LOSS OF TDT

Elisabeth Paietta, Josef D. Schwarzmeier

First Medical Dep., Univ. of Vienna, Medical School
Lazarettgasse 14
A-1090 Vienna, Austria

SUMMARY

Phenotypic changes in a case of blast crisis of chronic myelo-genous leukemia (CML-BC) were characterized by serial terminal transferase determinations simultaneously related to morphological, cytochemical, and cytogenetic data.

INTRODUCTION

Elevated levels of terminal deoxynucleotidyl transferase (TdT) activity have been observed in approximately one third of all pa-tients with CML-BC (Sarin et al., 1976; Kung et al., 1978; Janossy et al., 1978; Modak et al., 1980). The presence of TdT activity in CML-BC has proved to be of therapeutic significance in that the TdT containing cells are highly sensitive to hydrocortisone and its derivatives (Marks et al., 1978; Janossy et al., 1979). The loss of TdT in CML blast cells has recently been found parallelled by the emergence of therapy-resistant cells (Hutton and Coleman, 1976; Ross et al., 1979).

We present a case of CML-BC which was initially characterized by the predominance of a TdT positive cell population. The elimi-nation of this cell clone by chemotherapy was followed by the rise of a cell type with morphological features of myeloid blasts. Later on, TdT positive cells reappeared. This time, the therapeutic re-sponse was short and at relapse, a TdT negative cell clone with lymphatic morphology and cytochemical characteristics emerged. This clone was resistant to any further treatment.

METHODS

Mononuclear cells were isolated from heparinized peripheral blood or bone marrow samples according to Boyum (1968).

For TdT determination the micromethod developed by Modak et al. (1980) was used. TdT activity was assayed with ^{3}H-deoxyguanosine triphosphate (specific activity 10-12 Ci/mmol) as substrate and a polymer of deoxyadenylic acid (chain length 12-18 residues) as primer. Results were calculated from the difference in incorporation in the absence and presence of adenosine triphosphate, a specific inhibitor of TdT (Mertelsmann et al., 1978), and expressed in units/10^8 cells (1 unit = 1 nmol ^{3}H-deoxyguanosine monophosphate incorporated in 1 hour at 37°C).

Karyotyping was done on chromosome preparations from mononuclear peripheral blood cells cultured without phytohemagglutinin for 24-72 hours. G banding was performed using a modification of Seabright's technique (Seabright, 1971).

CASE REPORT

In October 1979, six years after the first diagnosis of Philadelphia chromosome (Ph[1]) positive CML (46,XX,Ph[1]), the 38-year-old female patient presented with fever, generalized lymphadenopathy, hepato- and splenomegaly, a white blood cell count (WBC) of 27 000 per μl and greater than 90% of blast cells in the differential count. These blast cells were positive for the acid phosphatase stain (focal pattern) and expressed markedly increased TdT activity in both the peripheral blood and the bone marrow. Figure 1 summarizes the levels of TdT activity in the peripheral mononuclear cells, total leukocyte count (WBC) and the absolute number of lymphoid blasts at the onset and throughout the course of blast crisis of our patient. The lymphoid blast cells which initially dominated were 100% Ph[1] positive - t(9q+;22q-) - without any additional karyotype aberrations. The TdT activity in the peripheral mononuclear cells amounted to 29 units/10^8 cells and to 57 units/10^8 cells in the bone marrow. Treatment with vincristine (V) and prednisone (PRED) resulted in a rapid decrease in both WBC and TdT. After 4 months of repeated V and PRED administrations, the TdT in the peripheral blood stabilized at a level of 0.4 units/10^8 cells. Despite maintenance therapy with PRED and constantly low TdT levels, the WBC again started to rise and a population of large blasts with myeloid morphological features appeared, negative for cytochemical stains and 100% Ph[1] positive (46,XX,Ph[1]). Therapy with 1,4-dimethane sulfonoxybutane (busulfan) largely eliminated this cell population but subsequently, TdT levels increased in the peripheral blood (32 units/10^8 cells) and in the bone marrow (23 units/10^8

Table 1. Characterization of The Various Phenotypes of Blast Cells Emerging in The Peripheral Blood of The Patient During The Course of Blast Crisis of CML

PHENOTYPE	MORPHOLOGY	CYTOCHEMISTRY		TdT	CYTOGENETICS
		Pox	acid P'tase (focal)	(units/10^8 cells)	
1	lymphoid	neg	pos	29,2	Ph^1
2	myeloid	neg	neg	0,4	Ph^1
3	lymphoid	neg	pos	32,2	Ph^1
4	lymphoid	neg	pos	<0,2	Ph^1

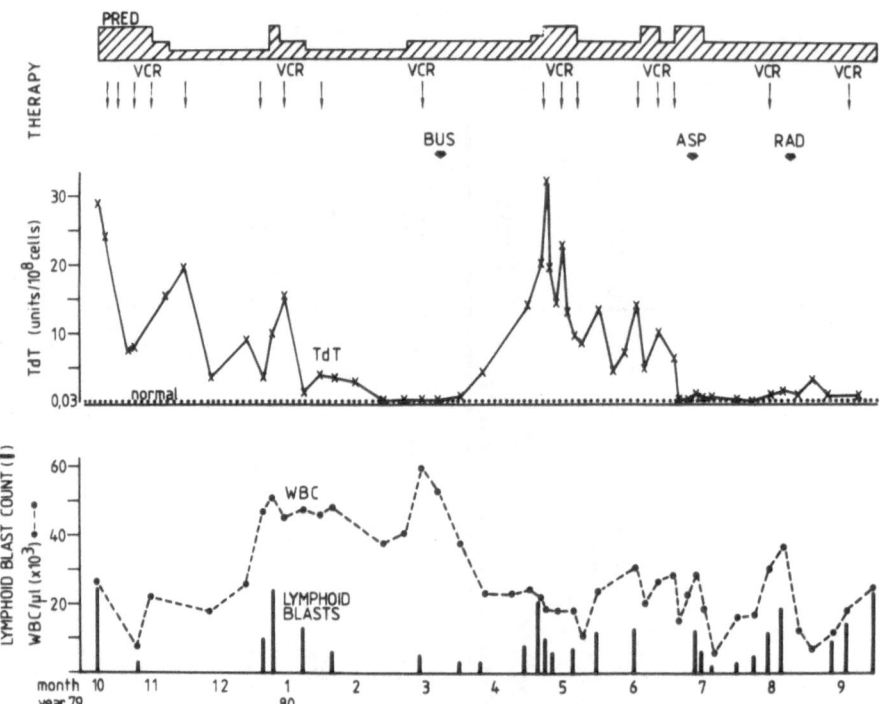

Fig. 1.   Therapy, TdT activity, WBC and the absolute number of lymphoid blasts in the peripheral blood of the patient from
time of blastic transformation of CML. Drugs are abbreviated as follows: PRED=prednisone (25,50,75 or 100 mg daily);
V=vincristine (arrows indicate daily injections of 1.8 to
2 mg V); BUS=busulfan (3x2 mg daily for 11 days); ASP=
L´asparaginase (200 units/kg body weight daily for 5 days);
for BUS and ASP, arrows indicate the date of first drug
application; RAD=radiation (200 rad daily for 6 days; arrow
indicates the first day of irradiation).

cells) enabling the early recognition of a second lymphatic blast
crisis which was dominated by small round, acid phosphatase positive
and Ph[1] positive blast cells. Administration of V and augmented dos-
ages of PRED again reduced the percentage of lymphoid blasts and the
TdT activity. Although the enzyme activity finally reached levels
below 0.2 units/10[8] cells, the WBC remained elevated between 15 and
30 000/µl accompanied by extensive splenomegaly. The blast cells
still present were of lymphoid origin according to morphology and
cytochemistry, with the karyotype 46,XX,Ph[1] but they lacked TdT
activity and did not respond to the therapy schedule chosen so far.
Treatment with L´asparaginase caused a transient drop in WBC and a

decrease of the spleen size but did not reduce the lymphoid blast count below 30%. The employment of any other cytostatic drug combination was precluded by severe thrombocytopenia. To reduce the size of the rapidly enlarging spleen, splenic irradiation was initiated. This gave only an unsatisfactory response. Under the constant rise of lymphoid blasts and massive splenomegaly, the patient died with symptoms of severe thrombocytopenia and anemia approximately one year after the first diagnosis of CML-BC.

DISCUSSION

We closely monitored the blast crisis in a case of CML using morphological, cytochemical, cytogenetic and biochemical (TdT) parameters and we were able to observe clonal changes during the course of therapy. The following table summarizes the phenotypes which emerged in the various phases of the blast crisis (Tab. 1). The first phenotype was lymphoid according to morphology and cytochemistry, expressed high levels of TdT activity and was Ph^1 positive; the second phenotype was myeloid according to morphology and Ph^1 positive; the third phenotype showed characteristics identical to those of the first one; and, finally, the fourth phenotype was again of lymphoid origin and Ph^1 positive but lacked TdT activity. Even though we found no significant chromosomal changes accompanying the phenotypic changes, the existence of two distinct Ph^1 positive, lymphoid blast cell populations must be postulated differing both in TdT activity and response to treatment. This failure of morphology, cytochemistry and cytogenetics to distinguish between these individual phenotypes characterized the TdT as a cell marker of not only important diagnostic but especially therapeutically prognostic value.

REFERENCES

Boyum, A., 1968, Isolation of mononuclear cells and granulocytes from human blood, Scand. J. Clin. Lab. Invest., 21 (Suppl. 97):77

Hutton, J.J., and Coleman, M.S., 1976, Terminal transferase measurement in the differential diagnosis of adult leukemias, Br. J. Hematol., 34:447

Janossy, G., Woodruff, R.K., Paxton, A., Greaves, M.F., Capellaro, D., Kirk, B., Innes, E.M., Eden, O.B., Levis, C., Catovsky, D., and Hoffbrand, A.V., 1978, Membrane marker and cell separation studies in Ph^1 positive leukemia, Blood, 51:861

Janossy, G., Woodruff, R.K., Pippard, M.J., Prentice, G., Hoffbrand, A.V., Paxton, A., Lister, T.A., Bunch, C., and Greaves, M.F., 1979, Relation of "lymphoid" phenotype and response to chemotherapy incorporating vincristine-prednisone in the acute phase of Ph[1] positive leukemia, Cancer, 43:426

Kung, P.C., Long, J.C., McCaffrey, R.P., Latliff, R.L., Harrison, T.A., and Baltimore, D., 1978, Terminal deoxynucleotidyl transferase in the diagnosis of leukemia and malignant lymphoma, Am. J. Med., 64:788

Marks, S.M., Baltimore, D., and McCaffrey, R., 1978, Terminal transferase as a predictor of initial responsiveness to vincristine and prednisone in blastic chronic myelogenous leukemia, N. Engl. J. Med., 298:812

Mertelsmann, R., Mertelsmann, I., Koziner, B., Moore, M.A.S., and Clarkson, B.D., 1978, Improved biochemical assay for terminal deoxynucleotidyl transferase in human blood cells: Results in 89 adult patients with lymphoid leukemias and malignant lymphomas in leukemic phase, Leuk. Res., 2:57

Modak, M.J., Mertelsmann, R., Koziner, B., Pahwa, R., Moore, M.A.S., Clarkson, B.D., and Good, R.A., 1980, A micromethod for determination of terminal deoxynucleotidyl transferase (TdT) in the diagnostic evaluation of acute leukemias, J. Cancer Res. Clin. Oncol., 98:91

Ross, D.D., Wiernik, P.H., Sarin, P.S., and Whang-Peng, J., 1979, Loss of terminal deoxynucleotidyl transferase (TdT) activity as a predictor of emergence of resistance to chemotherapy in a case of chronic myelogenous leukemia in blast crisis, Cancer, 44:1566

Sarin, P.S., Anderson, P.N., and Gallo, R.C., 1976, Terminal deoxynucleotidyl transferase activities in human blood leukocytes and lymphoblast cell lines: High levels in lymphoblast cell lines and in blast cells of some patients with chronic myelogenous leukemia in acute phase, Blood, 47:11

Seabright, M., 1971, Rapid banding technique for human chromosomes, Lancet, ii:971

ACKNOWLEDGEMENTS

We are indebted to the Institute of Cancer Research in Vienna, especially to the Drs. P. Fischer and O. Haas for their collaboration in performing the chromosomal analyses.

This work was financially supported by Grants of the Fonds zur Förderung der wissenschaftlichen Forschung in Österreich, Projects no. 2746, 3609, and 4106

ENRICHMENT OF TERMINAL DEOXYNUCLEOTIDYL TRANSFERASE ACTIVITY BY CELL SEPARATION

P. S. Sarin, M. Virmani and R. C. Gallo

Laboratory of Tumor Cell Biology
National Cancer Institute
Bethesda, Maryland 20205

SUMMARY

Terminal deoxynucleotidyl transferase is a unique DNA poly-merase that can carry out DNA synthesis on an initiator molecule in the absence of a template. The usefulness of this enzyme as a biological marker for following patients during treatment and remission has been suggested. The potential usefulness of this enzyme in predicting the onset of relapse before any morphological indications has been demonstrated in chronic myelogenous leukemia patients in blast phase of the disease. In order to be able to detect low levels of TdT activity especially during remission phase, we have used cell separation techniques which can enrich cell populations containing TdT activity. A number of cell separation techniques have been developed to separate different cell types. We have used the techniques of unit gravity sedimentation and free flow electrophoresis to achieve enrichment of TdT positive cell populations. Our results show that up to 20 fold enrichment of TdT activity in normal human bone marrow can be accomplished by using cell separation techniques. With the use of free flow elec-trophoresis, we have achieved enrichment of TdT positive cell populations from normal human bone marrow, cells from patients with acute lymphoblastic leukemia and chronic myelogenous leukemia in blast phase of the disease. No TdT positive cells were detected in patients with acute myelogenous leukemia. These cell separation techniques should prove to be useful in early detection of relapse in patients in remission.

INTRODUCTION

Terminal deoxynucleotidyl transferase (terminal transferase, TdT) is an enzyme that can catalyze the polymerization of deoxyribonucleotides on the 3'-OH ends of oligo- and polydeoxyribonucleotide initiators in the absence of a template. This enzyme was discovered in thymus tissue and was considered to be specific for this tissue[1]. High levels of this enzyme have since been detected in various forms of human leukemia[2], including acute lymphoblastic leukemia (ALL)[3,5,7-10], acute myelomonocytic leukemia (AMML)[4], acute undifferentiated leukemia[11], and chronic myelogenous leukemia in blast phase of the disease[6-8]. High levels of terminal transferase have also been detected in cell lines with T cell characteristics, such as Molt-4 and 8402, derived from leukocytes of patients with acute lymphoblastic leukemia[12,13].

Detection of terminal transferase in various forms of leukemia raises a question as to whether TdT is present in all cells or only in a select population of cells. In addition, in order to develop this enzyme as a biological marker for predicting the onset of relapse in patients during remission it is extremely important to be able to detect low levels of this enzyme activity. A number of cell separation techniques have recently been developed. One of the techniques separates cells into subpopulations based on net surface charge differences. The technique of free flow electrophoresis has been used in the past for separation of human[14] and mouse[15] bone marrow cells, and for the separation of cell membranes and organelles[16]. Separation of acute leukemic cells into different cell populations has also been achieved with the help of centrifugal elutriation[17]. We have utilized the technique of free flow electrophoresis for separation of human leukemic cells into TdT positive cell populations. With this technique it is possible to obtain up to 20 fold enrichment in TdT positive cells thus making the detection of low levels of enzyme simpler, and providing a tool for selection and identification of cell types containing this enzyme for biochemical and biological studies.

MATERIALS AND METHODS

Materials

Tritium labeled deoxyribonucleoside triphosphates were obtained from New England Nuclear, Boston, Mass. Oligo- and polydeoxyribonucleotides and deoxyribonucleoside triphosphates were obtained from P. L. Biochemicals and Sigma Chemicals.

Source of Cells

Cultured cells from T (8402) and B (8392) cell lines established from patients with acute lymphoblastic leukemia were grown
at Biotech Laboratories. Fresh human cells were obtained by
leukaphoresis from patients with leukemia, and were obtained from
M. D. Anderson Hospital, Houston, Texas and NIH Clinical Center.

Processing of Cells

Human leukemic cells (2×10^9 cells) were washed with RPMI-
1640 at 4°C. The process was repeated three times. The cells
were layered on top of isolymph (ficol/hypaque) and centrifuged at
400g at room temperature. The separated leukocytes at the interface
were removed, washed with phosphate buffered saline (PBS, pH 7.4)
twice and used for the cell separation studies.

Cell Separation by Free Flow Electrophoresis

Separation of human normal and leukemic cells was performed
with a free flow electrophoresis apparatus model FF5 (Bender and
Hobein, Munich, Germany) using conditions similar to the ones
already described[14-16]. Briefly, $2-5 \times 10^8$ cells were suspended
in 10 ml of chamber buffer (15 mM triethanolamine, pH 7.4, 10 mM
glucose and 4 mM potassium acetate) (TGK buffer), and applied
through the entry porthole of the electrophoresis chamber at a
dosage pump speed of 350. The Suction pump speed for the buffer
in the chamber was set at 140. The separation chamber was maintained at 5°C. Voltage was adjusted to 550 volts, and the current
was 150 amp. The electrode buffer consisted of 75 mM triethanolamine (pH 7.4) and 4 mM potassium acetate (TK buffer). At these
settings approximately 2×10^8 cells can be separated per hour.
Cell fractions were collected in the fraction collector maintained
at 4°. Cells were pelleted and washed with RPMI 1640. A portion
of the cell pellet was used for making cytospin slides and the
other was extracted in 0.5 ml buffer A (50 mM Tris-HCl (pH 7.5), 5
mM dithiothreitol and 20% glycerol) containing 0.3 M KCl and 0.3%
triton X100.

Cell Extraction

Cell pellets suspended in buffer A containing KCl and triton
X100 were sonicated at maximum output of a Branson sonifer (4 x 30
seconds) using a microtip. The extract was stirred in the cold
for two hrs. and centrifuged at 100,000 xg for one hr. The supernatant was analyzed for the presence of terminal transferase and
cellular DNA polymerase activity.

Enzyme Assays

Terminal transferase activity was assayed at 37°C for one hr. as described[6],[7],[13] in a standard reaction mixture (0.05 ml) which contained 50 mM Tris-HCl (pH 7.8), 50 mM KCl, 0.6 mM Mn^{2+}, 5 mM dithiothreitol, 100 µM of the labeled deoxyribonucleoside triphosphate, 2.5 µg of poly dA as the initiator and 5 µl of the enzyme fraction. The specific activity of $[^{3}H]dGTP$ used was 1030 dpm/pmol. The reaction was arrested by the addition of 50 µg of yeast tRNA and 2 ml of 10% trichloroacetic acid, collected on millipore filters, washed with 5% trichloroacetic acid containing 0.02M sodium pyrophosphate, dried and counted in a scintillation counter.

Cellular DNA polymerase activity was measured under conditions similar to those described for terminal transferase except $(dT)_{15}(dA)_{n}$ was used as the primer-template and $[^{3}H]$ TTP was used as the radiolabeled triphosphate. DNA polymerase activity was assayed both in the presence of 0.6 mM Mn^{2+} or 10 mM Mg^{2+}. Cell fractions containing enzyme activity were pooled and TdT purified by various column chromatographic techniques as described[7],[18].

RESULTS

The unique biochemical properties of TdT (see Table 1 for comparison of the properties of DNA polymerases) make it a novel enzyme for easy detection and characterization. To be effective as a biological marker, the enzyme should be detectable in low amounts so as to have a predictive value for early detection of relapse. Our studies on TdT positive patients with chronic myelogenous leukemia (CML) in blast crisis indicate that TdT positive CML patients respond to vincristine and prednisone therapy[20],[21] and that TdT levels begin to increase 5-6 months before any morphological indications of relapse[20].

In our earlier attempts to enrich TdT positive cell populations from human bone marrow, we have used the technique of unit gravity sedimentation in sucrose[19]. Various cell separation techniques are available and are summarized in Table 2. In order to be able to obtain a large number of cells for biochemical and biological studies, it is important to utilize a separation technique that can separate large quantitites of cells in a relatively short period. Free flow electrophoresis is one such technique by which it is possible to separate up to 2×10^{8} cells per hour. This method separates cells based on their net surface charge differences. This method could be used both for the detection of low levels of TdT as well as for obtaining large numbers of TdT positive cells for biochemical and biological studies. The protocol used for the free flow electrophoresis (FFE) of fresh human normal and leukemic cells and cell lines is outlined in Table 3.

Table 1: Characteristics of DNA Polymerases from Mammalian Cells and Type C RNA Tumor Viruses

Property	DNA Polymerase				
	α	β	γ	TdT	RT (Mammalian) Virus
1. Location	Cytoplasm	Nucleus	Cytoplasm & Nucleus	Nucleus	
2. Molecular Weight	150,000	40,000	90,000	32,000 (CT) 60,000 (Human)	70,000–100,000
3. Template/Initiator	Acitvated DNA	Activated DNA	$dT_{15} \cdot (A)_n$	$(dA)_{15-100}$	$(dT)_{15} \cdot (A)_n$
4. Divalent Cation	Mg^{++}	Mg^{++}	Mn^{++}	Mn^{++}/Mg^{++}	Mn^{++}
5. $dG_{15} \cdot (C)_n$/ $dG_{15} \cdot (OMeC)_n$	-	-	-	-	++
6. Response of antibody to:					
(a) α	++	+	-	-	-
(b) β	+	++	-	-	-
(c) γ	-	-	++	-	-
(d) TdT	-	-	-	++	-
(e) Reverse Transcriptase (RT)	-	-	-	-	++

Table 2:   Cell Separation Techniques

_____

1. Ficoll - Hypaque
2. Staput Gradients. Unit Gravity Sedimentation in Sucrose
    Gradients
3. Centrifugal Elutriation
4. Free Flow Electrophoresis
5. Electrophoresis in Sucrose Gradients
6. Affinity Chromatography
7. Fluorescence Activated Cell Sorter (FACS)

_____

Table 3:   Separation of Human Leukemic Cells and T/B Cell Lines
           by Free Flow Electrophoresis

_____

Fresh Leukemic Cells (Suspend in RPMI 1640)
|
Separate on Isolymph (Ficoll-Hypaque)
|
Interface Cells/or T or B Cells
|
1-5x10^8 Cells Resuspended in 10 ml. of Chamber Buffer
|
8 ml. Applied for Free Flow Electrophoresis
|
Fractions Count for Cell Number
|
Centrifuge at Low Speed
|
Make Slides from each Fraction for Morphological Study.
Assay for Terminal Transferase and Cellular DNA
Polymerase Activity

_____

Separation of T and B Cells

The cells from a patient with acute lymphoblastic leukemia have
been used to establish a T (8402) and a B cell line (8392). This
T cell line contains terminal transferase and is a useful cell line
to determine the distribution of cells positive for terminal transferase after free flow electrophoresis. We have used these cell
lines to check the distribution of T and B cells on free flow
electrophoresis. Figure 1 shows the separation of T cells (8402)
on free flow electrophoresis. As can be seen from this figure,
majority of the terminal transferase and cellular DNA polymerase
activity is contained in cells in fractions 20-30. These cells

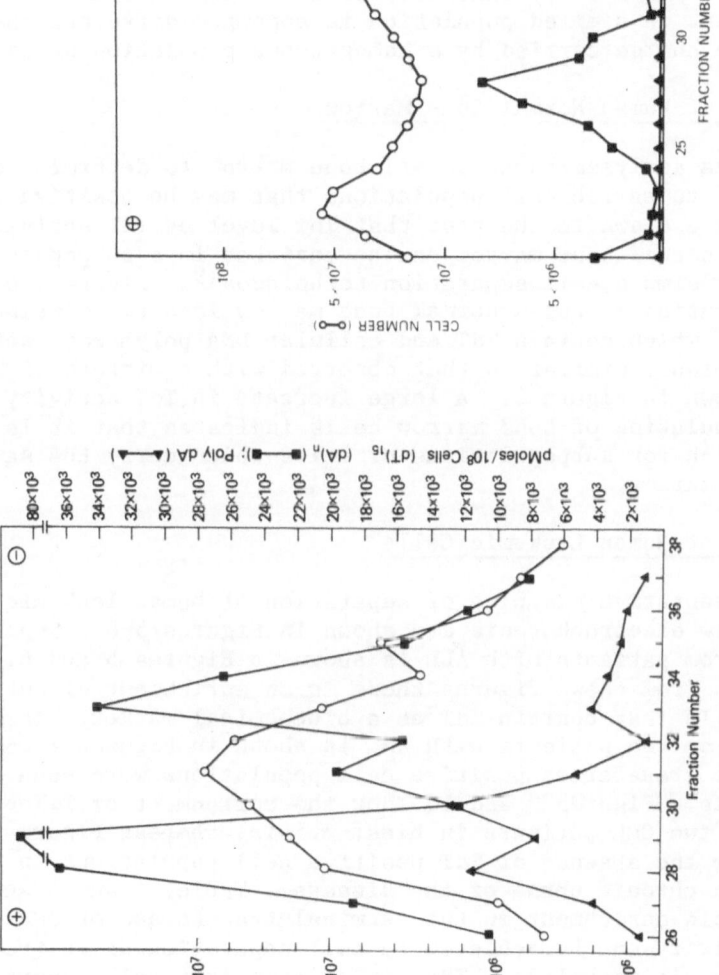

Fig. 1. Separation of T cells by free flow electrophoresis. Cell number (0---0); Poly dA (▲---▲); and (dT)₁₅·(dA)ₙ (■---■).

Fig. 2. Separation of B cells by free flow electrophoresis. Cell number (0---0); Poly dA (▲---▲); and (dT)₁₅·(dA)ₙ (■---■).

migrate toward the anode and carry a net negative surface charge.
Figure 2 shows the distribution of B cells (8392) on free flow
electrophoresis. A major portion of the cellular DNA polymerase
activity is distributed between fractions 25 and 30. Separation
of a mixture of T and B cells (a mixture of $2x10^8$ cells each of
8402 and 8392) by free flow electrophoresis is shown in Figure 3.
In this figure the leading and the trailing edge of the various
fractions contain terminal transferase and cellular DNA polymerase
activities. This is in contrast to the distribution of terminal
transferase and cellular DNA polymerase activities in T and B cells
shown in Figures 1 and 2. The distribution observed in Figure 3
suggests the possibility that net surface charge carried by the
cells present in a mixed population is somewhat different than the
net surface charge carried by a homogeneous population of cells.

Separation of Human Normal Bone Marrow

We have analyzed human normal bone marrow to determine if it
is possible to enrich cell populations that may be positive for
TdT. We have shown in the past that low level of TdT activity
present in normal bone marrow can be enriched in a subpopulation
of cells by simple cell separation techniques[19]. Figure 4 shows
the distribution of human normal bone marrow into two distinct cell
populations which contain TdT and cellular DNA polymerase activity.
This is somewhat similar to that observed with a mixture of T and
B cells shown in Figure 3. A large increase in TdT activity seen
in a subpopulation of bone marrow cells indicates that it is possible to enrich for subpopulations of cells which carry the same
biological marker.

Separation of Human Leukemic Cells

Representative examples of separation of human leukemic cells
by free flow electrophoresis are shown in Figures 5-8. Separation
of cells from patients with ALL is shown in Figures 5 and 6. As
can be seen from these figures there is an enrichment of subpopulation of cells that contain TdT as a biochemical marker. Analysis
of cells from two patients with AML is shown in Figures 7 and 8.
No terminal transferase positive cell populations were seen in these
AML patients. Figures 9 and 10 show the enrichment of TdT positive
cells from two CML patients in blast crisis, whereas Figures 11
and 12 show the absence of TdT positive cell populations in CML
patients in chronic phase of the disease. Tables 4 and 5 show that
up to 30 fold enrichment in the terminal transferase or DNA polymerase activity can be achieved by cell separation using free flow
electrophoresis technique. Thus it appears that cell separation
techniques may prove to be useful for the detection of biochemical
markers that may be expressed in these cells. This technique could
be very useful in following patients during remission phase for
early detection of biological markers that may signal the onset

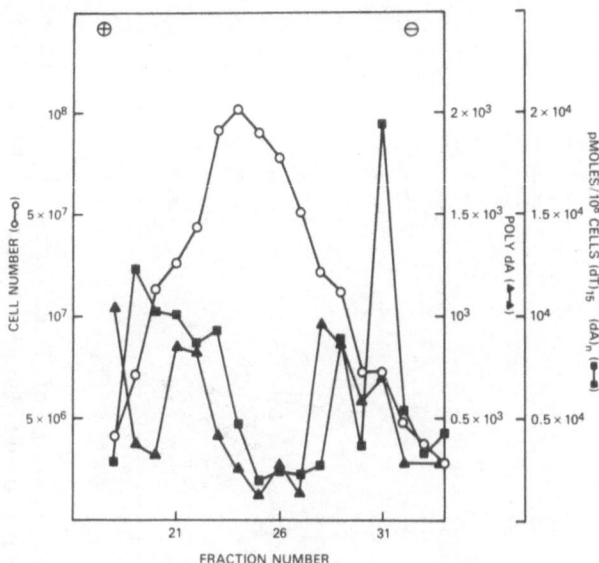

Fig. 3. Separation of a mixture of T and B Cells by free flow electrophoresis. A mixture of equal numbers of T and B cells was
used for this experiment. Cell number (0 —— 0); Poly dA
(▲ —— ▲); and $(dT)_{15} \cdot (dA_n$ (■ —— ■).

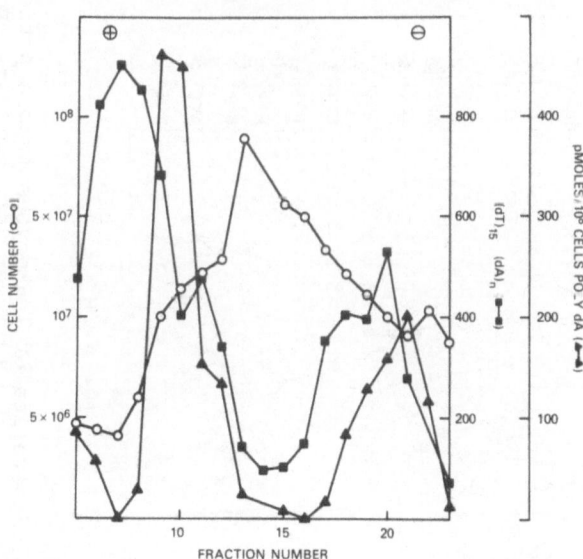

Fig. 4. Separation of human bone marrow cells by free flow electrophoresis. Cell number (0 —— 0); Poly dA (▲—— ▲); and
$(dT)_{15} \cdot (dA)_n$ (■ —— ■).

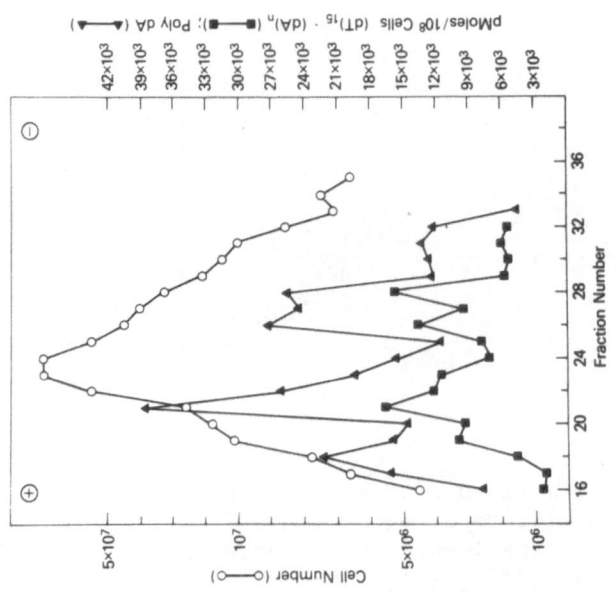

Fig. 6.   Separation of leukocytes from a patient with acute lymphoblastic leukemia (789) by free flow electrophoresis. Cell number (0---0); Poly dA (▲---▲); and $(dT)_{15} \cdot (dA)_n$ (■---■).

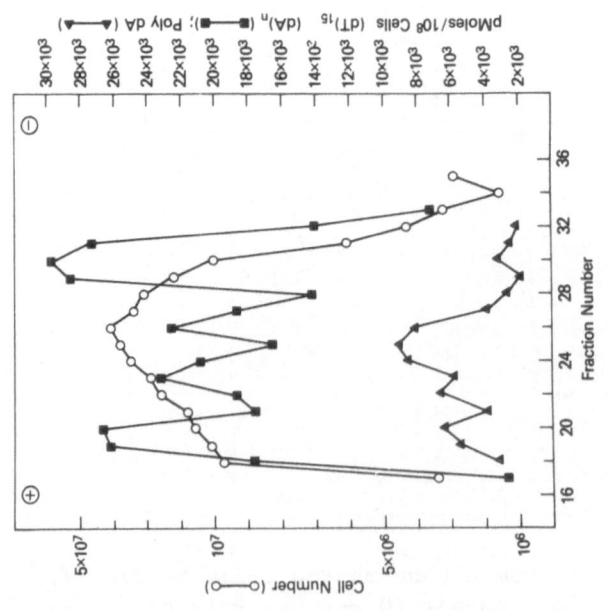

Fig. 5.   Separation of leukocytes from a patient with acute lymphoblastic leukemia (FG). Cell number (0---0); Poly dA (▲---▲)1 and $(dT)_{15} \cdot (dA)_n$ (■---■).

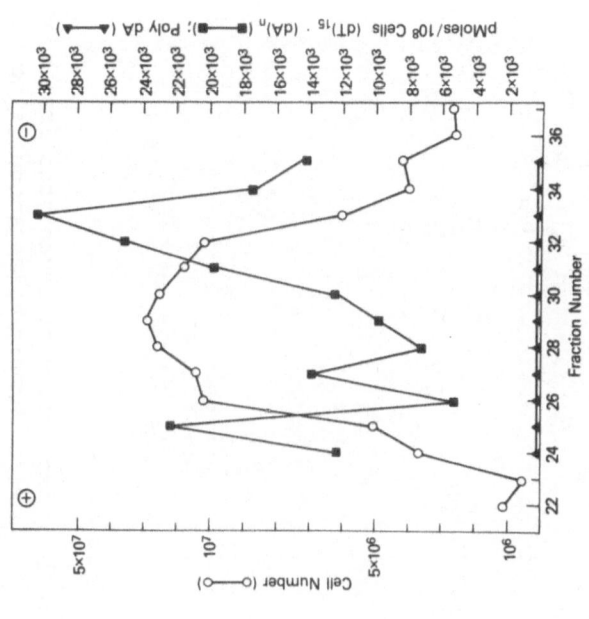

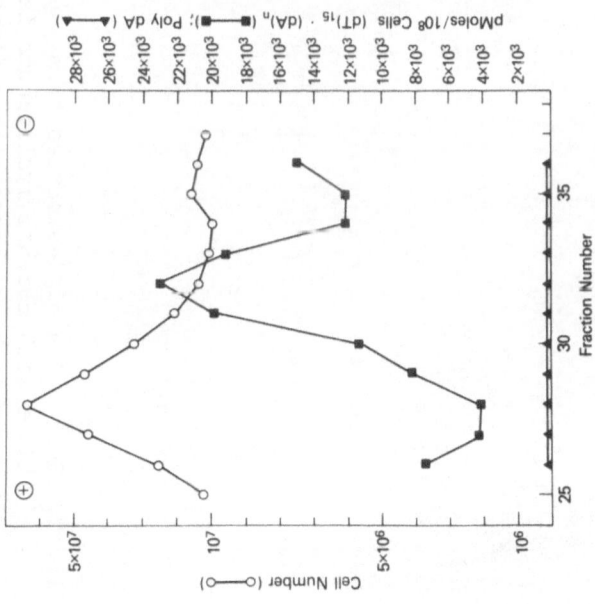

Fig. 7. Separation of leukocytes from a patient
with acute myelogenous leukemia (725) by
free flow electrophoresis. Cell number
(O——O); Poly dA (▲——▲); and
(dT)$_{15}$·(dA)$_n$ (■——■).

Fig. 8. Separation of leukocytes from a patient
with acute myelogenous leukemia (7101)
by free flow electrophoresis. Cell num—
ber (O——O); Poly dA (▲——▲);
and (dT)$_{15}$·(dA)$_n$ (■——■).

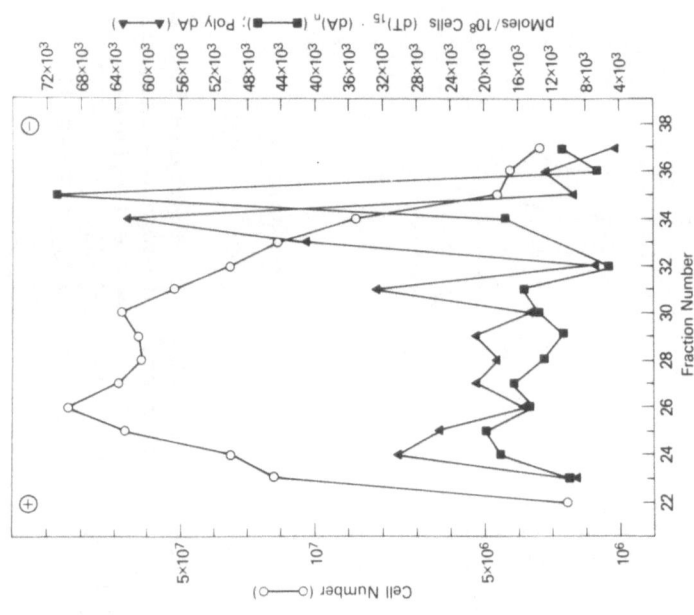

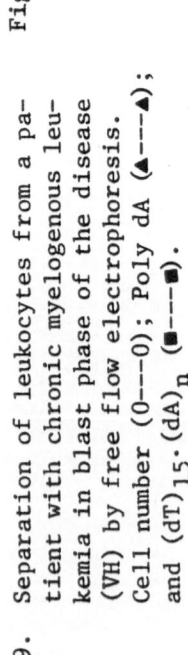

Fig. 10.   Separation of leukocytes from a patient with chronic myleogenous leukemia in blast phase of the disease (VH) by free flow electrophoresis. Cell number (0---0); Poly dA (▲---▲); and dT$_{15}$·(dA)$_n$ (■---■).

Fig. 9.   Separation of leukocytes from a patient with chronic myelogenous leukemia in blast phase of the disease (VH) by free flow electrophoresis. Cell number (0---0); Poly dA (▲---▲); and (dT)$_{15}$·(dA)$_n$ (■---■).

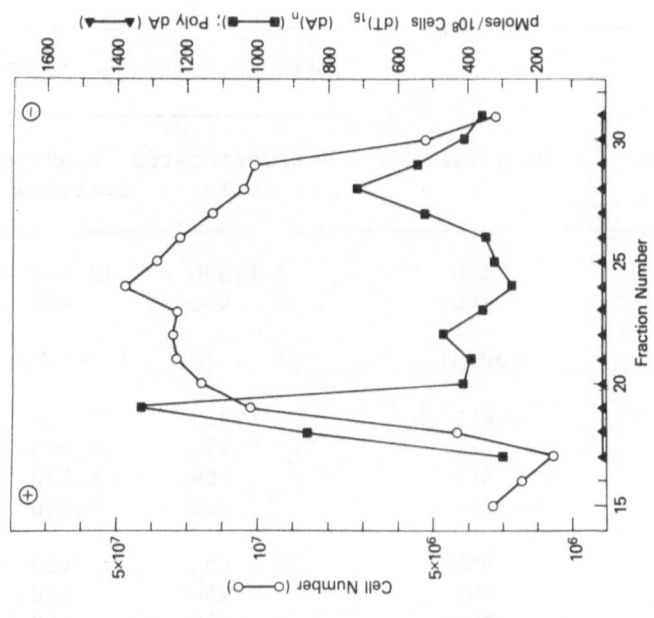

Fig. 12. Separation of leukocytes from a patient with chronic myelogenous leukemia (739) by free flow electrophoresis. Cell number (O——O); Poly dA (▲——▲); and $(dT)_{15} \cdot (dA)_n$ (■——■).

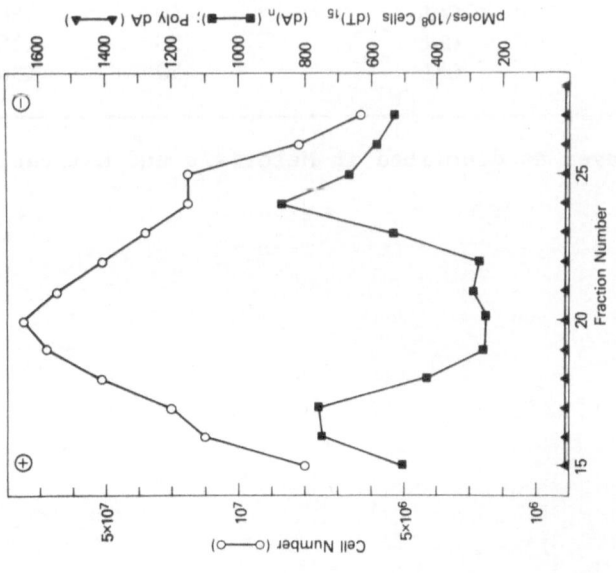

Fig. 11. Separation of leukocytes from a patient with chronic myelogenous leukemia (737) by free flow electrophoresis. Cell number (O——O); Poly dA (▲——▲); and $(dT)_{15} \cdot (dA)_n$ (■——■).

Table 4:   Terminal Transferase Activity in Human Normal and
Leukemic Cells Before and After Cell Separation

#	Source of Cells	Diagnosis	TdT[*] Activity (pmoles/10^8 cells)		
			Unfractionated Cells	Peak Fraction	Enrichment (Fold)
1.	T (8402)	ALL	1,390	12,900	9
2.	B (8392)	ALL	<50	<50	–
3.	Bone Marrow	Normal	70	500–1400	7–20
4.	719	ALL	1,830	22,500	12
5.	789	ALL	5,430	38,900	7
6.	792	ALL	460	8,530	30
7.	FG	ALL	250	7,540	18
8.	711	AML	<50	<50	–
9.	725	AML	<50	<50	–
10.	7101	AML	<50	<50	–
11.	VH	CML(BC)	710	23,400	33
12.	FM	CML(BC)	10,900	62,900	6
13.	7100	CML(BC)	<50	1,080	20
14.	737	CML	<50	<50	–
15.	739	CML	<50	<50	–
16.	799	CML	<50	<50	–

*TdT was assayed as described in Materials and Methods.

Table 5: DNA Polymerase Activity in Human Normal and Leukemic
Cells Before and After Cell Separation

| # | Source of Cells | Diagnosis | DNA Polymerase Activity* (pm/10^8 cells | | |
			Unfractionated Cells	Peak Fraction	Enrichment (Fold)
1.	T (8402)	ALL	6,380	82,600	13
2.	B (8392)	ALL	3,300	75,400	23
3.	Bone Marrow	Normal	130	900–3,000	6–25
4.	719	ALL	4,570	15,600	3
5.	789	ALL	5,430	13,400	3
6.	792	ALL	7,300	113,000	15
7.	FG	ALL	6,200	47,300	8
8.	711	AML	590	6,670	11
9.	725	AML	1,130	23,000	20
10.	7101	AML	2,200	30,500	14
11.	VH	CML(BC)	760	21,200	28
12.	FM	CML(BC)	18,670	79,700	4
13.	7100	CML(BC)	90	600	7
14.	737	CML	150	1,450	10
15.	739	CML	110	4,670	42
16.	799	CML	150	2,370	16

*DNA polymerase activity was assayed with $(dT)_{15} \cdot (dA)_n$ as
described in Materials and Methods.

of relapse. This early detection may prove useful in giving treatment earlier than otherwise may be possible by detection of relapse
by cell morphology alone.

Immunological Studies

We have studied the inhibition of TdT isolated from cells from
a number of leukemic patients after free flow electrophoresis by TdT
antibody. TdT antibody has been successfully produced in rabbits[22]
and mice[23]. We have been involved in the preparation of a monoclonal
antibody against TdT by using the technique of Kohler and Milstein[24].
We have isolated a number of clones that show a strong cross reaction
with terminal transferase. Surprisingly, antibodies produced from
some of these clones also cross react with cellular DNA polymerase γ
and some viral reverse transcriptases. As shown in Figure 13 the
hyperimmune antibody produced in goats or rabbits is specific for
TdT whereas the monoclonal antibody cross reacts to some extent with
reverse transcriptases from feline leukemia virus, gibbon ape leukemia virus, baboon endogenous virus, simian sarcoma virus and cellular DNA polymerase γ. This observation with the monoclonal antibody suggests that the monoclonal antibody is recognizing a common
antigenic site among type-C RNA tumor virus reverse transcriptases,
DNA polymerase γ and TdT. We are in the process of analyzing other
clones producing antibody for selection of a monospecific monoclonal
antibody against TdT. The availability of a monoclonal monospecific
antibody against TdT will be useful in identification and characterization of TdT positive cell populations by immunofluorescence or
by the use of fluorescence activated cell sorter.

DISCUSSION

Previous studies in evaluating human leukemic cells for leukemia specific molecules, including retrovirus related proteins and
nucleic acids, have only been successful in a limited number of
patients. This may be due to the fact that these leukemia specific
molecules are present only in a small population of leukocytes,
thereby making detection difficult and variable, or that the nega-
tives are truly negative. In the past, we have reported on the
enrichment of terminal transferase activity in human bone marrow
cells[19] by using the techniques of ficoll-hypaque followed by unit
gravity sedimentation in sucrose. A number of cell separation
techniques have recently been developed (see Table 2). We have
utilized the techniques of free flow electrophoresis for separation
of human leukemic cells into different cell populations for detection of terminal transferase. This technique of free flow electrophoresis has been used in the past for separation of human[14] and
mouse[15] bone marrow cells, and for the separation of cell membranes
and organelles[16]. Separation of acute leukemic cells into different
cell populations has also been achieved with the help of centrifugal
elutriation[17].

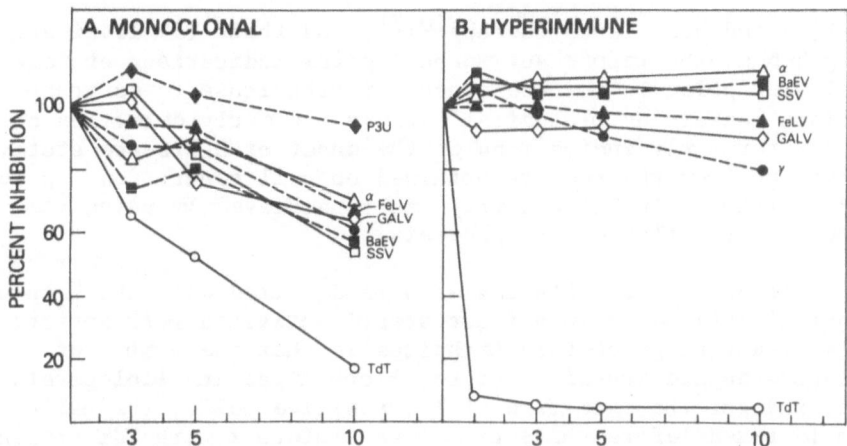

Fig. 13.   Inhibitory effect of monoclonal and hyperimmune TdT antibody on TdT and viral and cellular DNA polymerase. Antibody assays were carried out as described[25-27]. Viral
and cellular DNA polymerases (5 μl) were incubated with
nonimmune sera IgG and TdT antisera IgG at 4°C for 16 hr
in a reaction mixture containing 50 mM Tris-HCl (pH 7.5),
50 mM KCl, 5 mM dithiothreitol, 0.01% triton X100, and
160 μg/ml of bovine serum albumin. The standard reaction
mixture (0.05 ml) was then incubated at 37°C for 60 min
and contained 50 mM Tris-HCl (pH 7.5), 50 mM KCl, 5 mM
dithiothreitol, 0.5 mM Mn^{2+}, 0.01% triton X100, 80 μg/ml
bovine serum albumin, poly dA or $(dT)_{15}(A)_n$ (10 μg/ml)
and 100 μM of $[^3H]$dGTP or $[^3H]$dTTP. A. Effect of monoclonal TdT antibody on various DNA polymerases. (a) TdT
(O——O); (b) DNA polymerase α (Δ——Δ); (c) DNA polymerase γ (●---●); (d) feline leukemia virus (FeLV)
(▲---▲); (e) gibbon ape leukemia virus (GALV) (◆——◇);
(f) baboon endogenous virus (BaEV) (◪——◪); (g) simian
sarcoma virus (SSV) (☐——☐); and (h) control myeloma
line (P3U) supernatant (◆--◆). B. Effect of hyperimmune TdT antibody on TdT and other viral and cellular
DNA polymerases. Symbols for the various enzymes are
the same as in A.

     With the use of cell separation techniques it is possible to
enrich TdT positive cell populations for early detection of relapse
and for biochemical and biological studies. In our studies on the
followup of TdT positive CML patients during treatment and remission,
we have observed that the TdT positive CML patients respond to

vincristine and prednisone therapy[20,21], and their TdT level start
going up 5-6 months before any morphological indications of relapse[20].
These studies point to the usefulness of techniques to enrich cell
populations containing biological markers for early detection of
relapse or for early indications of the onset of a disease state.
As is evident from the results obtained on cell separation, up to
30 fold enrichment in TdT activity can be achieved by using the
technique of free flow electrophoresis.

A small number of cells can also be detected with the help of
cell surface antibodies in a fluorescence activated cell sorter.
The only disadvantage of this technique is that the number of
cells separated are insufficient for biochemical and biological
studies, and for the detection of TdT positive cells, the cells
need to be fixed before the antibody can interact with TdT present
in the nucleus of the cells. With the availability of monoclonal
antibodies against tumor cell specific antigens it should be possible to provide early indications of the onset of relapse. Free
flow electrophoresis is one technique which could be very useful in
enrichment of small number of cell populations containing a biological marker.

REFERENCES

1. F. J. Bollum, Terminal deoxynucleotidyl transferase, in: "The
    Enzymes" P. D. Boyer, ed., Academic Press, N.Y., p. 145
    (1974).
2. P. S. Sarin, Terminal transferase as a biological marker for
    human leukemia, in: "Recent Studies in Cancer Research"
    R. Gallo, ed., CRC Press, Cleveland, p. 131 (1977).
3. R. McCaffrey, D. F. Smoler and D. Baltimore, Terminal deoxy-
    nucleotidyl transferase in a case of childhood acute
    lymphoblastic leukemia. Proc. Nat. Acad. Sci. U.S.A.
    70:521 (1973).
4. M. S. Coleman, J. J. Hutton, P. D. Simone, and F. J. Bollum,
    Terminal deoxynucletidyl transferase in human leukemia.
    Proc. Nat. Acad. Sci. U.S.A. 71:4404 (1974).
5. S. Khan, J. Minowada, E. Henderson and I. Tabowski, Terminal
    deoxynucleotidyl transferase activity and blast cell char-
    acteristics in adult acute leukemias. Leukemia Res. 4:209
    (1980).
6. P. S. Sarin and R. C. Gallo, Terminal deoxynucleotidyl trans-
    ferase in chronic myelogenous leukemia. J. Biol. Chem.
    249:8051 (1974).
7. P. S. Sarin, P. N. Anderson and R. C. Gallo, Terminal deoxy-
    nucleotidyl transferase activities in human blood leukocytes
    and lymphoblast cell lines. High levels in lymphoblast
    cell lines and in blast cells of some patients with chronic
    myelogenous leukemia in acute phase. Blood 47:11 (1976).

8. P. S. Sarin and R. C. Gallo, Terminal deoxynucleotidyl trans-
    ferase as a biological marker in human leukemia, in: "Modern
    Trends in Human Leukemia" R. Neth, ed., J. F. Lehmann's
    Verlag, Munich, Vol. 2, p. 491 (1976).

9. R. McCaffrey, T. Harrison, R. Parkman and D. Baltimore, Termi-
    nal deoxynucleotidyl transferase activity in human leukemic
    cells and in normal human thymocytes. N. Engl. J. Med.
    292:775 (1975).

10. M. S. Coleman, M. F. Greenwood, J. J. Hutton, F. J., Bollum,
    B. Lampkin and P. Holland, Serial observations on terminal
    deoxynucleotidyl transerase activity and lymphoblast surface
    marker in acute lymphoblastic leukemia. Cancer Res. 36:120
    (1976).

11. S. L. Marcus, S. W. Smith, C. L. Jarowski and M. J. Modak,
    Terminal deoxyribonucleotidyl transferase activity in
    acute undifferentiated leukemia. Biochem. Biophys. Res.
    Commun. 70:37 (1976).

12. B. Srivastava and J. Minowada, Terminal deoxynucleotidyl
    transferase in a cell line (Molt-4) derived from the peri-
    pheral blood of a patient with acute lymphoblastic leukemia.
    Biochem. Biophys. Res. Commun. 51:529 (1973).

13. P. S. Sarin and R. C. Gallo, Characterization of terminal
    deoxynucleotidyl transferase in a cell line (8402) derived
    from a patient with acute lymphoblastic leukemia. Biochem.
    Biophys. Res. Commun. 76:673 (1975).

14. J. C. F. Schubert, F. Walther, E. Holzberg, G. Pascher and K.
    Zeiller, Preparative electrophoretic separation of normal
    and neoplastic bone marrow cells. Klin. Wschr. 51:327
    (1973).

15. K. Zeiller, J. C. F. Schubert, F. Walther and K. Hanning, Elec-
    trophoretic distribution analysis of in vivo colony forming
    cells in mouse bone marrow. Hoppe-Seyler's Z. Physiol. Chem.
    353:95 (1972).

16. K. Hanning and H. G. Heidrich, The use of continuous prepara-
    tive free flow electrophoresis for dissociating cell frac-
    tions and isolation of membranous components, in: "Methods
    in Enzymology" S. Fleischer and L. Packer, L. eds., Academic
    Press, New York, Vol. 31, p. 746 (1974).

17. H. D. Preisler, I. Walezak, J. Renick and Y. M. Rustum, Sepa-
    ration of leukemic cells into proliferative and quiescent
    subpopulations by centrifugal elutriation. Cancer Res.
    37:3876 (1977).

18. P. S. Sarin, M. Virmani and B. Friedman, Terminal transferase
    in acute lymphoblastic leukemia in gibbons. Biochem.
    Biophys. Acta. 608:62 (1980).

19. R. D. Barr, P. S. Sarin and S. Perry, Terminal transferase in
    human bone marrow lymphocytes. Lancet 1:508 (1976).

20. M. M. Oken, P. S. Sarin, R. C. Gallo, et al. Terminal trans-
    ferase levels in chronic myelogenous leukemia in blast
    crisis and in remission. Leukemia Res. 2:173 (1978).

21. D. D. Ross, P. H. Wiernik, P. S. Sarin and J. Whang-Peng, Loss
     of terminal deoxynucleotidyl transferase activity as a pre-
     dictor of emergence of resistance to chemotherapy in a case
     of chronic myelogenous leukemia. Cancer 44:1566 (1979).
22. F. J. Bollum, Antibody to terminal deoxynucleotidyl transfer-
     ase. Proc. Nat. Acad. Sci. U.S.A. 72:4119 (1975).
23. P. C. Kung, P. D. Gottlieb and D. Baltimore, Terminal deoxy-
     nucleotidyl transferase. Serological studies and radioimmuno-
     assys. J. Biol. Chem. 251:2399 (1975).
24. G. Kohler and C. Milstein, Continuous cultures of fused cells
     secreting antibody of predefined specificity. Nature
     256:495 (1975).
25. P. S. Sarin, J. Donlon, B. Friedman and R. C. Gallo, Character-
     ization of an RNA directed DNA polymerase from a cell line
     derived from a radiation induced lymphoma in mice. Biochem.
     Biophys. Acta 564:235 (1979).
26. P. S. Sarin, B. Friedman and R. C. Gallo, Purification and char-
     acterization of baboon endogenous virus DNA polymerase.
     Biochem. Biophys. Acta 479:198 (1977).
27. P. S. Sarin and R. C. Gallo, Purification and characterization
     of gibbon ape leukemia virus DNA polymerase. Biochem.
     Biophys. Acta 454:212 (1976).

TdT-POSITIVE AND TdT-NEGATIVE HUMAN LEUKEMIC CELLS: SPECIFIC DENSITY AND MORPHOLOGY

M. Marini°, G.P. Bagnara°, G. Biagini^,
M. Baccarani¶, P. Rosito§

°Istituto di Istologia ed Embriologia Generale
^Istituto di Microscopia Elettronica Clinica
¶Istituto di Ematologia "Lorenzo e Ariosto Seragnoli"
§Clinica Pediatrica
Facoltà di Medicina, 40100 Bologna, Italy

ABSTRACT

Peripheral blood cells from 9 E- and 6 E+ SIg- ALL patients and from 11 CML patients in TdT+ blastic crisis were subjected to discontinuous albumin gradient separation according to Dicke's method, and TdT assayed in the six fractions recovered therefrom. The major results were: (i) in E- ALL and in CML-BC, TdT+ cells could be recovered either within a narrow range of specific densities or were spread over most of the gradient, which might suggest differences in cell maturation among the patients; (ii) in some instances, most leukemic blasts were found in TdT-negative or faintly positive fractions; (iii) in most, but not all, E- ALL and BC, the majority of TdT+ cells was found in low density fractions; (iv) all E+ ALL had high density TdT+ cells. Cell fractions of most patients were also examined at the electron microscope, and correlations between morphology and marker characterization were tentatively drawn.

INTRODUCTION

High levels of terminal deoxynucleotidyl transferase (TdT) have been found to occur in white blood cells and marrow of patients affected by a number of leukemias[1-4], including most patients with non-B acute lymphoid leukemia (ALL) and a number of patients undergoing the blast crisis (BC) of chronic myeloid leukemia (CML). Since this enzyme provides a useful marker for a better characterization of human leukemias[5], in the past three years patients entering the

357

hematological branch of the Bologna University Hospital with a diagnosis of ALL and CML-BC were subjected to the evaluation of TdT in
white blood cells by the radiochemical assay. The results of this
screening are here reported.

　　At the same time, since the TdT+ leukemic populations apparently
arise within bone marrow lymphoid precursors which are thought to be
physiologically endowed with such a marker, we decided to compare the
cell density distribution of normal TdT+ bone marrow cells with that
of TdT+ leukemic blast cells, in the attempt to get a better understanding of the nature of the leukemic populations. Upon the observation of striking differences in the specific density of TdT+ cells
among patients, we examined the ultrastructural features of cell
fractions expressing or not expressing the marker in the different
leukemias.

MATERIALS AND METHODS

Cell Preparation
　　For TdT determination in the unfractionated cell suspension,
white blood cells were collected upon Dextran sedimentation of red
cells and washed thrice with PBS. They were subsequentially disrupted
by sonication and the 100,000 x g supernate used as crude enzyme
source in the radiochemical assay.

　　When cell fractionation was to be performed, cells were resuspended in 17% BSA solution and layered on the top of the gradient.
Cell fractions, recovered from the interfaces or the bottom of the
gradient, were treated in the same way as the unfractionated cell
suspension.

　　Within the same patient, no difference was found among marrow's
and blood's density distribution pattern of TdT+ cells.

Discontinuous Albumin Gradients
　　Albumin gradients were made after Dicke[6]. Albumin concentrations
and specific densities of the layers were: 17% - 1.0525 g/cm^3; 19% -
1.0580 g/cm^3; 21% - 1.0637 g/cm^3; 23% - 1.0682 g/cm^3; 25% - 1.0734
g/cm^3; 27% - 1.0780 g/cm^3. Fractions were labeled from the top, F_1
corresponding to the interface between the 17 and the 19% albumin
solution, and F_6 to cells sedimenting at the bottom of the tube.

TdT Assay
　　Radiochemical assays were performed according to the method of
Coleman[7], except that poly(dA)$_{12-18}$ (PL Biochemicals) was substituted for poly(dA)$_{50}$, which was not commercially available. Whenever
possible, assays were run in triplicate and to one control tube
50μM ATP, a specific TdT inhibitor[8], was added.

RESULTS

Normal Bone Marrow

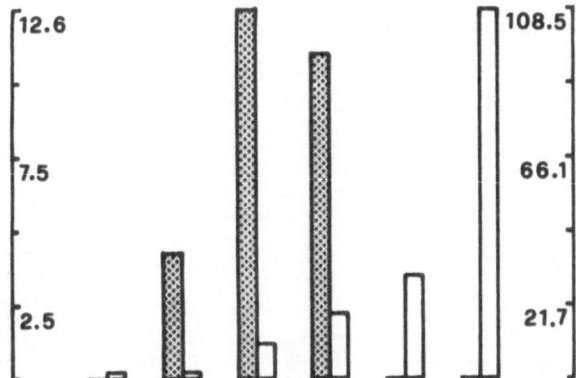

Fig.1. Discontinuous albumin gradient separation of normal pediatric bone marrow. In this, as well as in figs. 2, 5 and 6, columns are the various fractions recovered from the gradient, lowest density F_1 fraction being on the left hand. For each fraction, shaded areas are TdT Units/10^8 nucleated cells; open areas are the actual number of cells recovered in the fraction (x 10^6).

Unseparated normal pediatric bone marrow cells averaged 0.3 - 0.5 TdT U/10^8 nucleated cells. Fig.1 illustrates the pooled results from density gradient separations of 5 normal donors. TdT+ cells appear to be concentrated in fractions 2-4, with a ∿30-fold concentration respect to the unseparated cell suspension in F_3.

Their density distribution overlaps that described for murine bone marrow CFUs[9], while CFUc from both human and murine sources have been reported to concentrate in fractions 3 and 4[10]. Normal bone marrow TdT+ cells appear therefore to share the same specific density with hemopoietic progenitor cells. Similar results have been described by Pazmiño et al.[11] in murine bone marrow cells, although their gradient separation technique was somehow different from ours. For comparison purposes, it should be mentioned that TdT+ normal thymocytes distribute over a higher specific density range (fractions 3-5)[12], which therefore overlaps in part with that of bone marrow TdT+ cells.

Chronic Myeloid Leukemia - Chronic Phase

It is generally accepted that the blastic transformation, which sooner or later follows the "stable" phase of CML, occurs in the same transformed stem cells which, during that stable phase, had retained the capability to differentiate and mature along the myeloid lineage.

Since in the blastic transformation leukemic cells may express the same markers as in cALL, including TdT[13], we used albumin gradients as a means for concentrating blast cells during the chronic

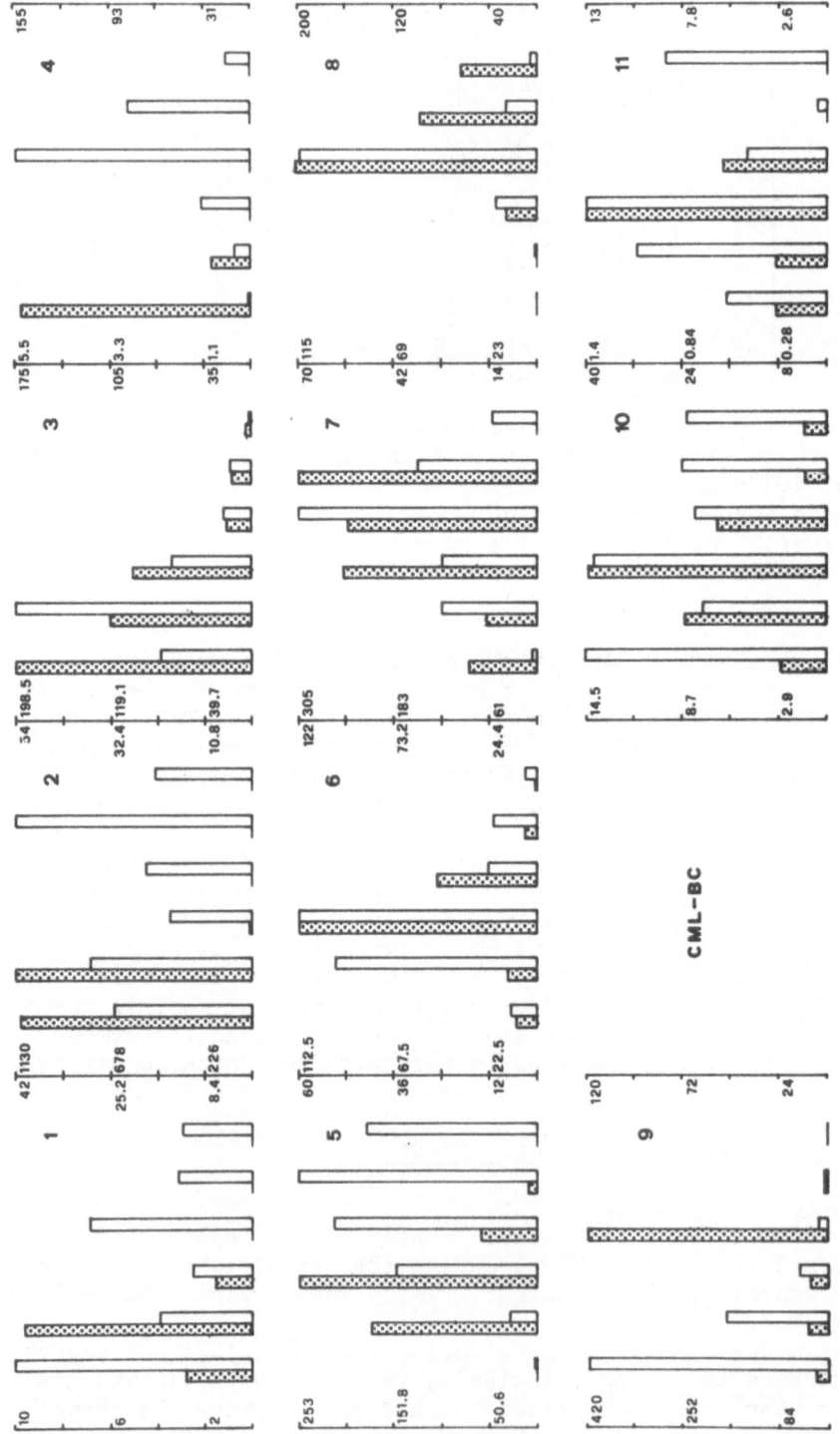

Fig. 2. Discontinuous albumin gradient separation in 11 cases of blastic transformation. Patient no. 11 had agnogeneic myeloid metaplasia, all the others had CML-BC. In patients nos. 10 and 11 the separation was performed in lymphonodes, in the others in peripheral blood cells.

phase, in the attempt ot see whether TdT+ cells were already present.
The study was performed in 10 untreated Phl+ CML patients at the onset
of the disease. No TdT activity was found: it is noteworthy that one
of them underwent a TdT+ blastic crisis 8 months later. This finding
suggests that TdT+ cells arise, at least in a sizeable number, just
at the onset of the blastic transformation and accumulate quickly.

Chronic Myeloid Leukemia - Blastic Phase

TdT vas evaluated in 31 patients during the blastic crisis of
CML and in 3 patients undergoing the blastic transformation following other myeloproliferative disorders.

14 CML-BC patients had TdT+ blood cells; one more patient was
found to have TdT+ cells in the involved lymphonodes. TdT+ cells
were found also in lymphonodes of one patient with agnogeneic myeloid
metaplasia in blastic proliferation.

All cases were a lymphoid cytotype could be recognized, either
alone or mixed with other blastic populations, were TdT+, whilst all
cases with frankly myeloid or monocytic blast cells were TdT-. 5 of
14 cases with mixed populations of unclassifiable and myeloid and/or
monocytic blast cells were TdT+.

3 TdT+ and 2 TdT- patients were Philadelphia negative. No correlation between TdT-positivity and length of chronic phase has been
found.

Blast cells from 11 TdT+ and 7 TdT- patients were separated in
discontinuous albumin gradients. In TdT- patients, some of whom had
a mixed (U+My) blastic population, the gradient-operated concentration step did not unmask TdT+ cells, which can therefore be assumed
to be totally absent. Results of TdT+ patients are reported in fig.2.
A marked variability in the specific density of TdT+ cells can be
observed. In most patients, including those where lymphonode suspensions were studied, the "TdT peak fraction" was in the lower density
range, while in the other maximal TdT activity was found in high density fractions. Thus, in this disease, the TdT+ leukemic population
appears a very heterogeneous one; it is interesting to note that i)
in 4 patients most TdT+ cells were recovered in F_1 and in F_2, at a
lower density than normal TdT+ bone marrow cells and that ii) TdT+
cells' sedimentation pattern is independent from that of myeloid-
committed stem cells, which in CML have always a low specific density[14,15].

Cell fractions from patients 2, 3, 5, 6 and 7 were examined at
the electron microscope. In all patients, lymphoid-like cells were
generally more concentrated in those fractions where maximal TdT activity was found. Low density TdT-positive fractions were characterized by very undifferentiated lymphoid-like blasts, with dispersed
chromatin. In patients 6 and 7, higher density TdT-positive fractions
had a majority of lymphoid-like cells with a more condensed chromatin.

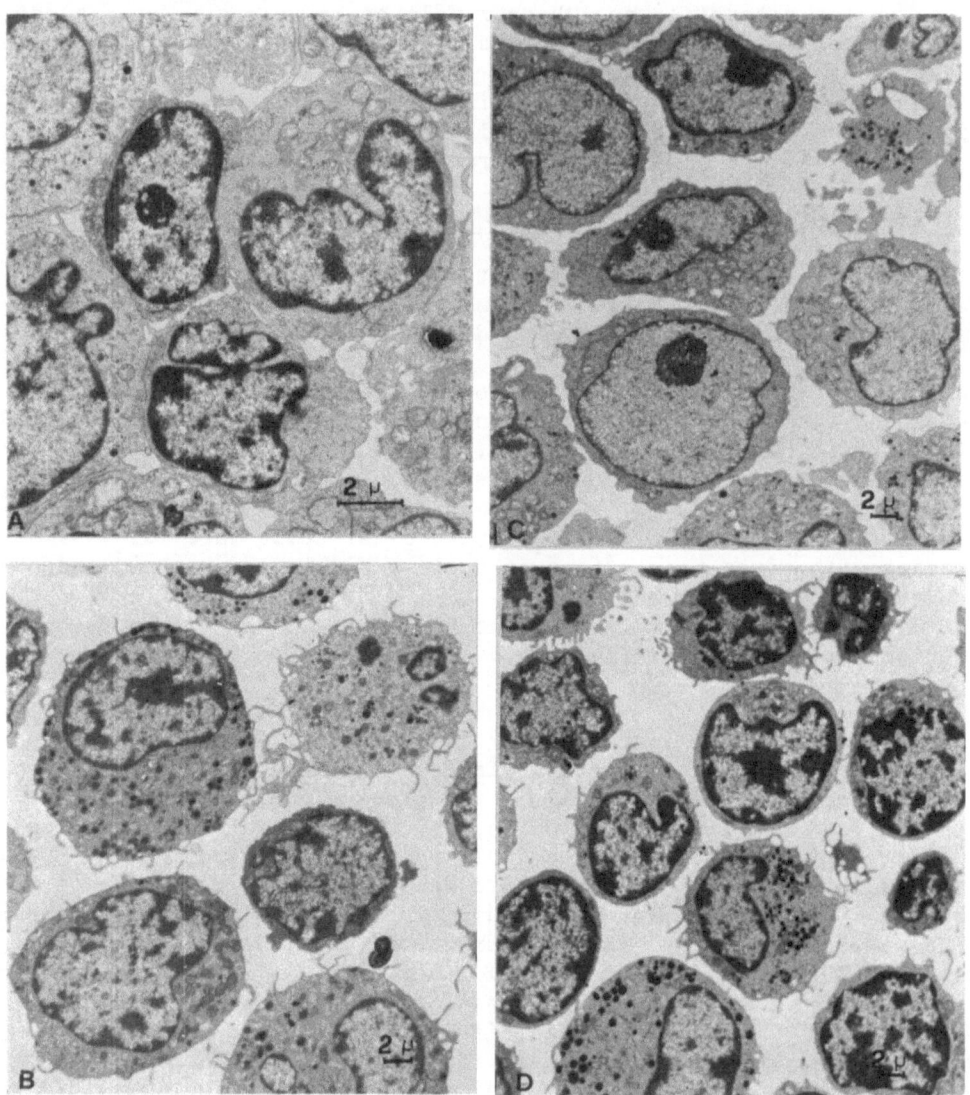

Fig. 3. Electron micrographs of cell fractions from CML–BC patients.
A. TdT+ F_1 from patient no.2. B. TdT– F_3 from the same patient. Note the predominance of very undifferentiated lymphoid-like cells in
F_1 and the presence of a mixed population of more mature lymphoid
cells and of myeloid elements in F_3. C. TdT– F_1 from patient no.6.
D. TdT+ F_3 from the same patient. In the TdT+ fraction lymphoid
cells with condensed chromatin are more numerous.

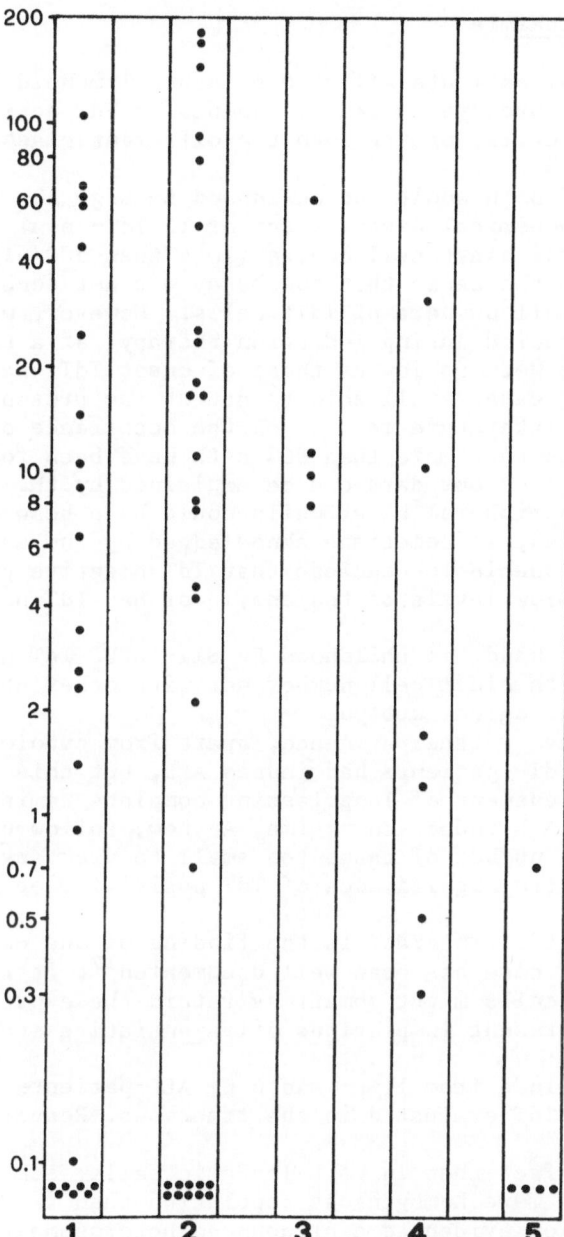

Fig. 4. TdT U/10[8] WBC in 67 patients affected by acute lymphoid
leukemia.
1. Childhood E- Sig- ALL.   2. Adult E- Sig- ALL.
3. Childhood E+ Sig- ALL.   4. Adult E+ Sig- ALL.
5. Adult E- Sig+ ALL.

Acute Lymphoid Leukemia

A total of 67 patients affected by acute lymphoid leukemia were
studied during a three-years period. Values of TdT activity in the
peripheral blood cells, broken into the different classes of patients,
are shown in fig.4.

About 1/3 of both adult and childhood E- SIg- ALL were found
TdT- at the radiochemical assay. 9 out of 10 TdT- adult patients had
very low peripheral blast cell counts (less than 500/µl), which might
be interpreted in the sense that the assay was not sensitive enough
to detect very small numbers of TdT+ cells. However, when TdT+ patients were re-studied during induction therapy, at a time when their
blast cell counts were as low as those of onset TdT- patients, the
assay was in most cases still able to detect the presence of TdT.
From a review of literature reports on the occurrence of this marker
in ALL, it results that more than 90% cALL have been found TdT+.
This discrepancy from our data can be explained by the fact that patients with few peripheral blast cells could have been disregarded
in previous studies, as sometimes aknowledged by the authors[16]. It
is therefore reasonable to conclude that TdT-negative patients had
indeed extremely low levels of the enzyme or had TdT-negative blast
cells.

On the other hand, in childhood E- SIg- ALL, TdT negativity was
not correlated with blast cell number nor with other characteristi cs,
such as the morphological subtype.

We do not have further evidence, apart from cytology and cytochemistry, that TdT- patients had indeed ALL, but this conclusion is
supported by achievement of long-lasting complete remission in most
of them, with an ALL induction regime. Anyhow, follow-up times are
too short and the number of cases too small to draw any conclusion
about the prognostic significance of TdT positivity or negativity in
cALL.

Another point of interest is the finding of one case of E- SIg+
TdT+ ALL; another case has been well documented[17]; it is not unlikely
that transformed cells might sometimes retain the expression of a
marker which is present in previous differentiative stages[13].

Peripheral blood from 11 E- and 6 E+ ALL patients was gradient
fractionated and TdT evaluated in the fractions. Results are shown in
figs. 5 and 6.

Despite the fact that in cALL leukemic cells seem to constitute
a morphologically more homogeneous population than in CML-BC, gradient fractionation evidenced a pronounced heterogeneity in their
specific density. Electron microscopy observation of cell fractions
confirmed that the gradient had achieved a separation along a sort of
differentiation gradient, the less differentiated and more "atypical"
blast cells being found at lower specific densities. The peak of TdT
activity did not follow in most cases that of cell recovery and could
be found, in different patients, at different specific densities. In

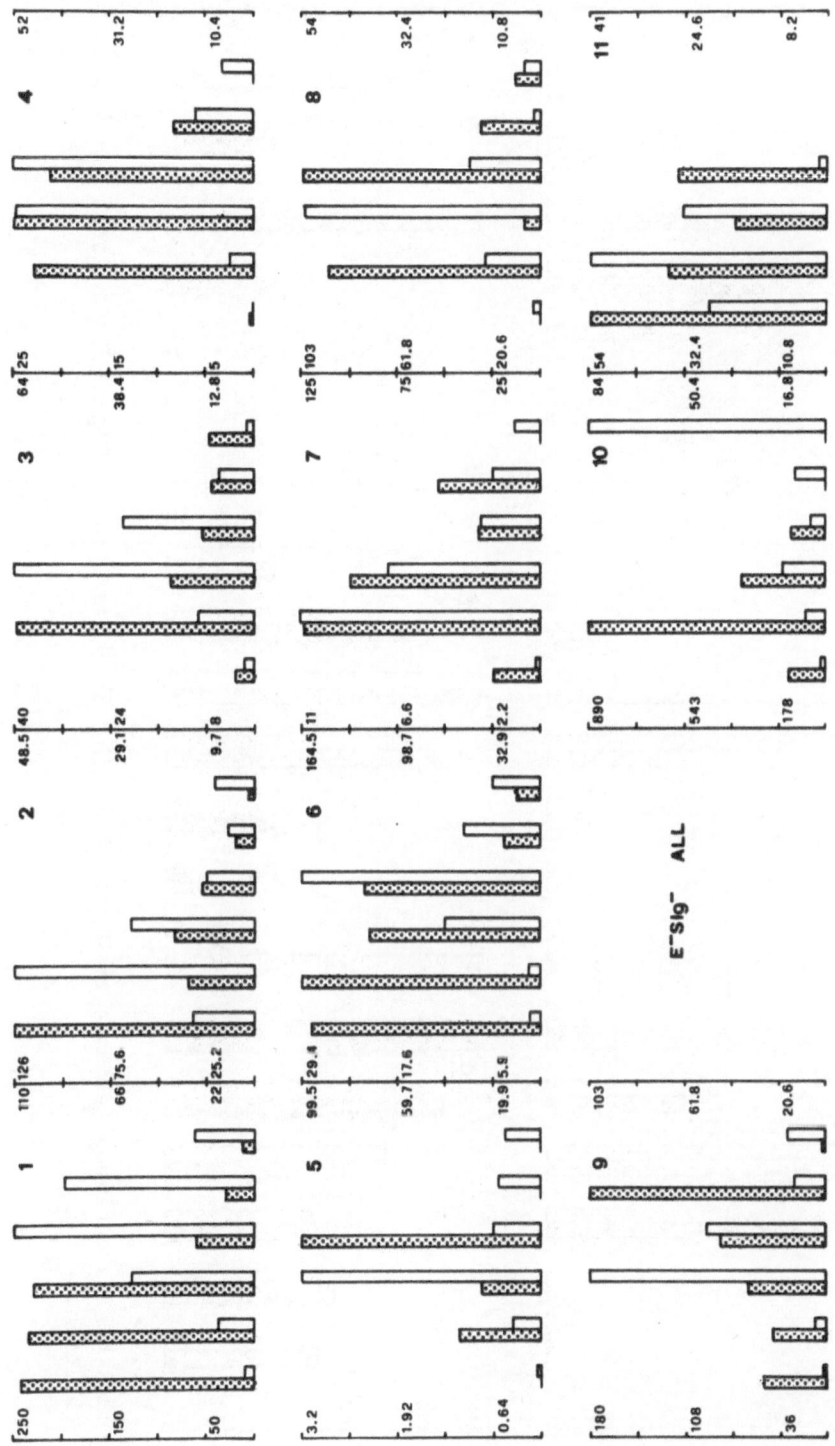

Fig. 5. Discontinuous albumin gradient separation in 11 cases of E- SIg- ALL. Patients 1 - 5 were children. Patients nos. 10 and 11 were studied during relapse.

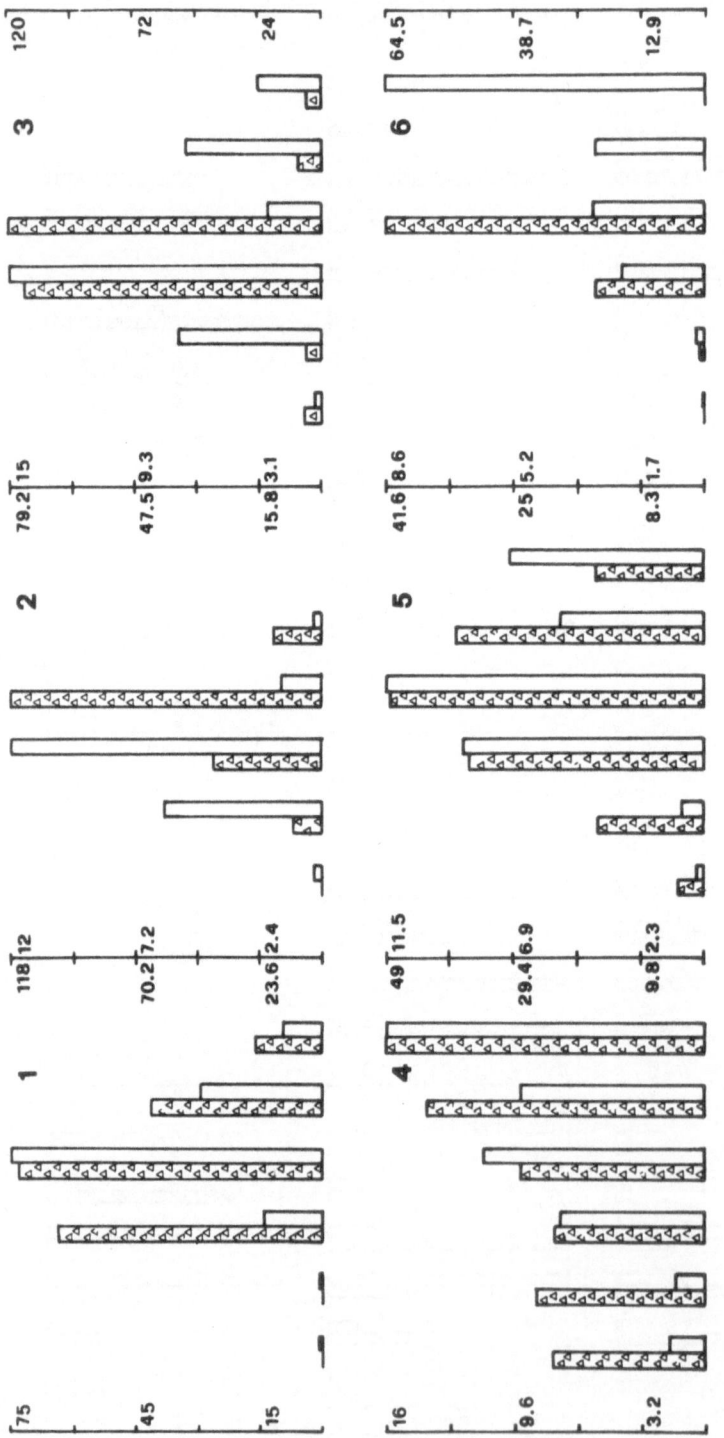

Fig. 6. Discontinuous albumin gradient separation in 9 cases of E+ SIg− ALL. Patients 1 − 4 were children.

one case (no.8), two TdT+ subpopulations were well separated.

We have observed that, in individual patients studied several times during relapses and induction therapy, TdT activity correlated well with blast cell count[16], and interpreted this observation in the sense that the average TdT content of leukemic cells is generally constant. From this interpretation follows that, in ALL patients, fractions with high specific TdT activity should be made up by a majority of TdT+ cells, while those with low specific activity should be made up by few TdT+ and many TdT- cells (rather than by cells with a lower specific TdT content). Therefore, since low density fractions are constituted by leukemic blasts and are virtually devoid of normal or mature cells, in some patients there should be a considerable number of TdT- leukemic cells, which, because of their highly undifferentiated features, are more likely to be precursors, instead of progeny, of TdT+ cells. The presence of a quota of TdT- cells in blood smears of TdT+ patients had already been observed by immunofluorescence, but no characterization of these cells had been attempted[5].

Gradient fractionation of E+ ALL cells gave a more homogeneous scenary, as in all patients most leukemic cells and the peak of TdT activity were found in high density fractions.

Electron microscopic examination of cell fractions from 6 E- and 3 E+ patients was unable to detect morphological differences among TdT+ and TdT- fractions of the same patient. Irrespective of where the TdT activity peak was found, E- patients'blasts were often characterized by bizzarre and irregularly shaped nuclei and the overall impression, within the single patient, was that of a reduced tendency toward differentiation. E+ patients' blasts displayed somewhat more regular nuclei, with condensed chromatin and the overall impression was that of a certain heterogeneity in cytotypes.

CONCLUSIONS

In both CML-BC and cALL patients we have found a wide variability in the specific density of the TdT+ leukemic population. In many patients TdT+ cells were found in low density fractions, in others in high density ones or through the whole gradient. Blast cells were generally undifferentiated and some displayed anomalous features. As a matter of fact, both diseases are thought to be derived by the leukemic transformation of a common progenitor cell, which, during the chronic phase of CML, is still able to progress into the differentiation path[13]. TdT+ blastic crisis might therefore derive from an "early" block in differentiation, upon which even leukemic myelopoiesis cannot progress. The observations that in most patients TdT+ blasts had a very low specific density (as compared with normal TdT+ bone marrow cells), that light TdT- blasts (pre-TdT?) were probably present and that low density fractions were made up by very undifferentiated blasts all corroborate to the hypothesis that the leukemic

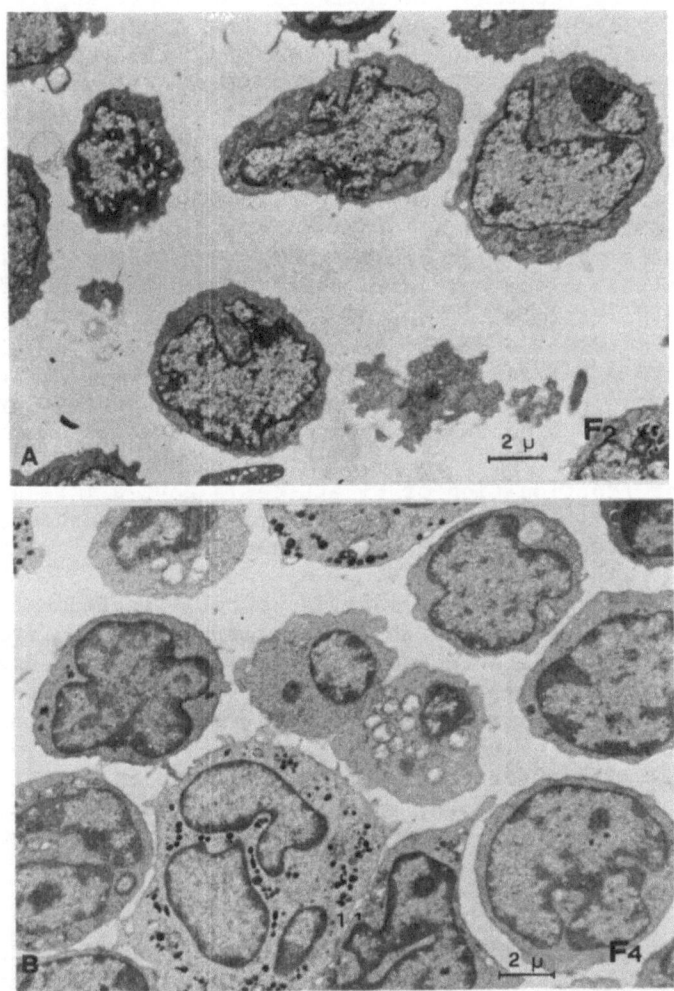

Fig. 7. Electron micrographs of TdT+ low density fraction (A) from
an E- ALL patient and of TdT+ high density fraction (B) from an E+
ALL patient. Note the difference in nuclear shape and in chromatin
condensation. Low density fractions were made up almost exclusively
by blast cells, while in heavier fractions other cell populations
were present.

transformation had occurred in <u>very</u> early progenitor cells, both in CML and in ALL.

On the other hand, E+ ALLs were characterized by TdT+ blast cells with more differentiated features and with the same specific density as normal cortical thymocytes. These observations agree with the concept that in this disease the leukemic transformation occurred in precursors cells which were already committed to the T-cell line.

Further developments of these studies will rely on the inequivocal identification of individual TdT+ cells by the use of immunostaining and by the contemporary detection of other cell markers, in order to see whether the heterogeneity in TdT+ cells we observed in cALL and in CML-BC could be due to the simultaneous presence of more than one leukemic population or to an anomalous progression in the maturative pathway.

REFERENCES

1. R. A. McCaffrey, D. Smoler, and D. Baltimore, Terminal deoxynucleotidyl transferase in a case of childhood acute lymphoblastic leukemia. <u>Proc. Natl. Acad. Sci. USA</u>. 70:521 (1973).

2. M. S. Coleman, J.J. Hutton, P. DeSimone, and F. J. Bollum, Terminal deoxynucleotidyl transferase in human leukemia. <u>Proc. Natl. Acad. Sci. USA</u>. 71:4404 (1974).

3. P. Sarin and R. C. Gallo, Terminal deoxynucleotidyl transferase in chronic myelogeneous leukemia. <u>J. Biol. Chem</u>. 249:8051 (1974).

4. R. Mertelsmann, I. Mertelsmann, B. Kozimer, M. A. S. Moore, and B. D. Clarkson, Improved biochemical assay for terminal deoxynucleotidyl transferase in human blood cells: results in 89 adult patients with lymphoid leukemias and malignant lymphomas in leukemic phase. <u>Leukemia Res</u>. 2:57 (1978).

5. F. J. Bollum, Terminal deoxynucleotidyl transferase as a Hemopoietic cell marker. <u>Blood</u>. 54:1203 (1979).

6. K. A. Dicke, Radiobiol Inst. TNO, Rijswijk, The Netherland (thesis). (1970).

7. M. S. Coleman, A critical comparison of commonly used procedures for the assay of terminal deoxynucleotidyl transferase in crude tissue extracts. <u>Nucl. Acid. Res</u>. 4:4305 (1977).

8. R. B. Bhalla, M. K. Schwartz, and J. M. Mukund, Selective inhibition of terminal deoxynucleotidyl transferase (TdT) by adenosine ribonucleoside triphosphate (ATP) and its application in the detection of TdT in human leukemia. <u>Bioch. Biophys. Res. Commun</u>. 76:1056 (1977).

9. J. S. Haskill, T. A. Mc Neill, and M. A. S. Moore, Density distribution analysis of <u>in vivo</u> and <u>in vitro</u> Colony Forming Cells in bone marrow. <u>J. Cell Physiol</u>. 75:167 (1970).

10. M. A. S. Moore and N. Williams, Analysis of proliferation and

differentiation of foetal granulocyte-macrophage progenitor cells in hemopoietic tissue. Cell. Tissue Kinet. 6:461 (1973).

11. N. H. Pazmiño, R. N. McEvan, and J. N. Ihle, Distribution of terminal deoxynucleotidyl transferase in bovine serum albumin gradient. Fractionated thymocytes and bone marrow cells of normal and leukemic mice. J. Immun. 119:494 (1977).

12. R. McCaffrey, T. A. Harrison, R. Parkmann, and D. Baltimore, Terminal deoxynucleotidyl transferase activity in human leukemic cells and in normal human thymocytes. N. EngL. J. Med. 292:775 (1975).

13. M. Greaves and G. Janossy, Patterns of gene expression and the cellular origins of human leukemias. Bioch. Biophys. Acta 516:193 (1978).

14. M. A. S. Moore, in: "Advances in Acute Leukemia", p.160, F. J. Cleton, D. Crowther, and J. S. Malpas eds., North Holland Publishing Company, Amsterdam (1975).

15. G. Biagini, G. P. Bagnara, M. Baccarani, P. Preda, L. Valvassori, L. Bonsi, M. A. Brunelli, G. Astaldi, and R. Laschi, Ultrastructure and culture growth of peripheral blood cells from discontinuous albumin density gradient fractions obtained in the blastic phase of chronic granulocytic leukemia. In: Electron Microscopy 1980, vol.2, p.344 (1980).

16. M. Baccarani, M. Marini, G. P. Bagnara, M. Gobbi, F. Saviotti, S. Tura, and M. A. Brunelli, TdT in adult acute lymphoblastic leukemia: relationship with blast cell count. Hematologica (in press).

17. M. T. Shaw, J. M. Dwyer, H. S. Allandeen, and H. A. Weitzmann, Terminal deoxynucleotidyl transferase activity in B Acute Lymphoblastic Leukemia. J. Am. Soc. Hemat. 51:181 (1978).

This work was supported by CNR Finalized Projects, contracts no. 79.01850.04 and 80.0166.96.

NORMAL MONONUCLEAR BLOOD CELLS IN DIFFUSION

CHAMBERS: OCCURRENCE OF cALLA AND TdT

Gundula Jäger and Barbara Lau

Abteilung für Experimentelle Hämatologie der
Gesellschaft für Strahlen- und Umweltforschung
8000 München 2, West Germany

INTRODUCTION

It is known, that normal peripheral mononuclear cells
show remarkable growth and blastic transformation in the
diffusion chamber culture system. The rise in cell number
is mainly caused by lymphoid cells.

The purpose of this study was to examine if cALL and
TdT positivity is inducible in normal mononuclear blood
cells which contain neither TdT nor cALL antigen during a
13 day diffusion chamber culture.

MATERIALS AND METHODS

Peripheral mononuclear blood cells of two healthy individuals were separated on a Ficoll-Hypaque gradient. The
diffusion chamber culture technique was performed as described elsewhere (Benestad et al.,1975). 5×10^5 cells
were filled into each chamber. Cells were harvested on
days 3, 6, 9, and 13 after implantation. Details concerning this technique were published previously (Lau et al.,
1979).

Immunofluorescence

The cALL-antigen was demonstrated using a monoclonal anti-
cALL antibody (Ritz et al., 1980). The antibody was conjugated with FITC. 10^6 cells were incubated with the antibody (dilution 1:100) for 30' at 4°C. After two washing pro-

371

cedures the number of positive cells was counted in a Zeiss
fluorescence microscope with epi-illumination. Cells were
classified as positive if the fluorescence intensity was
clearly above the background fluorescence. Controls were
performed using normal rabbit serum conjugated with FITC.

Unlabeled antibody enzyme method

TdT was demonstrated by the PAP-method (Sternberger
et al., 1970). Rabbit anti-calf-TdT was obtained from PL-
Biochemicals. Details of the TdT test are described elsewhere (Jäger, 1981).

RESULTS

The results from both culture periods are summarized
in Fig. 1. Phenotypes of the cells in both experiments were
identical at the beginning of culture: neither cALL antigen nor TdT were detectable. In case 1 a high percentage
of cALL positive staining cells was observed at day 3 (69%).
Two different types of staining could be clearly distinguished: a very low number of cells showing bright ring-
like staining (3 %), and a relatively high percentage (66%)
of cells with a diffuse and weak staining pattern. These
weakly stained cells disappeared till day 6, whereas cells
with circular staining pattern increased up to 8 % until
day 9. The first TdT positive cells occurred on day 3 (9%),
the maximum of the TdT+ cells was observed on day 9 (90 %),
and the percentage declined to 15 % on day 13. The staining pattern was nuclear as well as cytoplasmatic.

The second culture showed a well comparable behavior.
The number of cALL+ cells was nearly identical (60 %) but
these cells did not appear before day 6. Also in this case
we observed the above-mentioned types of fluorescence: a
low number of ring-like stained cells and a large percentage of cells with a diffuse and weak fluorescence. The
weakly fluorescent cells disappeared till day 9, and the
number of cells showing ring-like staining increased till
day 13 to a maximum of 9 %. TdT positive cells first appeared on day 3 (17 %) and increased up to 89 % on day 13.
There was a random pattern of cytoplasmic and nuclear staining with anti-TdT during the whole culture period. The
shift in time of the peaks of cALL positivity as well as
the reaction with anti-TdT towards a later harvesting day
in case 2 is in congruence with the growing behavior. Case
1 showed the cell minimum already on day 3 and started
growing on day 6, whereas in case 2 no significant growth
was observed until day 9. Controls performed in both test

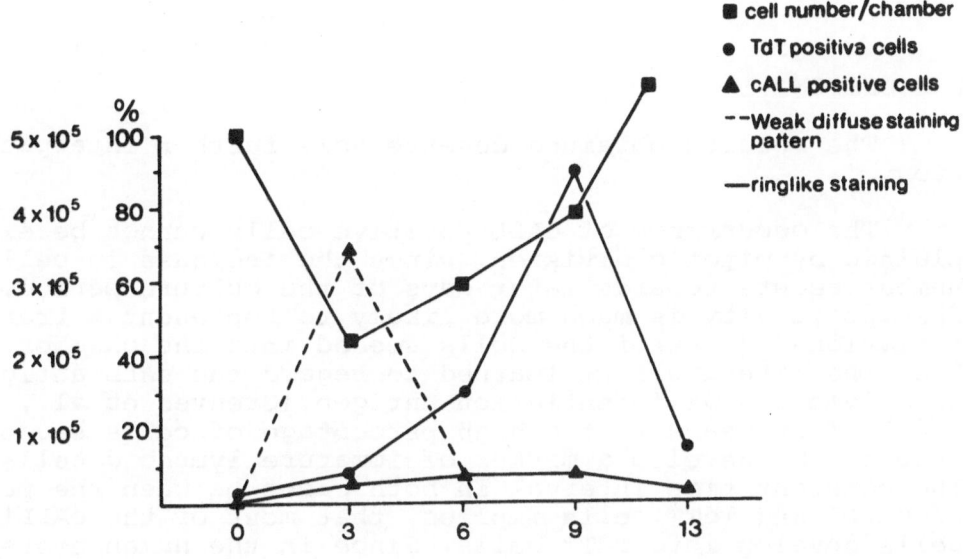

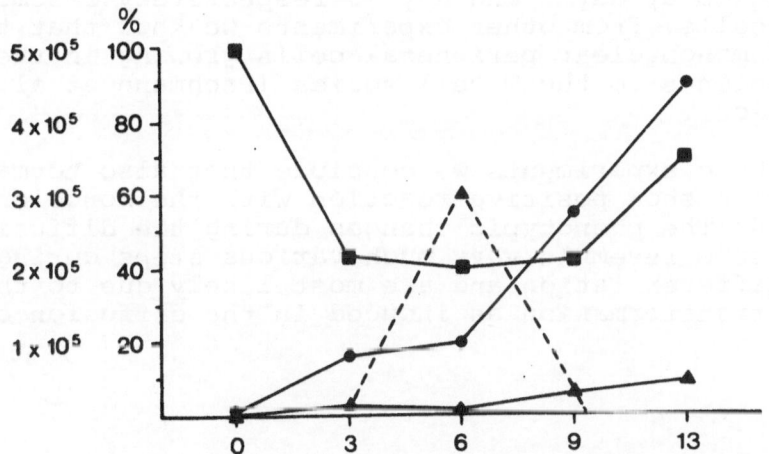

Course of cALL and TdT positivity during a
13-day diffusion chamber culture. ■ cell num-
ber/chamber, ● TdT positive cells, ▲ cALL positive cells, -- weak diffuse staining,
── ring-like staining.

systems with rabbit normal serum did never exceed 3 %.
Autofluorescence could not be observed at any harvestin
day.

DISCUSSION

The results obtained deserve some further interpre
tion.

The occurrence of cALL positive cells cannot be ex
plained by mitotic division, since the increase in cell
number occurs towards later days of the culture period.
This positivity is much more likely to represent a tran
formational event of the cells seeded into the chamber.
From the literature we learned to regard the cALL antig
as a lymphoid differentiation antigen (Greaves et al.,
1980). That means that a high percentage of cells seede
into the DC develop a marker of immature lymphoid cells
The constant time interval in both cases between the pe
of cALL$^+$ and TdT$^+$ cells implies, that most of the cALL$^+$
cells develop into TdT$^+$ cells. Since in the human syste
TdT positivity of a cell is regarded as an early T-diff
rentiation marker (Silverstone et al., 1976) most of th
harvested cells at day 9 and day 13 respectively resemb
immature T-cells. From other experiments we know that t
majority of mononuclear peripheral cells growing in the
DC system belongs to the T-cell series (Pachmann et al.
in preparation).

From these experiments we conclude that also norma
lymphoid cells show positive reaction with the monoclon
antibody J.5. The phenotypic changes during the diffusi
chamber culture resemble very much various steps during
lymphatic differentiation and are most likely due to th
process of transformation as induced in the diffusion c
ber system.

REFERENCES

Benestad, H. B., and Reikvam, A., 1975, Diffusion chamb
    culturing of hematopoietic cells: methodological i
    vestigations and improvement of the technique,
    Exp.Hematol., 3:249.
Boecker, W. R., Boyum, A., Carsten, A. L., and Cronkite
    E. P., 1971, Human bone marrow (HBM) and blood (HP
    stem cell kinetics. Blood, 38:8.

Greaves, M. F., Delia, D., Janossy, G., Rapson, N., Chessells, J., Woods, M., and Prentice, G., 1980, Acute lymphoblastic leukemia-associated antigen. IV.Expression on non-leukemic lymphoid cells. Leuk.Res., 1:15.

Jäger, G., 1981, A simple test for terminal deoxynucleotidyl transferase using the peroxidase-anti peroxidase technique. BLUT, 42:259.

Lau, B., Jäger, G., Thiel, E., Rodt, H., Huhn, D., Pachmann, K., Netzel, B., Böning, L., Thierfelder, S., and Dörmer, P., 1979, Growth of the Reh cell line in diffusion chambers. Scand.J.Hematol., 23:285.

Ritz, J., Pesando, J.M., Notis-McConarky, J., Lazarus, H., and Schlossmann, S. F., 1980, A monoclonal antibody to human acute lymphoblastic leukemia antigen. Nature, 283:583.

Silverstone, A., Cantor, H., Goldstein, G., and Baltimore, D., 1976, Terminal deoxynucleotidyl transferase is found in prothymocytes. J.Exp.Med., 144:543.

Sternberger, L. A., Hardy Jr., P.H., Cuculis, J.J., and Meyer, H. G., 1970, The unlabeled antibody enzyme method of immunohistochemistry. J.Histochem.Cytochem., 18:315.

ENZYMATIC PROPERTIES IN PERICELLULAR MEMBRANES OF LEUKEMIC CELLS

Gabriele Losa

Laboratory of Cellular Pathology
Ticino Institute of Pathology
Locarno, Switzerland

INTRODUCTION

Many of the physiological events taking place at the cell
surface, as for example ion and amino acid transport between the
external and inside cell compartment, membrane rearrangement, transfer of hormonal and other stimuli, are carried out and regulated by
enzymes tightly associated to the plasma membrane. Thanks to their
location such enzymes are considered specific markers of the plasma-
lemmal component and reliable indicators of its physiological state,
as shown by Beaufay et al.[1]. Therefore changes in their activities
could account for the biochemical alterations occurring in the plasma
membrane of peripheral blood cells during the malignant proliferation. The aim of this study was to distinguish the various types of
acute leukemias by analyzing biochemically the enzyme characteristics
of the plasma membrane in cell populations isolated from the peripheral blood of leukemic patients.

MATERIAL AND METHODS

Mononuclear cells were isolated by gradient of Ficoll[2] from
heparinized peripheral blood of patients with acute leukemia. After
centrifugation (800 g / 30 min. / 20° C), cells were harvested from
the Ficoll-serum interface and washed three times with a large volume
of ice-cold saline solution. T cells were isolated from mononuclear
cells using nylon fiber columns[3]. Viability of isolated cells determined by tripan blue exclusion excedeed 90%.

ouabain sensitive (Na-K)-adenosine triphosphatase was calculated as
the difference between the total and the ouabain insensitive (Na-K)-
ATPase activity. 0.2 ml of cell suspension were preincubated for 10
min. at 37° C in a total reaction volume of 0.5 ml containing 1 mM
EDTA, 25 mM Tris-HCl buffer at pH 7.4, 0.02% of Na-deoxycholate with
or without 2 mM ouabain. Pretreatement of cells with deoxycholate at
this concentration allowed the maximal activity to be reached. Thereafter 0.5 ml of an aqueous solution containing 10 mM $MgCl_2$, 6 mM
ATP-Tris, 100 mM NaCl, 40 mM KCl , 2 mM EDTA and 100 mM Tris adjusted
to pH 7.5 was added to the preincubation mixture. The incubation was
performed at 37° C during 30 min. and the reaction was stopped with
1 ml of ice-cold TCA 10%. An aliquot of supernatant recovered after
centrifugation was determined for the inorganic phosphate by the
method of Fiske and Subbarow[11]. Blank assays, either substrate on
enzyme free, were run in parallel with all enzyme determination.
Enzyme activity was expressed as nanomole of substrate liberated or
converted per hour and per 10^6 cells.
Terminal deoxynucleotidyl transferase was measured in cytosol fractions obtained by ultracentrifugation (100000 g / 30 min. / 4° C) of
cell homogenates. Assays were performed at 37° C in a total reaction
volume of 50 μl containing 0.2 M of K-Cacodylate pH 6.8, 5 mM $MnCl_2$,
2.5% bovine serum albumine, H^3-dGTP / dGTP (1 mCi / ml, 0.5 mM), 10
μl of p (dA)$_{12-18}$ (10 OD u / ml) and 10 μl of the enzyme solution.
After incubation (15 min., 37° C) 20 μl volumes were added to a tube
containing 20 μl of yeast RNA and 3 ml of ice-cold 5% trichloroacetic
acid TCA. The precipitates were sedimented by centrifugation (1000
g / 10 min. / 4° C), rinsed with 5% TCA and dissolved in 0.3 ml of
0.1 N NaOH[12]. The "blank" assays contained the same reagents except
for the substitution of the initiator with 10 μl of water. The total
activity was taken by placing into a counting vial 5 μl of the remaining unprecipitated reaction mixture. All samples were counted for
10 minutes with 10 ml of Aqualuma.

RESULTS

     The specific activities of several plasma membrane enzymes,
i.e. 5'-nucleotidase (5'-AMPase), alkaline 5'-phosphodiesterase
(PDAase), alkaline phosphomonoesterase (PNPPase), γ-glutamyltranspep-
tidase (GLUPAase), total and ouabain-dependent adenosine triphosphatase (ATPase), were determined in cells isolated from peripheral
blood of leukemic patients, of normal donors and in normal T cells
(table 1). With respect to their normal counterparts very low activities were generally recorded in cells of both non B / non T lymphoblastic and T leukemias. Differences in the activity of some enzymes
were sufficient to allow a clear-cut differentiation of the two leukemic cell populations. Indeed in lymphoblastic leukemias high

Phenotypic characterization of cells

T cells were identified by rosette formation with modified sheep erythrocytes[4] or by staining with fluorescent coniugated monoclonal antibodies. B cells were analyzed by direct immunofluorescence after staining of surface immunoglobulins (sIg) with TRITC-labelled $F(ab')_2$ fragments of polyvalent antiimmunoglobulin[5]. Monocytes were identified by reacting an aliquot of isolated cells for the unspecific α-naphtylacetate esterase. The presence of the Immune associated antigen was assessed by indirect immunofluorescence after staining cells with monoclonal anti-Ia antibodies followed by staining with an FITC labelled anti-mouse IgG antiserum.

Enzyme assays

5'-nucleotidase was assayed with the method of Beaufay et al.[6] Incubations were performed in duplicate at 37° C during 20 minutes in a total reaction volume of 4 ml. 0,2 ml of cell suspension were added to a mixture containing 50 mM Tris-HCl buffer at pH 7.5,8 mM $MgCl_2$ and 2 mM 5'-AMP. The reaction was stopped with 30% ice-cold TCA and inorganic phosphate removed by centrifugation was measured colorimetrically at 820 nm by the method of Chen[7].
Alkaline 5'-nucleotide-phosphodiesterase was carried out in a total reaction volume of 1 ml containing 0.2 ml of cell suspension, 2 mM Zn-acetate, 0.1 mM glycine-KOH buffered at pH 9.6 and 1.5 mM of p-nitrophenylthymidine-5'-monophosphate. Incubations were performed in duplicate at 37° C. After 20 min. the reaction was stopped by addition of 3 ml of NaOH 0.1 N. After the centrifugation (2200 rpm / 10 min. / 4° C) the absorbance of the liberated product was measured at 400 nm spectrophotometrically. Activity was calculated by relating the optical density values to a standard curve of p-nitrophenol[6].
Alkaline phosphomonoesterase was estimated by measuring at 400 nm the liberation of p-nitrophenol from p-nitrophenylphosphate[6]. 0.2 ml of cell suspension were incubated 20 min. at 37° C in a volume of 1 ml containing 5 mM of p-nitrophenylphosphate as substrate, 5 mM Mg-acetate, 70 mM KCl and 50 mM glycine-KOH buffer at pH 9.0. The reaction was stopped by addition of 3 ml of ice-cold NaOH 0.1 N and the tubes were centrifuged at the previously mentioned conditions.
γ-glutamyltranspeptidase was determined according to Meister et al[8] with slight modification. Incubations were performed in duplicate at 37° C in a total reaction volume of 1 ml containing 60 mM glycylglycine, 0.3 M NaCl, 20 mM Tris-HCl buffer pH 8.0 and 2.5 mM of γ-glutamylparanitroanilide. The reaction was stopped by addition of 2 ml of ice-cold 1.5 N acetic acid. After centrifugation the absorbance of the liberated p-nitroaniline was measured at 410 nm.
(Na-K-Mg) Adenosine-triphosphatase and (Na-K) Adenosine-triphosphatase assayed as previously described[9,10]. The activity of the

Table 1. Plasma membrane enzyme activities
of human leukemic cells (nmole / hr / 10^6 cells)

INTACT CELLS	5'-AMPase	5'-PDEase	PNPPase	GLUPase	(Na-K)Mg -ATPase		
					TOTAL	OUABAIN	% INHIBITION
ALL B⁻/ T⁻ (N = 10)	44.5±24.0 (238.1±95.1)	0.74±0.9	4.7 ±2.3	0.7 ±0.9	37.8±7.6	7.1±6.1	19
ALL T⁺ (N = 5)	4.8±3.5	1.4±1.3	4.7 ±1.7	3.1±2.7	25.5±12.4	8.3±2.6	32
AML (N = 16)	12.5±7.7	6.1±3.3	11.4±7.3	11.0±6.9	73.9±45.6	18.5±15.3	25
MONONUCLEAR CELLS	20.3±8.6	2.0±0.7	23.4±18.6	19.8±7.6	73.6±27.7	20.1±8.3	27
T CELLS	19.4±6.2	2.1±0.8	7.3±3.4	10.6±2.4	51.8±20.5	33.9±4.7	65

5'-AMPase activity was measured with a mean level of about 45 nmole / hr / 10^6 cells. In some cases this activity reached a mean value of 230 nmole / hr / 10^6 cells, suggesting an earlier differentiation stage of the corresponding leukemic cells. In this group 5'-PDAase and GLUPAase were practically undetectable. In contrast T leukemic cells showed a very low activity of the 5'-nucleotidase[13] and a 50% reduced inhibitory effect of the ouabain on the adenosine triphosphatase when compared to the activity measured in normal T cells.
With regard to acute myeloid leukemia, cells isolated by Ficoll gradient were characterized by a quite different profile of enzyme activity distribution. Compared to the control mononuclear cell population, moderate decreased activities were measured for 5'-AMPase, PNPPase and GLUPAase. The behaviour of the other purine enzyme alkaline 5'-phosphodiesterase was noteworthy: the activity was significant high in about 40% of the cases while in the remainder cases was practically undetectable or wery low (range: 0-2.5 nmole / hr / 10^6 cells). The absence of activity of this membrane enzyme was found inversely related to the expression of another surface marker, the Immune associated antigen (Ia-like), whereas the absence of Ia correlated with the presence of a large PDAase activity (range: 3.8-16.8 nmole / hr / 10^6 cells) as reported in table 2.
Surprisingly the activities of the other plasma membrane enzymes were not essentially affected with respect to both phenotypes Ia⁺ / PDA-

Table 2. Enzyme profiles in acute myeloid leukemia.

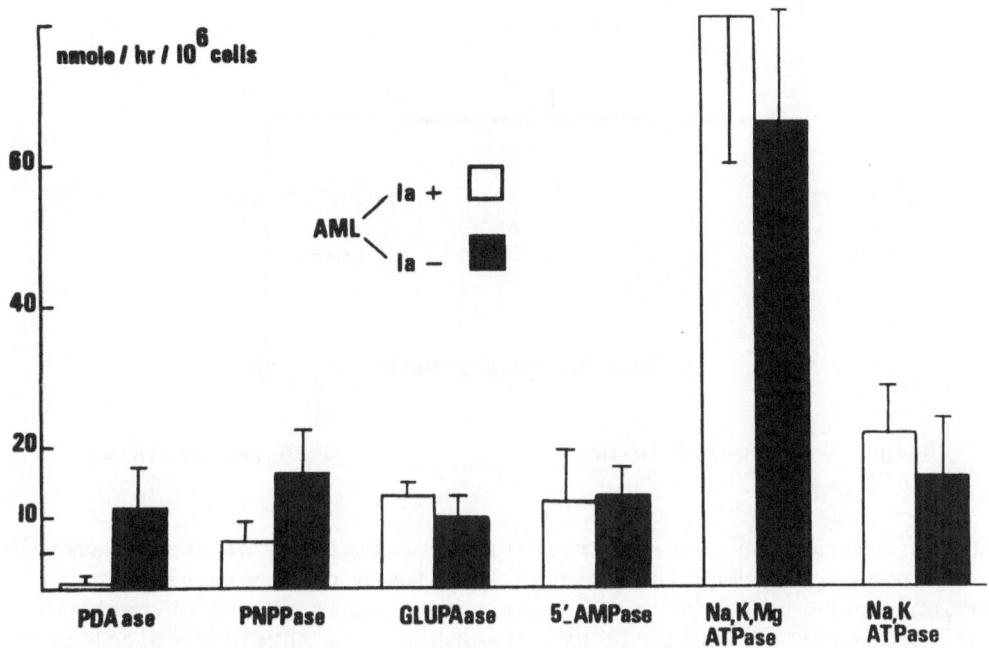

ase⁻ or Ia⁻ / PDAase⁺. Acute myeloid leukemic cells with the former
phenotype could constitute a subpopulation presenting some lymphoid
features, and the scarcity of the PDAase could prove as a further
peculair trait of the lymphoid character. Indeed similar membrane
properties characterized lymphoblastic cells which revealed a positivity for Ia and for the terminal deoxynucleotidyl transferase
(TdT) but lacked alkaline phosphodiesterase activity (table 1).
Consistently with these properties, TdT should be detectable in that
subgroup of AML cells with the above defined phenotype Ia⁺ / PDA-
ase⁻, rather than in the other subgroup. To this aim TdT activity
was biochemically assessed in cytosol fractions from cells of both
groups of AML as summarized in table 3. Our findings revealed a
positive low level of TdT in about 20% of the cases tested but with
an activity range (0-30 picomole / min / mg protein) and activity
distribution similar for both subgroups of acute myeloid leukemia.

CONCLUSIONS

Mononuclear cells isolated from patients with acute leukemia
were found to possess an heterogenous enzymatic composition of their

Table 3.   Terminal deoxynucleotidyl transferase
in AML subgroups.

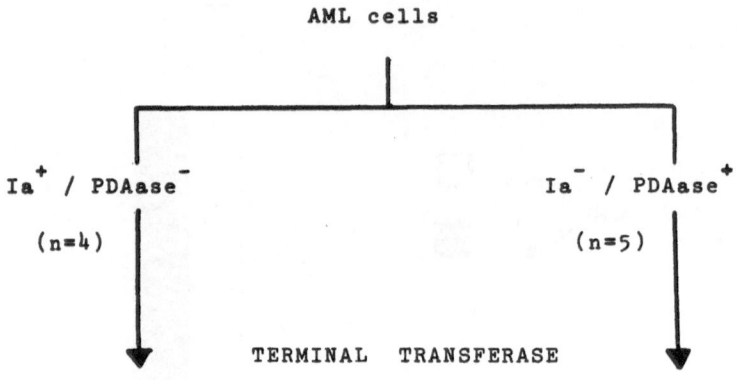

AML cells

Ia$^+$ / PDAase$^-$ Ia$^-$ / PDAase$^+$

(n=4)   (n=5)

TERMINAL   TRANSFERASE

Range: 0-27 pmole/min/mg   0-20 pmole/min/mg

plasma membrane. However characteristic variations of enzyme activities in the different leukemic cells allowed to distinguish the various acute leukemias on a quantitative basis. These enzymatic variations may be explained by changes of the physicochemical properties of the plasmalemma raised by the malignant proliferation. As a consequence, a rearrangement of the membrane enzymes and other constituents may occur, leading to an altered expression of their activities.

Other authors[14] reported differences in morphological and marker characteristics of AML blasts and concluded that these leukemic populations constitute a mixture of lymphoblastic and myeloblastic cells. At glance our data are in agreement with this interpretation of blast heterogeneity in acute myeloid leukemia.

However separate differentiation pathways occurring in a single clone may amplify or suppress the expression of phenotypic markers and enzymes of the plasma membrane or of other organells. If so, the heterogeneity may not be simply due to a mixture of blasts of different origin but could be inherent to the altered distribution of the constituents within the cell membrane.

References

1.   H. Beaufay, A. Amar-Costesec, D. Thinès-Sempoux, M. Wibo, M. Robbi, J. Berthet, Analytical study of microsomes and isolated subcellular membranes from rat liver. III. J Cell Biol. 61:213-231 (1974).

2. A. Böyum, Isolation of mononuclear cells and granulocytes from human blood. Scand. J Clin Lab. Invest. 21, suppl. 27:77-88 (1968).

3. MF. Greaves, G. Janossy, P. Curtis, Purification of human T lymphocytes using nylon fiber columns. In vitro methods in cell-mediated and tumor immunity. Edited by BR Bloom, JR David. New York, Academic Press, pp 217-230 (1976).

4. ME. Kaplan, C. Clark, an improved rosetting assay for detection of human T lymphocytes. J Immun. Methods, 5:131-135 (1974).

5. RJ. Winchester, Techniques of surface immunofluorescence applied to the analysis of the lymphocytes. In vitro methods in cell-mediated and tumor immunity. Edited by BR Bloom, JR David. New York, Academic Press, pp 395-402 (1976).

6. H. Beaufay, A. Amar-Costesec, E. Feytmans, D. Thinès-Sempoux, M. Wibo, M. Robbi, J. Berthet, Analytical study of microsomes and isolated subcellular membranes from rat liver. I. Biochemical methods. J Cell Biol. 61:188-200 (1974).

7. P.S. Chen, T.Y. Toribara, H. Warner; Microdetermination of phosphorus. Analytical Chemistry. 28:1756-1759 (1956).

8. A. Meister, SS. Tate, LL. Ross, Membrane-bound Gamma-glutamyltranspeptidase. The enzymes of biological membranes. Vol. 3. Edited by E. Martonosi,New York, Plenum Press, pp 315-330 (1976).

9. J. Dornand, C. Mani, M. Mousseron-Canet, B. Pau, Propriétés d'une ATPase Ca^{2+} ou Ma^{2+} dépendante des membranes plasmiques de lymphocytes. Effet de la concanavalin A sur les ATPases membranaires. Biochimie 56:1425-1432 (1974).

10. E. Costantino-Ceccarini, PM Novikoff, PH Atkinson, AB. Novikoff, Further characterization of HeLa S_3 plasma membrane ghosts. J Cell Biol, 77:448-463 (1978).

11. CH. Fiske, Y. Subbarow, The colorimetric determination of phosphorus. J Biol Chem, 66:375-379 (1925).

12. NL. Edwards, GW. Gelfand, L. Burk, HM. Dosch, IH. Fox, Distribution of 5'-nucleotidase in human lymphoid tissues. Proc Nat Acad Sci, USA 76:3474-3476 (1979).

13. E. Beutler, W. Kuhl, An assay for Terminal deoxynucleotidyl transferase in leukocytes and bone marrow. J Clin Pathol. 70:733-737 (1978).

14. A.V. Hoffbrand, G. Janossy, Enzyme and membrane markers in leukemia: recent developments. J Clin Pathol. 34:254-262, (1981).